Lecture Notes in Energy

Volume 57

Lecture Notes in Energy (LNE) is a series that reports on new developments in the study of energy: from science and engineering to the analysis of energy policy. The series' scope includes but is not limited to, renewable and green energy, nuclear, fossil fuels and carbon capture, energy systems, energy storage and harvesting, batteries and fuel cells, power systems, energy efficiency, energy in buildings, energy policy, as well as energy-related topics in economics, management and transportation. Books published in LNE are original and timely and bridge between advanced textbooks and the forefront of research. Readers of LNE include postgraduate students and non-specialist researchers wishing to gain an accessible introduction to a field of research as well as professionals and researchers with a need for an up-to-date reference book on a well-defined topic. The series publishes single and multi-authored volumes as well as advanced textbooks.

More information about this series at http://www.springer.com/series/8874

Miriam Rabaçal · Ana F. Ferreira
Carla A.M. Silva · Mário Costa
Editors

Biorefineries

Targeting Energy, High Value Products and Waste Valorisation

Springer

Editors
Miriam Rabaçal
Instituto Superior Técnico
Universidade de Lisboa
Lisboa
Portugal

Ana F. Ferreira
Universidade de Lisboa
Lisboa
Portugal

Carla A.M. Silva
Faculdade de Ciências
Universidade de Lisboa
Lisboa
Portugal

Mário Costa
Instituto Superior Técnico
Universidade de Lisboa
Lisboa
Portugal

ISSN 2195-1284 ISSN 2195-1292 (electronic)
Lecture Notes in Energy
ISBN 978-3-319-83905-9 ISBN 978-3-319-48288-0 (eBook)
DOI 10.1007/978-3-319-48288-0

Printed on acid-free paper

This Springer imprint is published by Springer Nature
The registered company is Springer International Publishing AG
The registered company address is: Gewerbestrasse 11, 6330 Cham, Switzerland

Preface

Energy is inherent to all human activities and its higher consumption per capita correlates well with the higher living standards. Coal, petroleum, and natural gas are energy reserves formed million years ago and are the main raw material for our needs of energy, materials, and chemicals. Recent years have witnessed significant research activity in converting biomass into useful liquid fuels in an attempt to replace totally or partially the oil consumption. Besides bioenergy, researchers are looking for the biorefinery concept perceiving agroindustrial value-chains. In this concept, multiple products could be potentially obtained from biomass including products from its own processing wastes: bioplastics, bioenergy, and biochemicals.

Shifting from fossil fuels to biomass-based fuels requires a deep knowledge of the biomass characteristics (e.g., composition and energy content) and appropriate conversion technology processes (e.g., Fischer–Tropsch, gasification, pyrolysis, combustion, and fermentation) in such a way that the same products obtained from a crude oil refinery could be produced. Both experimental and simulation endeavors are crucial to help understanding the feasibility of an industrial-scale biorefinery-based system. The material contained in this book allows a number of fundamental questions to be tackled such as understanding the best combination of biomass-biorefinery system (with what thermochemical and/or biological and/or catalytic processes in between) for each specific case study and contains tips for evaluating its sustainability. It provides an explanation of the existing methodologies, tools, and metrics to address this pertinent issue.

Chapter 1 presents the definition and examples of biorefinery systems along with basic concepts. Chapter 2 describes the main types of biomass used in the energy field and their composition and lists and discusses the main parameters used to evaluate their energy potential. Chapter 3 focuses on pyrolysis, fermentation, and catalytic conversion technologies with real experimental data. Chapter 4 deals with simulation of the main thermochemical (combustion, pyrolysis, and gasification), biochemical (fermentation), and chemical (fractionation, lignin depolymerization, and platform molecules) processes for biomass processing. Chapter 5 concentrates on the sustainability analysis of biorefinery systems in terms of environmental and socioeconomic indicators, discussing metrics and uncertainty of the concept, configuration and operation. Chapter 6 introduces a systematic methodology for designing integrated biorefineries using process systems engineering tools, which include market analysis, techno-economic assessment, cost accounting, energy integration analysis, life cycle assessment, supply chain analysis, as well as a multi-criteria decision-making framework to put forward the most effective biorefinery strategies that fulfil the needs of the forest industry. Finally, Chap. 7 gives a brief overview of the current conventional and advanced biomass-based biorefineries in the world. While the conventional biorefineries use mature and commercial technology, the advanced biorefineries (e.g., lignocellulosic-based biofuel biorefineries and microalgae-based biorefineries) have different degrees of maturity technology-readiness level and, regardless the process technology, only a few of them have reached the commercial scale although the profitability remains a quest.

Lisboa, Portugal

Miriam Rabaçal
Ana F. Ferreira
Carla A.M. Silva
Mário Costa

Acknowledgements

The editors are thankful to Fundação para a Ciência e a Tecnologia (FCT), through IDMEC under LAETA, project UID/EMS/50022/2013 and IDL under IDL-UID/GEO/50019/2013.

The authors of Chap. 6 are thankful for financial support from the Program on Energy Research and Development (PERD) and the Forest Innovation Program (FIB) of the Canadian Forest Service from Natural Resources.

The author of Chap. 1 is thankful to FCT for the Post-Doctoral financial support through grant no. SFRH/BPD/95098/2013.

Contents

Nomenclature

ABC	Annual basis carbon
ABE	Acetone/butanol/ethanol
ALCA	Attributional life cycle analysis
ASTM	American Society for Testing and Materials
BioLPG	Bio-derived liquid petroleum gas
BioSNG	Bio-derived synthetic natural gas
BTG	Biomass to gas
C	Carbon
CBP	Consolidated bioprocessing
CEN	European Committee for Standardization
CFB	Circulating fluid bed
CIRAIG	Interuniversity Research Centre for the Life Cycle of Products, Processes and Services
CLCA	Consequential life cycle assessment
CNRS	Centre National de la Recherche Scientifique
CO_2 eq	CO_2 equivalent
CSR	Corporate social responsibility
DALY	Disability-adjusted life year
DHMF	Dihydroxymethyl furan
DMF	Dimethylfuran
EBITDA	Earnings Before Interest, Taxes, Depreciation and Amortization
EC	European Commission
EIA	Environmental impact assessment
EMS	Environmental Management System
EPA	US Environmental Protection Agency
ESIA	Environmental social impact assessment
EtOH	Ethanol
FCC	Fluid catalytic cracking
FDCA	Furandicarboxylic acid
FU	Functional unit

GHG	Greenhouse gas
GRI	Global reporting initiative
GTI	Gas Technology Institute
GVL	γ-valerolactone
HMF	Hydroxymethylfurfural
IEA	International Energy Agency
IFBR	Implementing Integrated Forest Biorefinery
ILO	International Labour Organization
iLUC	Indirect land use change
IRR	Investment return rate
ISO	International Standards Organization
JRC	Joint Research Centre
KIT	Karlsruhe Institute of Technology
LA	Levulinic acid
LBA	Large-block analysis
LCA	Life cycle assessment
LCC	Life cycle costing
LCI	Life cycle inventory
LCIA	Life cycle impact assessment
LHS	Latin hypercube sampling
LUC	Land use change
MCDM	Multi-criteria decision-making
MF	Metric of flexibility
MILP	Mixed integer linear programing
MINLP	Mixed integer nonlinear programing
MIT	Massachusetts Institute of Technology
MR	Metric of robustness
MTHF	Levomefolic acid
NEV	Net economic value
NGO	National government organization
NGO	Non-government organization
NLP	Nonlinear problem
NMVOC	Non-methane volatile organic compound
NREL	National Renewable Energy Laboratory
NRR	Non-renewable energy resources
OECD	Organization for Economic Co-operation and Development
PA	Pentanoic acid
PDF	Potentially disappeared fraction of species
PLA	Polylactic acid
PM	Particulate matter
PNNL	Pacific Northwest National Laboratory
PSE	Process systems engineering
PUFAs	Polyunsaturated fatty acids
R&D	Research and development
RED	Renewables Energy Directive

ROCE	Return on capital employed
RTFO	Renewable Transport Fuel organization
SC	Supply chain
SEA	Strategic environmental assessment
SHCF	Separate hydrolysis with co-fermentation
SIA	Social impact assessment
S-LCA	Social life cycle assessment
SRC	Standardized regression coefficients
SSF	Simultaneous saccharification and fermentation
TANs	Total acid numbers
UCL	University College London
UPM	The Biofore Company
USDOE	United States Department of Energy
VOC	Volatile organic compound
VTT	Technical Research Centre of Finland
ZSM-5	Zeolite Socony Mobil–5

Editors and Contributors

About the Editors

Miriam Rabaçal, Ph.D. is a post-doc researcher at Instituto Superior Técnico (IST), Lisboa, Portugal, and was a collaborator with the IDMEC/IST Research Institute from 2010 until she became an integrated member in 2016. She received an M.Sc. degree in Mechanical Engineering from IST in 2010. During her master thesis project, she held an Initiation to Research grant while working on small-scale biomass pellets combustion. After a 4 months' internship in Karunya University, Coimbatore, India, she returned to IST to pursue her Ph.D. studies supported by a FCT Ph.D. grant on large eddy simulations of coal and biomass combustion. During her Ph.D. studies, she spent 6 months in Imperial College London, 18 months in Duisburg Essen University, and 3 months in TUB Freiberg for collaborative research on her Ph.D. topic. She received her Ph.D. degree in Mechanical Engineering in 2016. She participated in more than five international and national research projects in the areas of high-performance computing, combustion, and energy. She has more than 10 papers in international peer-reviewed journals, one book chapter, and about 30 communications in international scientific meetings.

Ana F. Ferreira, Ph.D. is a post-doctoral in the area of environment and energy at the Mechanical Engineering Department of Instituto Superior Técnico (IST). She graduated and obtained her Master's degree in Chemistry, in the environmental area of analysis of atmospheric aerosols and atmospheric exposure, at Faculty of Science of University of Lisbon in 2005 and 2008, respectively. In 2013 completed her Ph.D. in Environmental Engineering at IST of University of Lisbon. Currently she is working on alternative biomass and biofuels. She has more than 15 papers in international peer-reviewed journals, more than one book chapter, and more than 30 papers in international conferences. She has participated in more than five national and international projects, including a Cost Action, in the area of environment and energy. She has been awarded third place in YEAR 2012—The Young European Arena of Research during the Transportation Research Arena (TRA2012) in Athens, in 2012. Her areas of interest include biomass, biomass conversion (e.g., combustion and pyrolysis), pollutant emissions, biofuels, energy and environment and life cycle assessment.

Dr. Carla A.M. Silva, Ph.D. received the Mechanical Engineering degree (five-year course) in 2000 and the Ph.D. degree in Mechanical Engineering in 2005, at Instituto Superior Técnico (IST), University of Lisbon, Portugal. She went for a post-doc at both IST and University of Michigan working on CO_2 mitigation in road vehicles. She has been a senior researcher at the Institute of Mechanical Engineering IDMEC since 2008. She is now Assistant Professor at the Department of Geographic, Geophysics and Energy Engineering. She teaches, in the Integrated Master of Energy and Environment Engineering, the courses Sustainable Mobility, Bioenergy, Combustion Technologies, Life Cycle Assessment and Electrical Circuits. She has supervised more than 50 M.Sc. and Ph.D. students. She has one book, two book chapters, more than 50 papers in international peer-reviewed journals and more than 60 papers in international conferences. She has been awarded seven prizes, including the Best Mechanical Engineering Internship and the 3M Prize for Innovation. Her areas of interest include new fuels; biomass; alternative road vehicle simulation; energy and environment impacts; life cycle assessment; indicators for sustainable mobility.

Mário Costa, Ph.D. is Full Professor in the area of environment and energy at the Mechanical Engineering Department of Instituto Superior Técnico (IST). He graduated in Chemical Engineering at University of Coimbra in 1984, obtained his Ph. D. in Mechanical Engineering at Imperial College London in 1992 and his Habilitation in Mechanical Engineering at Technical University of Lisbon in 2009. Currently, he teaches the courses Thermodynamics, Combustion, Renewables Energies, and Integrated Energy Systems. He has supervised more than 100 M.Sc. and Ph.D. students. He has participated in more than 50 national and international projects in the area of energy and environment and has (co-)authored one book, more than 100 papers in international peer-reviewed journals, and more than 150 papers in international conferences. He was the recipient of the Caleb Brett Award of the Institute of Energy in 1991, of the Sugden Award of the British Section of the Combustion Institute in 1991, of the *Prémio Científico UTL/Santander Totta* in 2010 and of a *Menção Honrosa* Universidade de Lisboa/Santander in 2016.

Contributors

Dieudonné R. Batsy Department of Chemical Engineering, Polytechnique Montreal, Montreal, QC, Canada

Marzouk Benali Natural Resources Canada, CanmetENERGY, Industrial Systems Optimization, Varennes, QC, Canada

A.V. Bridgwater Bioenergy Research Group, European Bioenergy Research Institute, Aston University, Birmingham, UK

Paula Costa LNEG, Lisboa, Portugal

Rocio A. Diaz-Chavez Instituto Centre for Environmental Policy, ICEPT, Imperial College London, South Kensington, London, UK

Javier Dufour Systems Analysis Unit, IMDEA Energy Institute, Móstoles, Spain; Chemical and Environmental Engineering Group, Rey Juan Carlos University, Móstoles, Spain

M. Fantini CONSORZIO L.E.A.P.—Laboratorio Energia Ambiente Piacenza, Piacenza, Italy

Ana F. Ferreira IDMEC, Instituto Superior Técnico, Universidade de Lisboa, Lisboa, Portugal

Krist V. Gernaey Department of Chemical and Biochemical Engineering, CAPEC-PROCESS Research Center, DTU Chemical Engineering, Lyngby, Denmark

Banafsheh Gilani Natural Resources Canada, CanmetENERGY, Industrial Systems Optimization, Varennes, QC, Canada

Francisco Gírio LNEG, Lisboa, Portugal

Luisa Gouveia Bioenergy Unit, Laboratório Nacional de Energia e Geologia, Lisboa, Portugal

Behrang Mansoornejad Natural Resources Canada, CanmetENERGY, Industrial Systems Optimization, Varennes, QC, Canada

Susana Marques LNEG, Lisboa, Portugal

Patrícia Moura LNEG, Lisboa, Portugal

Ana Cristina Oliveira LNEG, Lisboa, Portugal

Paula C. Passarinho Bioenergy Unit, Laboratório Nacional de Energia e Geologia, Lisboa, Portugal

Jens Peters Systems Analysis Unit, IMDEA Energy Institute, Móstoles, Spain; Helmholtz Institute Ulm (HIU), Karlsruhe Institute of Technology (KIT), Karlsruhe, Germany

Filomena Pinto LNEG, Lisboa, Portugal

Remus M. Prunescu Department of Chemical and Biochemical Engineering, CAPEC-PROCESS Research Center, DTU Chemical Engineering, Lyngby, Denmark

Alberto Reis LNEG, Lisboa, Portugal

Shabnam Sanaei Department of Chemical Engineering, Polytechnique Montreal, Montreal, QC, Canada

Abel Sanz Systems Analysis Unit, IMDEA Energy Institute, Móstoles, Spain

Juan Carlos Serrano-Ruiz Universidad Loyola Andalucía, Sevilla, Spain

Carla A.M. Silva Faculdade de Ciências, Instituto Dom Luiz, Universidade de Lisboa, Lisboa, Portugal

Gürkan Sin Department of Chemical and Biochemical Engineering, CAPEC-PROCESS Research Center, DTU Chemical Engineering, Lyngby, Denmark

Paul R. Stuart Department of Chemical Engineering, Polytechnique Montreal, Montreal, QC, Canada

Ana Susmozas Systems Analysis Unit, IMDEA Energy Institute, Móstoles, Spain

Chapter 1
Biorefinery Concept

Ana F. Ferreira

Abstract Biomass valorization is an important issue which could be significant in the reduction of the global dependence on fossil fuels. This chapter will focus on the overview of pathways for conventional and alternative biomass valorization, including transformation to valuable materials and energy. The efficient and flexible use of biomass and the innovative technologies will be discussed as a biorefinery concept. The "raw" biomass, its transformation, and the further conversion into energy and coproducts are considered, aiming to achieve sustainable proposals and maximum valorization. The development and application of the best possible technologies for all processes (e.g., combustion, pyrolysis, gasification, fermentation), including also pretreatment are essential in a biorefinery. The biorefineries can be classified considering the feedstock or the technology: lignocellulosic and marine biorefineries, biochemical and thermochemical biorefineries and advanced biorefinery. The sustainability and the economic factor of the biorefinery are extremely important, and should be evaluated to understand the energy and environmental issues, and the associated costs of any conversion system. Therefore, several sustainability assessment tools have been developed. It is expected that this chapter will contribute to improve understanding the biorefinery concept, the intense and sustainable use of biomass.

1.1 Introduction

The concern of the society regarding environmental issues and sustainability has increased in the last decades. The growing consumption of fossil fuels, in its majority those derived from crude oil, has been questioned especially in the context of its long-term environmental, energy and material sustainability, being at the center of global climate change policies discussion worldwide, which aim to

A.F. Ferreira (✉)
IDMEC, Instituto Superior Técnico, Universidade de Lisboa,
Av. Rovisco Pais 1, 1049-001, Lisboa, Portugal
e-mail: filipa.ferreira@tecnico.ulisboa.pt

M. Rabaçal et al. (eds.), *Biorefineries*, Lecture Notes in Energy 57,
DOI 10.1007/978-3-319-48288-0_1

implement new and alternative solutions to respond to these concerns. It is fundamental to increase and develop renewable energy generation, aiming to improve energy production and use efficiency, increase the security of supply, and enhance primary energy sources diversification. New energy sources have been regarded as a potential commodity to reduce fossil fuel dependence and to mitigate the negative environmental effects (European Commission 2005).

Biomass is considered to be one of future's key renewable resources, being the most important source of energy for three quarters of the world's population living in developing countries. Globally, it already supplies around 14% of the world's primary energy consumption (Star-colibri 2011).

The use of biomass as a resource for energy and fuel production will be limited by the maximum production rates and the supply of biomass, rather than the demand for energy and fuel. Significant barriers to the use of biomass are the relatively low energy content and the seasonality and discrete geographic availability of biomass feedstocks. (Basu 2013; FitzPatrick et al. 2010) In addition, sustainable biomass production is a crucial issue, especially concerning a possible fertile land competition with food and feed industries (Cherubini 2010; Harmsen and Powell 2010).

In order to optimize and exploit all of the elements of biomass, reuse secondary products and wastes of the conversion process into valuable products, as also to produce energy which may help to power the process itself, should be developed—this should be considered within an integrated biorefinery strategy (i.e., efficient and flexible use of biomass).

Basically, the concept of an integrated biorefinery is similar to a petroleum refinery, where oil is refined into many marketable products including chemicals, energy, and fuels. Though, the main difference is that biorefineries are based on the use of renewable materials as a feedstock, namely, biomatter, while the petroleum refineries are based on the use of nonrenewable materials such as fossil fuels (Biernat and Grzelak 2015).

The term biorefinery is derived from the raw material feedstock which is renewable biomass and also from the conversion processes often applied in the treatment and processing of the raw materials. The biorefinery approach involves multistep processes in which the first step, following feedstock selection, usually involves the treatment of biomass for further processing (pretreatment). Afterwards, the biomass is subjected to biological and/or chemical treatments (Strezov and Evans 2014).

There are different definitions for the term "biorefinery," however all of them have the same propose. For the National Renewable Energy Laboratory (NREL), a simple biorefinery concept has been devised that is built on three different "platforms" to promote different product routes, biochemical, thermochemical, and microorganism platforms (King et al. 2010). A biorefinery is a facility that integrates biomass conversion processes and equipment to produce fuels, power, chemicals, and value-added products productsfrom biomass. Following the U.S. Department of Energy, the biorefinery is an overall concept of a processing plant where biomass feedstocks are converted and extracted into a spectrum of valuable products. The IEA (IEA Bioenergy Task 42 2009) has the similar definition to the biorefinery: "Biorefinery is the sustainable processing of biomass into a spectrum of marketable products," as an integrated biobased industries using a variety of technologies to make products such as chemicals, biofuels, food and feed ingredients, biomaterials, fibers, and heat and power, aiming at maximizing the added value along the three pillars of sustainability environment, Economy and society (Sonnenberg et al. 2009). In product-driven biorefineries the biomass is fractionized into a portfolio of biobased products with maximal added value and overall environmental benefits, after which the process residues are used for power and/or heat production, used internally, and/or sold to the national electrical grids. Some residues or coproducts after appropriate treatment processes may also be considered as valuable outputs. These high value-added materials have considerable importance from the industrial and economic point of view, and their appropriate exploitation is a key step in the development of an economy based on recycling and renewable resources—the biobased economy or simply bioeconomy.

A crucial factor for biomass utilization is the cost of the input materials, which today can range from "negative costs" for waste materials (credit for waste disposal), to the more expensive and specialized crops (Grigg and Read 2001). Meaning that, the spectrum of biomass resources can ensure the diversity of the raw material exploitation.

There is a competition for different biomass uses for bioenergy (heat, electricity, transportation fuelsfuels), food (vegetables, meat, among others), biomaterials (paper, construction material, chemicals, cotton, ribber, fertilizer, among others), and feed. The new way in bioeconomy is from competition to integration. Bioeconomy means the part of the economy using biological resources (biomass), or bioprocesses, for the production of value-added products, such as food, feed, materials, fuels, chemicals, biobased products and bioenergy (Sonnenberg et al. 2009). Agriculture, forestry, fisheries, and aquaculture, but also the biotechnological use and conversion of biomass, in addition to biogenic waste materials and residual materials: these are the central starting points for the bioeconomy's value chains and value adding networks, which are interlinked in a multitude of ways.

The transition to a biobased economy has multiple drivers: need to develop an environmentally and socially sustainable global economy; the anticipation that oil, gas, coal, and phosphorous will reach peak production in the not too distant future and that prices will climb; the desire of many countries to reduce an over dependency on fossil fuel imports, so the need for countries to diversify their energy sources; the global issue of climate change and the need to reduce atmospheric greenhouse gases (GHG) emissions; and the need to stimulate regional and rural development. A strategic factor of a successful biobased economy will be the development of biorefinery systems, allowing highly efficient and cost-effective processing of biological feedstock to a range of biobased products, aiming also to reduce GHG emissions and make efficient use of resources (IEA Bioenergy Task 42 report 2011).

Biorefineries should be highly energy efficient and make use of mostly zero-waste production processes, where those "waste" products are regarded as coproducts and may be reallocated for added value use or conversion processes.

Most biorefineries are closely integrated with traditional biomass processing industries. The purpose of the biorefinery is to optimize the use of resources and minimize waste, thus maximizing the benefits and profitability. Full-scale, highly efficient, integrated biorefineries allow competitive manufacturing of high value biobased products. Flexibility is the key. Flexibility has to be intended as the possibility to choose among different processes and, within a specific process, the possibility to select optimal operating conditions and proper technology in dependence on the specific characteristics of the available biomass stream. This allows the production of a broad spectrum of valuable, marketable products.

This chapter will give an overview of several types of biorefinery. Biorefineries can be classified considering the feedstock or the technology, such as: lignocellulosic biorefineries (uses nature dry raw material, such as cellulose-containing biomass and wastes), marine biorefineries (based on marine biomass), biochemical and thermochemical biorefineries (based on a mix of several technologies), and advanced biorefinery (multiple feedstocks, products, and platforms), among others (Cherubini 2010; Sonnenberg et al. 2009; Ververis et al. 2007).

Figure 1.1 shows a scheme of the overview of the biorefinery concept, in which a biorefinery that admits one of the many possible biomass inputs, is able to use several treatment and conversion processes aiming to produce a wide spectrum of products, including primary products and coproducts valorization.

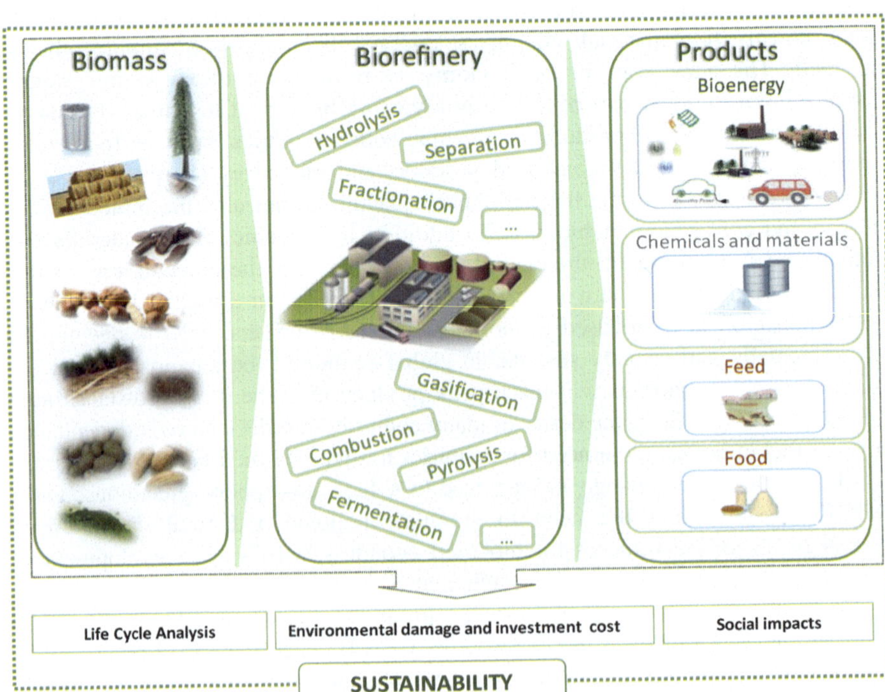

Fig. 1.1 Scheme of biorefinery concept

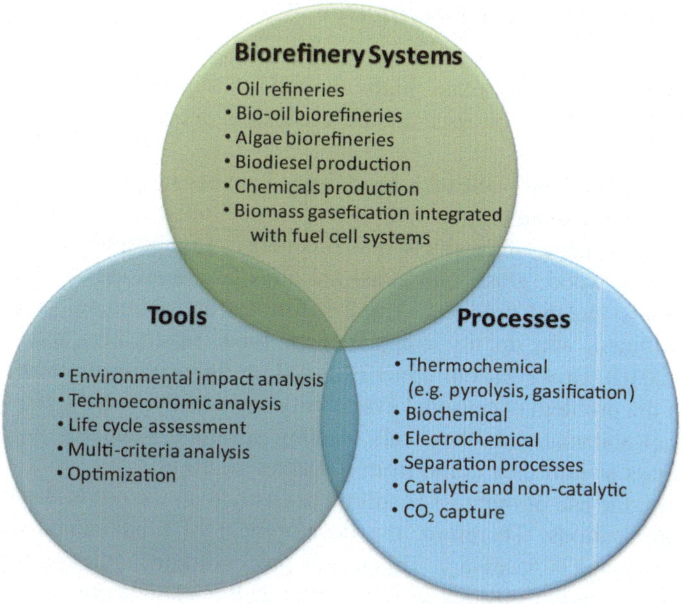

Fig. 1.2 Examples of some of the topics considered in this book

Figure 1.2 introduces some important topics treated in this book, such as some examples of biorefinery systems, the most relevant processes considered in these systems to obtain the primary products and coproducts, and some tools that are essential to evaluate the system viability and sustainability.

1.2 Biomass Feedstock Biorefinery

The choice of biomass and final products is fundamental in biorefinery design due to the large-scale production implications. Initial biomass availability, its potential use, and its characteristics need to be considered (Mabee et al. 2005). There are many options available, each with advantages and disadvantages. The biomass from forest, agriculture, aquaculture, and residues from industry and households can be used on the biorefinery, including wood, short-rotation woody crops, agricultural wastes, wood wastesWaste, bagasse, waste paper, sawdust, biosolids, grass, and organic residues (e.g., waste from food processing), aquatic plants and algae, animal wastes, among others (Demirbas 2005). Detailed and accurate characterization of biomass feedstocks, intermediates, and products is a necessity for any biomass conversion process, to understand how the individual biomass components and reaction products interact at each stage in the process. Based on biomass feedstock, the biorefineries can be classified as lignocellulosic or marine.

1.2.1 Lignocellulosic Biorefineries

Lignocellulose materials contain high amount of sugars, but in the form of polysaccharides, and can be used in the context of biorefinery (Fernando et al. 2006).

So, the lignocellulosic feedstock biorefinery consists in the refinery of the raw material into intermediate outputs (cellulose, hemicellulose and lignin) that will be processed into a range of products and bioenergy. Lignocellulosic biomass and residues such as wood, grass, and straw are abundant, nonfood raw materials for renewable fuels fueland products. These substrates are abundant, geographically widely distributed, and do not compete with food, freshwater, and fertile land (Cherubini 2010). However, biorefineries based on lignocellulosic feedstocks have to face with the problem of seasonal availability and mainly with the requirement of biobased fuel and materials/chemicals meeting specified standards independently on the biomass stream they come from (Clark et al. 2009).

Nowadays, the use of biomass is still more costly than the use of fossil resources for these applications. Therefore, the development of fundamental and applied research in this area is necessary. This means processing technologies that can deal with multiple biomass feedstock streams either within a single process or through a combination of several integrated ones. This allows the production of a broad spectrum of valuable, marketable products. A scheme of Lignocellulosic biorefineries is represented in Fig. 1.3.

Many pilot and demonstration plants have been developed based on lignocellulosic matter biorefinery, and several commercial projects are under development. A most extensive review will be shown in Chap. 7. However here are some examples of pilot plants and commercial projects.

- There are some pilot plants in Europe, Australia, Canada, and USA such as LEUNA (Germany), a 2-platform (C5 and C6 sugars, lignin) biorefinery for the production of biobased synthesized building blocks and polymers from lignocellulosic residues (wood, straw) (http://de.total.com/en-us/search/site/LEUNA);

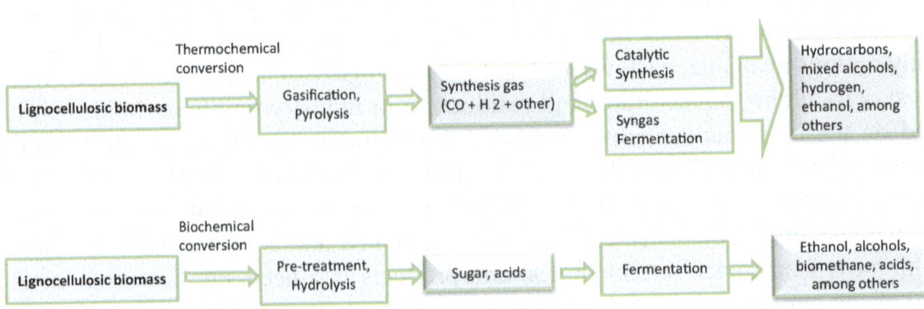

Fig. 1.3 Scheme of two types of the lignocellulosic biorefinery

- Bioprocess Pilot Facility—BPF (The Netherlands), a sugars and lignin platform biorefinery for the production of biobased products and bioenergy from lignocellulosic biomass (www.bpf.eu);
- BDI bioCRACK Pilot Plant (Austria), a one-platform (pyrolysis oil) refinery for the production of diesel fuel, pyrolysis oil, and bio-char from solid biomass (http://www.bdi-bioenergy.com/de-biomass_to_liquid-24.html);
- Microbiogen—Lignocellulosic "Fuel and Feed" Biorefinery (Sydney, Australia), a 2-platform (lignin, C5 and C6 sugars) biorefinery using wood chipschips to produce bioethanol, green coal, and yeast;
- GreenField Specialty Alcohols 2G Ethanol Pilot Plant (Ontario, Canada), a 2-platform (lignin, C5 and C6 sugars) biorefinery producing bioethanol, acetic acid and CO_2 from lignocellulosic biomass (http://www.gfsa.com/);
- Enchi Corporation (former Mascoma Corp.) (USA) Pilot plant CBP with little or no pretreatment—only mechanical disruption or hydrothermal wood chips, switchgrass, and other lignocellulosic biomass (http://www.enchicorp.com/).

Some of commercial lignocellulosic pretreatment technologies (e.g., Liberty™ Technology, POET/DSM, USA; Abengoa Bioenergy Biomass of Kansas, USA) are already available for the production of bioethanol and coproducts from a wide variety of woody material, wastes and other residues, contributing to the deployment of advanced biorefineries using raw materials other than readily available sugar and starch feedstocks. However, these advanced biorefineries are not mature yet but still under development.

Some of commercial scale are also available in Europe:

- CELLULAC (Ireland), a 3-platform (C5, C6 sugars and lignin) for the production of chemicals and fuels from lactose whey permeate and lignocellulosic biomass (http://cellulac.com/sf/);
- Matrica SPA (Italy), under construction, where the complex is expected to employ directly 680 people with a total investment of 500 million Euro. A 1-platform (bio-oil) biorefinery for the production of chemicals (bio-lubricants), bio-polymers, bio-fillers from oil-seed (http://www.matrica.it/);
- Dupont (Nevada, Iowa) and POET-DSM (Emmetsburg, Iowa) in USA; Iogen Corporation in Canada; GranBio (Alagoas) and Raízen/Iogen (Piracicaba) in Brazil, are giving the first steps as major players to commercialize cellulosic ethanol. (IEA Bioenergy Task 42 report, 2014).

1.2.2 Marine Biorefineries

As the name indicates, this biorefinery is based on marine biomass such as aquatic plants, macroalgae (e.g., seaweed) and microalgae. This type of biomass has some advantages, such as no competition for arable land, high areal productivities, and production of a wide range of biobased products and energy, however the cultivation and processing are still at its beginning.

Microalgae biofuels are also likely to have a much lower impact on the environment and on the world's food supply than conventional biofuel-producing crops. When compared with plants biofuel properties, microalgae biomass has a high caloric value, low viscosity, and low density. These characteristics make the microalgae more suitable for biofuels production than lignocellulosic materials, as well as their inherently high lipid content, semi-steady-state production, and suitability to grow in a variety of climates (Gouveia 2011). The main advantages of microalgae are: a high photon conversion efficiency and growth at high rates; a high CO_2 sequestration capacity; they utilize nitrogen and phosphorous from a variety of wastewater sources (e.g., concentrated animal feed operations and industrial and municipal wastewaters) providing the additional benefit of wastewater bioremediation; they do not compete with food production since they use marginal areas which are unsuitable for agricultural purposes (e.g., desert and seashore lands); production is not seasonal; cultures can be induced to produce a high concentration of oil, starch and biomass; they can be cultured without the use of fertilizers and pesticides, resulting in less waste and pollution; and they produce value-added products or coproducts (e.g., proteins, polysaccharides, pigments, biopolymers, animal feed, fertilizers…) (Campbell et al. 2011; Gouveia 2011).

The characteristics of microalgae described above make this biomass a great potential to use in biorefineries (Fig. 1.4). However, the industrial viability of microalgae-based biofuels depends upon the economical aspects which are fundamental to the process. Furthermore, whatever advances might arise in terms of technological innovations, the market will not exhibit an enthusiasm for funding capital-intensive energy projects unless the risk–return ratio is acceptable (Ferreira et al. 2013a; Singh and Gu 2010). Consequently, the technologies concerning this type of biomass are still in development, and must be further investigated to make microalgae-based products, energy, and environmentally relevant.

The marine biorefineries are not already developed, however there are a few examples of this type of biorefineries at pilot plant or demonstration stage:

- AlgaePARC (The Netherlands) Pilot Plant, a multi-platform biorefinery for the production of proteins, lipids, carbohydrates, and pigments from microalgae (http://www.algaeparc.com/project/2/algaeparc-biorefinery);
- Ecoduna Algae Biorefinery (Austria) Demonstration Plant, a 3-platform biorefinery producing biofuels, electricity and heat, omega-3/6 fatty acids and fertilizer from microalgae (http://www.ecoduna.com/);
- Solvent Rescue Ltd (previously Solray Energy Ltd) (New zealand), Pilot plant for supercritical water processing of algae to bio-crude oil. Development work on woody feedstocks (http://www.solventrescue.co.nz/);
- Aquafarming (Maris, Leuven, FeyeCon) Demonstration Plant (http://www.maris-projects.nl/) (IEA Bioenergy Task 42 report 2014).

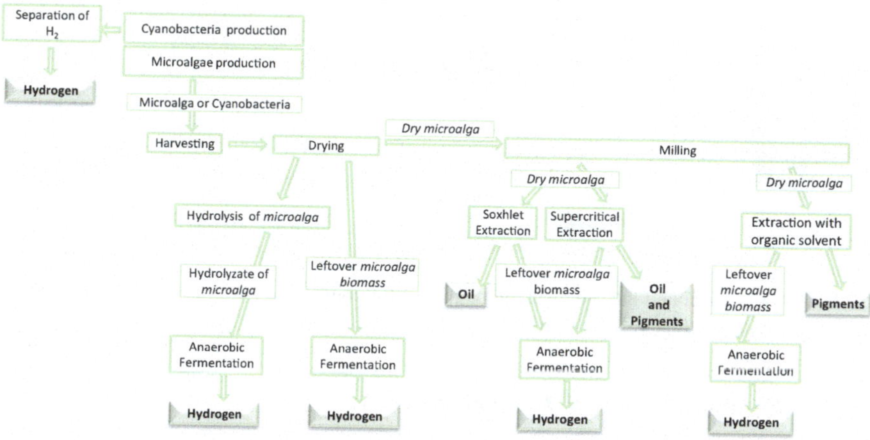

Fig. 1.4 Example of the marine biorefinery. Based on Ferreira et al. (2012, 2013a, b), Pacheco et al. 2015)

1.3 Biomass Conversion Processes

Nowadays, the biobased products have a great importance for several industries; however there are still several technical, strategic, and commercial challenges that need to be overcome before any large-scale commercialization of the industry can succeed. Then, the development and application of the best possible technologies for all processes (e.g. chemical conversion, gasification, fermentation), and also for pretreatment and storage, should be concerned for any projected biorefinery (Löffler et al. 2010).

Processes like combustion, gasification and pyrolysis and biological conversion to sugar and volatile fatty acids could be considered in order to take into account the variability in the biomass chemical compositionchemical composition (moisture, content and quality of both inorganic and hemicellulose fractions), as well as the need to provide a wide selection of energy carriers, end products and secondary raw materials for the fulfillment of the market needs. Thermochemical conversion of biomass in practical systems results from a strong interaction between chemicals and physical processes at the levels of both the single particle and the reaction environment. Feedstock restrictions for thermochemical conversion mostly pertain to particle size, moisture, and ash content (Küçük and Demirbaş 1997; Strezov and Evans 2014).

In terms of conversion processes biorefineries could be divided in two distinct pathways, as can be seen in Fig. 1.5: biochemical conversion and thermochemical conversion.

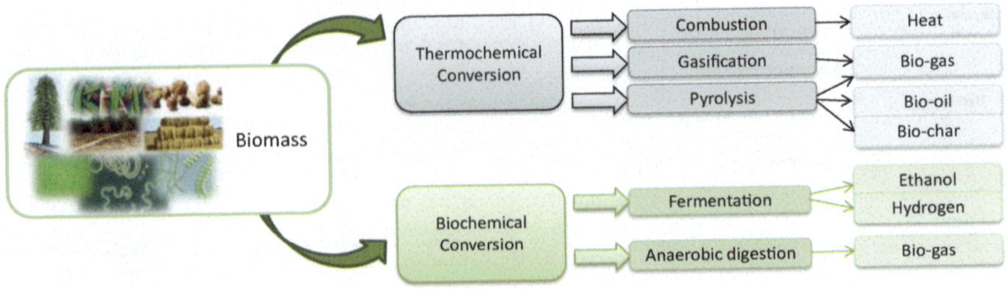

Fig. 1.5 Biochemical and thermochemical pathways for converting biomass to biobased products (e.g. of the main products of each process)

1.3.1 Thermochemical Conversion

Thermochemical conversion is characterized by higher temperatures and faster conversion rates. The three main pathways are: combustion (complete oxidation), gasification (partial oxidation), and pyrolysis (thermal degradation without oxygen). Other example of thermochemical conversion is the Fischer–Tropsch synthesis and liquefaction.

Thermochemical conversion involves controlled heating or oxidation of biomass synthetic gas as an intermediate product, which can be upgraded to valuable products (Demirbas 2004; Tanger et al. 2013).

Thermochemical-based refinery processes are generally consisting of the following interconnected unit operations: pretreatment (i.e., drying, size reduction), feeding, conversion, product clean up and conditioning, and product end use. Thermochemical conversion technologies convert biomass and its residues to fuels, chemicals, and power. The products by thermochemical conversion of biomass and their relative amounts depend on process conditions such as temperature, pressure, feed rate, time of heating, particle size of biomass, and any quenching processes that are applied. The thermochemical conversion is represented in Fig. 1.6.

Combustion of biomass is the thermochemical conversion technique most studied and established for generating heat and power. Combustion processes are responsible for over 97% of the world's bio-energy production (Demirbas, 2004). Combustion is an exothermic reaction between oxygen and the hydrocarbon in

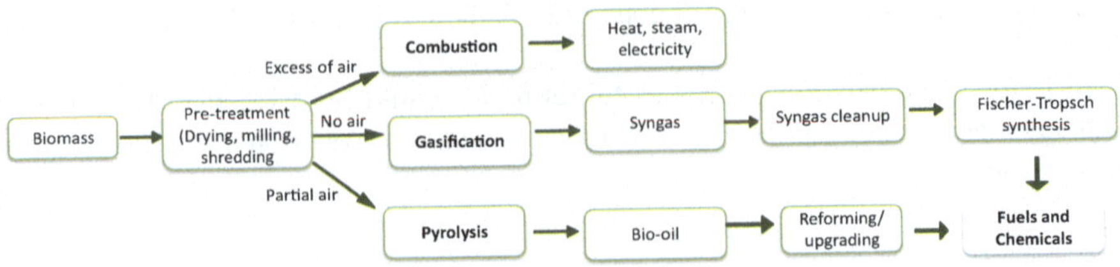

Fig. 1.6 The overall scheme of the thermochemical conversion process

biomass. (Liñán and Williams 1993). Here, the biomass is converted to H_2O and CO_2, where the main direct source of H_2O is drying of biomass and main indirect source of H_2O is oxidation of volatiles. Heat and electricity are two principal forms of energy derived from biomass (Basu 2013). However, the detailed chemical kinetics of the reactions that take place during biomass combustion are complex. Unfortunately, this technique is still associated with high emissions of particulate matter (PM), from which PMs smaller than 1 μm are regarded as a major indicator for the health relevance of ambient air pollution (Fernandes and Costa 2012). The imperfect combustion results in the release of intermediates including environmental air pollutants such as volatile organic compounds (VOC), large oxygenated organic carbon species, CH_4, CO, and PM. Additionally, fuel impurities, such as sulfur and nitrogen, are associated with emission of SO_X and NO_X. (Tanger et al. 2013).

Considering the high level of maturity of combustion technology, technology developments will only produce incremental improvements. Furthermore, combustion produces heat which is not an easy energy carrier to store. If heat is the desired energy carrier, the efficiency of conversion can be quite high, but storage is difficult. If electricity is produced, the efficiency of the process is relatively low, and it can only be stored in batteries, which is not a fully mature technology.

Combustion is widely utilized and commercially available for small-, medium- and large-scale applications. Large-scale co-firing of bio-oil has been carried out, for example, at EnSyn, but few other cases of application exist.

Gasification involves a chemical reaction in an oxygen-deficient environment. Gasification is the exothermic partial oxidation of biomass, with about one-third of the oxygen necessary for complete combustion, produces a mixture of CO_2 and hydrogen, known as syngas. The gas can be cleaned and used directly as a stationary biofuel or can be a chemical feedstock through biological fermentation or catalytic upgradingcatalytic upgrading via the Fischer–Tropsch process for the production of fuels or chemicals (alcohols, organic acids, ammonia, methanol and so on) (Cherubini, 2010; Tanger et al. 2013). The gasification process is faced with some challenges such as the development and commercialization of biomass gasification due to "tars" formation (Milne and Evans 1998). Tars and other contaminants formed during gasification must be removed prior to fuel synthesis; these are both a fouling challenge and a potential source of persistent environmental pollutants (Basu 2013; Foust et al. 2009).

Gasification of biomass has had little commercial impact owing to the competition from other conversion techniques. There has, however, been renewed interest in this process, yet economically viable examples are rare (Bridgwater 1995).

Pyrolysis is a process of heating biomass in the absence of oxygen at a relatively low temperature. Pyrolysis is a promising bioconversion technique for energy recovery, waste management, and converting biomass into useful energy products that has attracted considerable attention during the past decades due to its bioenergy production capability (Liew et al. 2014). Within a pyrolysis process, the raw material is converted into different reactive intermediate products: solid (bio-char), liquid bio-oils (heavy molecular weight compounds that condense when cooled down), and gaseous products (low molecular weight gases) (Fernández et al. 2011). Both gas and pyrolysis oil can be used as fuels that are cleaner and more efficient than the solid biomass, but can also be chemically converted to other valuable fuels and chemicals. Depending on the pyrolysis temperature, the char fraction contains inorganic materials to varying degrees, any unconverted organic solid and carbonaceous residuescarbonaceous residue produced from thermal decomposition of the organic components. Bio-char offers numerous benefits when applied to soils and it potentially delivers a net reduction of atmospheric carbon dioxide, achieved across the combined cultivation and processing regime overall as a function of time. The oil fraction is a complex mixture of organic chemicals (Silva et al. 2016) The pyrolysis method has been used for commercial production of a wide range of fuels, solvents, chemicals, and other products from biomass feedstock. Conventional pyrolysis consists of the slow, irreversible, thermal decomposition of the organic components in biomass. Slow pyrolysis has traditionally been used for the production of char. Short residence time pyrolysis (fast, flash, rapid) of biomass at moderate temperatures has generally been used to obtain high yield of liquid products (Yaman 2004). This technology is not already maturating but in contrast with combustion technology the resulted biofuels can be stored if fuel quality and storage conditions are appropriate.

The use of pyrolysis and the properties of the bio-oil produced are still in development, but it is thought that it can reduce the costs of gasification compared with feeding solid biomass directly into the gasifier (Bridgwater 1995).

A thermo-catalytic conversion can also be considered, including liquefaction and Fischer–Tropsch synthesis. The liquefaction process occurs in the presence of a catalyst and at a still lower temperature. In this process the biomass is converted into liquid. Liquefaction of solid biomass into liquid fuel can be done through pyrolysis, gasification as well as through hydrothermal process. In this case, when there is contact of the biomass with water at elevated temperatures (300–350 °C) with high pressures (12–20 MPa) for a period of time, the biomass is converted into oil (Basu 2013).

The Fischer–Tropsch synthesis is a process used to convert syngas obtained by gasification into liquid transport fuels. This process is widely recognized, but there is a possibility of catalyst shortages in large-scale productions if catalyst regeneration is not improved. This technology is commonly found in the commercial generation of electricity and synthetic fuels from conventional fossil fuels (The Royal Society 2008).

1.3.2 Biochemical Conversion

In biochemical conversion, biomass molecules are broken down into smaller molecules by bacteria or enzymes. In biochemical conversion technology, these biocatalysts, in addition to heat and other chemicals, convert the carbohydrates of the biomass (hemicellulose and cellulose) into sugar. These sugars are intermediate products that can be fermented or chemically catalyzed, using biocatalysts, into a range of advanced biofuels and value-added chemicals such as ethanol and other fuels, chemicals, heat, and/or electricity.

Unlike thermochemical conversion processes, biochemical processes occur at lower temperatures and have lower reaction rates. The biochemical process consists of the following crucial steps: feedstock supply, pretreatment, hydrolysis, biological conversion, and product recovery. The most common biological conversions are fermentation and anaerobic digestion, however it can be considered the enzymatic conversion. The overall scheme of the biochemical conversion process is shown in Fig. 1.7.

Fermentation uses microorganisms or/and enzymes to convert fermentable substrate into recoverable products (alcohols or organic acids). With this process ethanol (the most required fermentation product), butanol, hydrogen, methanol, butyrate acid, and acetate acid can be produced. The fermentation of lignocellulose into cellulosic ethanol has been substantially developed in the past few decades. Lignocellulosic ethanol plants as a whole are at the large demonstration stage, with the Beta Renewables plant becoming operational in 2013, and several others under construction. Enzymatic hydrolysis is being used in these demonstration scale plants. There are around 6–7 small-scale demonstration plants currently operational in Europe with capacities of 1–6 million liters per year (ML/year) and 2–3 pilot plants, Beta Renewables, SEKAB, Clariant, and Inbicon (Bacovsky 2014). The US has a similar number of demonstration plants of the same scale, but at a more advanced stage of development, with four plants under construction, Abengoa, Bluefire, Beta Renewables, Zeachem, Fiberight, Poet-DSM, Mascoma, and Dupont (Council 2013; DOE 2014; Sheridan 2013). In Brazil, one of the key actors is GranBio which plans to bring a 90 ML/year plant into operation in 2014 based on Beta Renewables technology (Bacovsky 2014).

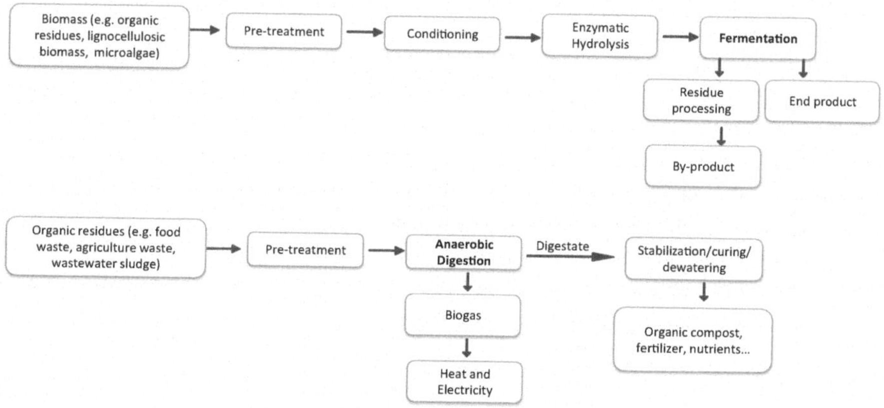

Fig. 1.7 The overall scheme of the biochemical conversion process (fermentation and anaerobic digestion processes)

Fermentation of C5 and C6 sugars to butanol is commercial using the Acetone–Butanol–Ethanol (ABE) process, although the process yields are typically found to be uneconomic for fuel production. Other fermentation pathways for producing only butanol are at the demonstration stage. Most developers are currently focusing on demonstrating butanol production based on sugar and starch feedstocks, with an aim to move to lignocellulosic feedstocks in the longer term, using technologies developed and demonstrated for lignocellulosic ethanol. For example, Gevo have licensed organisms from Cargill that would allow them to use lignocellulosic feedstocks (Alpena Biorefinery 2013; European Biofuels Technology Platform 2013).

Anaerobic digest occurs in controlled reactors or digesters and uses bacterial breakdown of biodegradable organic material. This process occurs in the absence of oxygen over a temperature range from about 30 to 65 °C. The main product of this process is biogas (methane and carbon dioxide and solid residue), which can be upgraded up to 97% methane content and can be used to replace natural gas (Cherubini 2010). An anaerobic digestion for the production of biogas is a well-established commercial technology. Small-scale biogas digesters have been used throughout many developing countries, most notably China and India, but also Nepal, South Korea, Brazil, and Thailand. However, this technology has some limitations in terms of conversion efficiency and productivity of lignocellulose (Consortium 2014).

1.3.3 Advanced Biorefinery

The integrated thermo-biorefinery or advanced biorefinery is considered a biorefinery of the future. The concept of advanced biorefineries is similar to a conventional biorefinery, however, in this case multiple feedstocks, products, and platforms are considered (Fig. 1.8). This type of biorefinery integrates all technologies mentioned in previous subsections.

In advanced biorefinery, different biomass feedstocks and innovative technologies are used. In this case, significant investments in development and new ways to reduce costs and achieve competitiveness with fossil fuels fossil fuelsare essential (Office 2013). An integrated biorefinery produces various products, which include electricity produced from thermochemical and bioproducts from the combination of sugar and other existing conversion technology platforms.

The example of an advanced biorefinery is Abengoa in USA (Office 2013). The current challenges and opportunities in the world and the pilot/demonstration plants will be mentioned with more detail in Chap. 7.

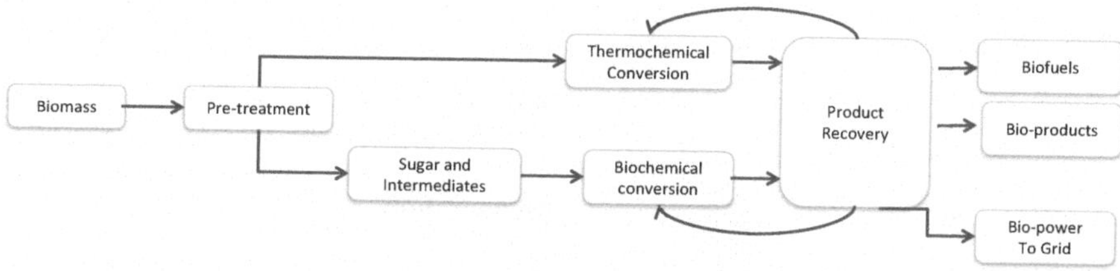

Fig. 1.8 Scheme of advanced biorefinery. Adapted from Office (2013)

1.4 Economic and Sustainability Analyses

Evaluating the economic and sustainability of the biorefinery is extremely important to understand the energy, environmental burdens, and costs of any production/conversion system giving insight into its sustainability (DG Tren—European Commission 2008).

Each process involved in the biorefinery has a relevant impact on the sustainability of biobased products. In any biorefinery, the recycling of energy/heat should be improved with influence on the energy consumption and emissions of the whole processes and respective products. The efficiency of the conversion processes on the biorefinery is essential to make it more sustainable and consequently more economic. Moreover, the use of coproducts and residues as added value products, or as potential fuels to be sold or to produce electricity and heat onsite, and the reutilization of residual heat, are topics that potentiates the increase of the system efficiency and economical gain. To make a biorefinery sustainable and efficient, a significant investment in terms of special technologies for biobased product production and in infrastructures and supply security is needed. Therefore, it is of extreme importance to minimize the costs and the cost of end products (Löffler et al. 2010). And this may be achieved with the proper coproduct allocation and valorization.

The optimization of biorefineries is becoming increasingly significant (Peters et al. 2015). It will allow the identification of bottlenecks and improvement of pathways in the biorefinery processes, improving the biomass conversion yield, carbon footprint, water footprint, fossil energy addiction, and net economic value (NEV). The selection of the most suitable processes, production pathways, and energy and material fluxes are some of the desired results of optimization methods applied to the refinery system. The selection of the most valuable or sustainable pathways within a biorefinery system is a challenge when optimizing biorefinery systems, namely because different products have different value, requirements, demands, and yield efficiency (Fig. 1.9). If in some cases the correct allocation is key, in other cases there may not be an optimal pathway allocation—flexibility depending on the fluctuating demand and value market may be one of the solutions.

The energy and emissions balance of an engineered process is crucial to verify if the processes and technologies used on biorefinery are environmentally friendly and

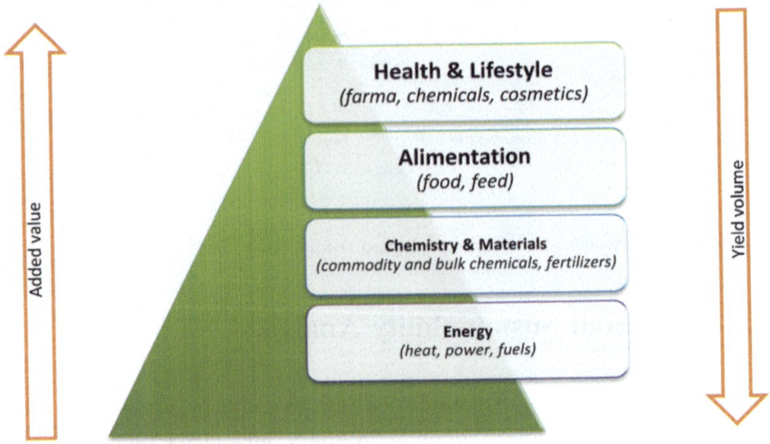

Fig. 1.9 Value pyramid of biomass use in a biorefinery concept. Adapted from Ree and Zeeland (2014)

sufficiently efficient, and identify the bottlenecks aiming to apply further improvements on the energy and CO_2 emission chain. As part of a sustainability assessment, the economic, energetic, and environmental impacts depend on these global process balances. The environmental carbon footprint based on life cycle methodology has shown to be an important tool and it is frequently used in studies of this research area (Pacheco et al. 2015).

The life cycle assessment methodology (LCA) analyzes a product during its lifetime from its production, to its utilization, and end of life, including its recycling process. It is an important methodology to estimate the energy balance and environmental impact of a system. This methodology is defined by the ISO 14040 principles (14040:2006, Life Cycle Assessment: Principles and Framework and 14044:2006, Life Cycle Assessment: Requirements and Guidelines).

In this methodology, each step of the processes should be considered, such as biomass feedstock (e.g., culture, harvesting, drying), conversion process (e.g., pyrolysis, fermentation) and use of the product.

Figure 1.10 exemplifies a biorefinery system with several life cycle steps, from its primary feedstock extraction to the final product achievement and coproducts or residues post-processing.

Several items should be addressed when dealing with bioenergy systems to ensure its sustainability, such as direct and indirect land use change, water footprint, and energy demand. The most adequate indicators and methodology to carry the social and environmental life cycle assessments should be selected according to their representativeness with a geographical and time approach. This will allow getting valid results to check land use changes and social impacts. All these indicators allow to determine the actual sustainability of the systems. Checking different

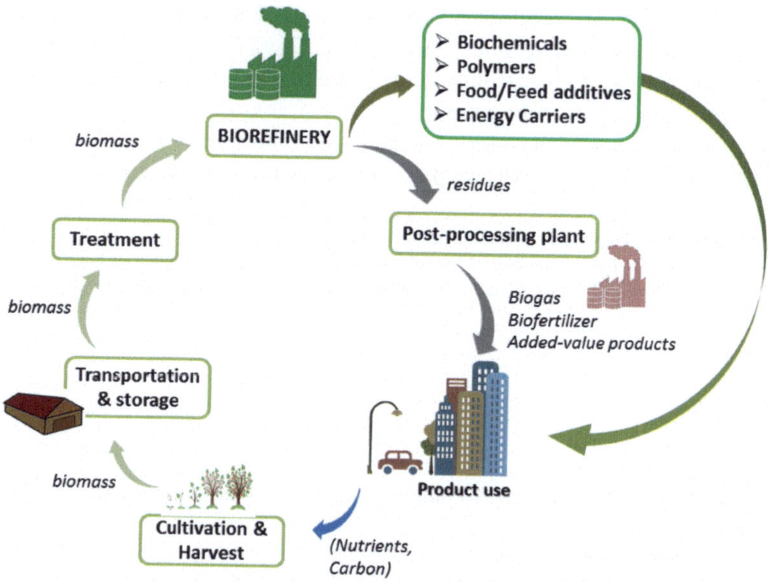

Fig. 1.10 Example of a biorefinery system and its integration within the life cycle of the input and output products. Adapted from https://www.chalmers.se/en/departments/bio/research/industrial-biotechnology/Biorefineries/Pages/default.aspx

biorefinery schemes will allow proposing a common methodology to be used for the assessments of each facilities and plants.

When it comes to life cycle, it is fundamental to refer the associated uncertainty. There can be two types of parameter uncertainty: measurement uncertainty, which is related to imperfections, assumptions or the inability to take an exact measurement when the actual inventory is being developed, and uncertainty related to the data quality of the inventories used. Identifications of sources of uncertainties in biorefinery assessment will therefore consist of a literature survey to determine the uncertainties existing in life cycle assessment of biorefinery, e.g., plantation of biomass and land use control (LUC) considerations, life of the plantation, different methods for harvesting, different processes of transformation and energy requirements, different allocation methods, and different end products. Definition of uncertainty in market demand will cause different biorefinery end product quantities needs as referred previously. A flexible biorefinery should respond to this market stimulation.

In the cost analysis, the economic viability and evaluation of the costs of each process can assess its economic feasibility. These types of analyses can be useful in determining which emerging technologies have the highest potential for success and to minimize the costs involved in whole processes. The socioeconomic factor should cover the impact created due to the biorefinery at a local and national level, associated job creations, land valorization, region incentives, associated secondary companies, and services contracted; and should also account with the capital investment, insurances, maintenance and relative services, human resources, energetic, and material requirements during the biorefineries lifetime, toward a viable or competitive leveled cost of the output.

In resume, in order to develop a sustainable biorefinery, the economic, environmental, and social impacts should be considered and analyzed.

Acknowledgements This work was supported by Fundação para a Ciência e a Tecnologia (FCT), through IDMEC, under LAETA-UID/EMS/50022/2013. The author is pleased to acknowledge the FCT for the Post-Doctoral financial support through grant no. SFRH/BPD/95098/2013.

References

Alpena Biorefinery (2013) Fact sheet

Bacovsky DL (2014) Status of advanced biofuels demonstration facilities in 2012. A report to iea bioenergy task 39 (February)

Basu P (2013) Biomass gasification, pyrolysis and torrefaction

Biernat K, Grzelak PL (2015) biorefinery systems as an element of sustainable development. Biofuels—Status and perspective, pp 427–458

Bridgwater AV (1995) The technical and economic feasibility of biomass gasification for power generation. Fuel 74(5):631–653

Campbell PK, Beer T, Batten D (2011) Life cycle assessment of biodiesel production from microalgae in ponds. Bioresour Technol 102(1):50–6. Elsevier Ltd. Retrieved 20 Mar 2014, from http://www.ncbi.nlm.nih.gov/pubmed/20594828

Cherubini F (2010) The biorefinery concept: using biomass instead of oil for producing energy and chemicals. Energy Convers Manag 51(7):1412–1421

Clark JH, Deswarte FEI, Farmer TJ (2009) The integration of green chemistry into future biorefineries. Biofuels Bioprod Biorefining

Consortium AU (2014) Advanced biofuel demonstration competition feasibility study annex 1 : technology status update, vol 2

Council AE (2013) Cellulosic Biofuels industry progress report: 2012–2013

Demirbas A (2004) Combustion characteristics of different biomass fuels. Prog Energy Combust Sci 30(2):219–230

Demirbas A (2005) Potential applications of renewable energy sources, biomass combustion problems in boiler power systems and combustion related environmental issues. Prog Energy Combust Sci 31(2):171–192

DG Tren—European Commission (2008) Sustainability criteria and certification systems for biomass production final report

DOE (2014) US department of energy

European Biofuels Technology Platform (2013) European biofuels technology platform—Biobutanol

European Commission (2005) Biomass action plan

Fernandes U, Costa M (2012) Particle emissions from a domestic pellets-fired boiler. Fuel Proc Technol 103:51–56

Fernández Y, Arenillas A, Menéndez JÁ, Nacional I (2011) Microwave heating applied to pyrolysis. Advances in induction and microwave heating of mineral and organic materials. Intechopen

Fernando S, Adhikari S, Chandrapal C, Murali N (2006) Biorefineries: current status, challenges, and future direction. Energy Fuels 20(4):1727–1737

Ferreira AF, Marques AC, Batista AP, Marques PASS, Gouveia L, Silva CM (2012) Biological hydrogen production by Anabaena sp.—yield, energy and CO_2 analysis including fermentative biomass recovery. Int J Hydrog Energy 37(1):179–190. Retrieved 9 Apr 2014, from http://linkinghub.elsevier.com/retrieve/pii/S0360319911021690

Ferreira AF, Ribeiro LA, Batista AP, Marques PASS, Nobre BP, Palavra AMF, da Silva PP et al (2013a) A biorefinery from Nannochloropsis sp. microalga—energy and CO_2 emission and economic analyses. Bioresour Technol 138:235–44. Elsevier Ltd. Retrieved 29 Mar 2014, from http://www.ncbi.nlm.nih.gov/pubmed/23619136

Ferreira AF, Ortigueira J, Alves L, Gouveia L, Moura P, Silva CM (2013b) Energy requirement and CO_2 emissions of bioH2 production from microalgal biomass. Biomass and Bioenergy 49:249–259. Retrieved 9 Apr 2014, from http://linkinghub.elsevier.com/retrieve/pii/S0961953412005314

FitzPatrick M, Champagne P, Cunningham MF, Whitney RA (2010) A biorefinery processing perspective: treatment of lignocellulosic materials for the production of value-added products. Bioresour Technol 101(23):8915–8922

Foust TD, Aden A, Dutta A, Phillips S (2009) An economic and environmental comparison of a biochemical and a thermochemical lignocellulosic ethanol conversion processes. Cellulose 16 (4):547–565

Gouveia L (2011) Micoralgae as a feedstock for biofuels. Springer briefs in microbiology

Grigg SV, Read AD (2001) A discussion on the various methods of application for landfill tax credit funding for environmental and community projects. Resour Conserv Recycl 32(3–4):389–409

Harmsen J, Powell JB (2010) Sustainable development in the process industries—Cases and impact

IEA Bioenergy Task 42 (2009) Biorefineries: adding value to the sustainable utilisation of biomass

IEA Bioenergy Task 42 Report (2011) Bio-based chemicals value added products from

IEA Bioenergy Task 42 Report (2014) Task42 biorefining—sustainable and synergetic processing of biomass into marketable food and feed ingredients, products (chemicals, materials) and energy (fuels, power, heat)

King D, Inderwildi OR, Wiliams A (2010) The future of industrial

Küçük MM, Demirbaş A (1997) Biomass conversion processes. Energy Convers Manag 38(2): 151–165. Retrieved from http://www.sciencedirect.com/science/article/pii/0196890496000313

Liew WH, Hassim MH, Ng DKS (2014) Review of evolution, technology and sustainability assessments of biofuel production. J Clean Prod 71:11–29

Liñán A, Williams FA (1993) Fundamental aspects of combustion. Oxford University Press

Löffler K, Gillman N, Leen RW, van Schäfer T, Faaij A, Plata LG (2010) The future of industrial biorefineries, vol 40

Mabee WE, Gregg DJ, Saddler JN (2005) Assessing the emerging biorefinery sector in Canada. Appl Biochem Biotechnol 121–124:765–778

Milne TA, Evans RJ (1998) Biomass gasifier “Tars”: their nature, formation, and conversion. Constraints (Nov)

Office BT (2013) Integrated biorefineries : key challenges , pp 1–4

Pacheco R, Ferreira AF, Pinto T, Nobre BP, Loureiro D, Moura P, Gouveia L et al (2015) The production of pigments and hydrogen through a Spirogyra sp. biorefinery. Energy Convers Manag 89:789–797

Peters JF, Petrakopoulou F, Dufour J (2015) Exergy analysis of synthetic biofuel production via fast pyrolysis and hydroupgrading. Energy 79(C):L325–336

Sheridan, C. (2013). Big oil turns on biofuels. Nat Biotech 31(10):870–873 (Nature Publishing Group, a division of Macmillan Publishers Limited. All Rights Reserved)

Silva CM, Ferreira AF, Dias AP, Costa M (2016) A comparison between microalgae virtual biorefinery arrangements for bio-oil production based on lab-scale results. J Clean Prod, vol 130

Singh J, Gu S (2010) Commercialization potential of microalgae for biofuels production. Renew Sustain Energy Rev 14(9):2596–2610. Elsevier Ltd. Retrieved 21 Mar 2014, from http://linkinghub.elsevier.com/retrieve/pii/S1364032110001619

Sonnenberg A, Baars J, Hendrickx P (2009) IEA bioenergy task 42 biorefinery

Star-colibri (Grant Agreement 241535) (2011) Joint European biorefinery vision for 2030

Strezov V, Evans TJ (2014) Biomass processing technologies. Retrieved from http://books.google.com/books?id=AePMAwAAQBAJ&pgis=1

Tanger P, Field JL, Jahn CE, Defoort MW, Leach JE (2013) Biomass for thermochemical conversion: targets and challenges. Front Plant Sci 4(July):218

The Royal Society (2008) Sustainable biofuels: prospects and challenges

Van Ree R, Van Zeeland A (2014) IEA bioenergy: task42 biorefining, pp 1–66. Retrieved from http://www.ieabioenergy.com/wp-content/uploads/2014/09/IEA-Bioenergy-Task42-Biorefining-Brochure-SEP2014_LR.pdf

Ververis C, Georghiou K, Danielidis D, Hatzinikolaou DG, Santas P, Santas R, Corleti V (2007) Cellulose, hemicelluloses, lignin and ash content of some organic materials and their suitability for use as paper pulp supplements. Bioresour Technol 98(2):296–301

Yaman S (2004) Pyrolysis of biomass to produce fuels and chemical feedstocks. Energy Convers Manage 45(5):651–671. Retrieved 19 Mar 2014, from http://linkinghub.elsevier.com/retrieve/pii/S0196890403001778

Author Biography

Ana F. Ferreira, Ph.D. is a postdoctoral researcher in the area of Environment and Energy at the Mechanical Engineering Department of Instituto Superior Técnico (IST). She graduated and obtained her Master's degree in Chemistry, in the environmental area of analysis of atmospheric aerosols and atmospheric exposure, at Faculty of Science of University of Lisbon in 2005 and 2008, respectively. In 2013, she completed her PhD in Environmental Engineering at IST of University of Lisbon. Currently she is a postdoctoral researcher at Institute of Mechanical Engineering of IST, on alternative biomass and biofuels. She has more than 15 papers in international peer-reviewed journals, more than one book chapter and more than 30 papers in international conferences. She has participated in more than five national and international projects, including a Cost Action, in the area of Environment and Energy. She has been awarded third place in the year 2012—The Young European Arena of Research during the Transportation Research Arena (TRA2012) in Athens, in 2012. Areas of interest are biomass, biomass conversion (e.g., combustion and pyrolysis), pollutant emissions, biofuels, energy and environment, and life cycle assessment.

Chapter 2
Biomass Availability, Potential and Characteristics

M. Fantini

Abstract Biomass refers to any organic materials that are derived from plants (including algae, trees and crops) and comprises material which has an organic origin. This leads to a wide range of possible sources of biomass each with their own composition and to different possible classifications. In the energy field, "biomass" refers to the organic matter (living or residue) that can be used as combustible material or energy source for power generation, heat or biofuels. After a description of the main types of biomass used in the energy field, their composition is given and the main parameters to evaluate their energy potential are listed and commented. In particular, the energy potential of virgin biomass, both terrestrial and aquatic, and waste biomass is investigated. An analysis on the availability of the biomass is outlined together with its worldwide productivity. Finally, a list of limitation of energy production from this resource related to historical, technical, economic, environmental and social reasons is described.

2.1 Biomass Definition and Classification

In a broad sense, biomass is defined as the whole organic matter produced by biological processes. In the energy field, "biomass" refers to the organic matter (living or residue) that can be used as combustible material or energy source for power generation, heat or biofuels. The plethora of biomass suggests a huge variety of classifications. The first distinction is based on differentiation upon animal and plant biomass. The use of animal biomass for energy purposes is always considered a secondary matter for two reasons:

M. Fantini (✉)
CONSORZIO L.E.A.P.—Laboratorio Energia Ambiente Piacenza, Piacenza, Italy
e-mail: martina.fantini@polimi.it

M. Rabaçal et al. (eds.), *Biorefineries*, Lecture Notes in Energy 57,
DOI 10.1007/978-3-319-48288-0_2

– Animal beings are heterotrophic organisms, feeding themselves directly or indirectly with organic matter produced by autotrophic organism (plant beings). Therefore the energy inside animal matters always derives from plant organisms;
– Animal biomass usable for power energy consists of remaining portions of other activities, mainly wastes and livestock excrements. That is, energy from animal biomass is a necessary outcome of other activities linked to the production of energy.

Differently from animal beings, autotrophic organisms as vegetables algae and some kinds of bacteria produce organic matters which they need from inorganic substances and are able to convert the energy from solar radiation into chemical energy. Plant organisms can be used not only as food but also for processes dedicated to energy production. The energy produced in such processes is sun-derived. This chapter is dedicated to vegetal biomass as, capturing solar energy, it can realize a renewable energy supply.

Common sources of biomass are:

- *Agricultural*: food grain, bagasse (crushed sugarcane), corn stalks, straw, seed hulls, nutshells, poultry and hogs;
- *Forest*: trees, wood waste, wood or bark, sawdust (SW), timber slash and mill scrap;
- *Municipal*: municipal solid waste (MSW), sewage sludge, refuse-derived fuel, food waste, waste paper and yard clippings;
- *Energy crops*: poplars, willows, switch grass, corn, soybean, canola and other plant oils;
- *Biological*: animal waste, aquatic species and biological waste.

2.1.1 Types of Biomass

Biomass refers to any organic materials that are derived from plants (including algae, trees and crops) (McKendry 2002) and comprises material which has an organic origin. Biomass does not include organic materials that over many millions of years have been transformed by geological processes into fossil fuels such as coal or petroleum. The definition of biomass for energy given in the European Directive 2009/28/EC is the following: "the biodegradable fraction of products, waste and residues from biological origin from agriculture (including vegetal and animal substances), forestry and related industries including fisheries and aquaculture, as well as the biodegradable fraction of industrial and municipal waste". This leads to a wide range of possible sources of biomass each with their own composition. Different classifications are possible and were published. European committee for standardization published two standards for classification and specification (EN 14961) and quality assurance (EN 15234) of biomass. Based on their origin, it classified biomass under four main categories (McKendry 2002; Williams et al. 2001):

- Woody plants. A woody plant is a vascular plant that has perennial stem that is above ground and covered by a layer of thickened bark. Woody plants contain wood, which is composed of structures of tightly bound fibres of cellulose and lignin (trees, bushes, shrubs ...);
- Herbaceous plants. An herbaceous plant is a plant that has leaves and stems that die down at the end of the growing season to the soil level. These plants do not contain wood; therefore, their structure is composed of more loosely bound fibres of lignin and cellulose. This also means that the lignin content for herbaceous plants is lower than for woody plants (grasses, stalk, straw, grains and cereals);
- Aquatic plants. Aquatic plants are plants, which grow underwater like kelp and algae. The moisture content of this type of biomass is usually high (e.g. kelp);
- Wastes. Wastes are all kinds of waste streams, like manure, sewage sludge, refused-derived fuel (RDF). Most of the wastes are extremely wet, like sewage sludge and moisture manure (sewage sludge, refuse-derived fuel and animal wastes).

Loosely speaking, biomass includes all plants and plant-derived materials, including livestock manures. Primary or virgin biomass comes directly from plants or animals. Waste or derived biomass comes from different biomass-derived products. Table 2.1 (Klass 1998) lists a range of biomass types grouping them into two broad group, virgin or waste:

1. Virgin biomass includes wood, plants, leaves (lignocellulose), and crops and vegetables (carbohydrates).
2. Waste biomass includes solid and liquid wastes (municipal solid waste); sewage, animal and human wastes; gases derived from landfilling (mainly methane); and agricultural wastes.

2.1.1.1 Virgin Biomass

Considering the incident solar radiation, or insolation, that strikes the earth's surface, at an average daily insolation worldwide of about 220 W/m^2, the annual insolation on about 0.01% of the earth's surface is approximately equal to all the primary energy consumed by humans each year (Klass 1998). The most widespread and practical process for capture of this energy as organic fuels is the growth of virgin biomass. Extremely large quantities of carbon are fixed each year in the form of terrestrial and aquatic biomass. Using the figures in Table 2.1, the energy content of standing biomass carbon; that is, the renewable, above-ground biomass reservoir that in theory could be harvested and used as an energy resource, is about 100 times the world's annual energy consumption.

Table 2.1 Estimated net photosynthetic production and carbon storage in biomass (Consonni 2006)

Ecosystem	Area		Average net production of carbon			Carbon stored in biomass			Stored/production rate
	10^6 km	%	t/(ha-year)	10^9 t/year	%	t/ha	10^9 t	%	Years
Tropical pluvial forest	17	11.4	9.9	16.83	32.8	202.5	344.25	41.6	20.5
Boreal forest	12	8.1	3.6	4.32	8.2	90	108	13	25
Tropical seasonal forest	7.5	5	7.2	5.4	10.2	157.5	118.13	14.3	21.9
Temperate decidual forest	7	4.7	5.4	3.78	7.2	135	94.5	11.4	25
Evergreen temperate forest	5	3.4	5.85	2.93	5.5	157.5	78.75	9.5	26.9
Total forest	48.5	32.6		33.26	62.9		743.63	89.8	22.4
Extreme desert, sand, rocks	24	16.1	0.01	0.02	0	0.1	0.24	0	10
Shrub deserts and semi-deserts	18	12.1	0.41	0.74	1.4	3.2	5.76	0.7	7.8
Savanna	15	10.1	4.05	6.08	11.5	18	27	3.3	4.4
Cultivated land	14	9.4	2.93	4.1	7.8	4.5	6.3	0.8	1.5
Temperate grasslands	9	6	2.7	2.43	4.6	7.2	6.48	0.8	2.7
Woods and shrubs	8.5	5.7	3.15	2.68	5.1	27	22.95	2.8	8.6
Tundra and mountain regions	8	5.4	0.63	0.5	1	2.7	2.16	0.3	4.3
Marshes and swamps	2	1.3	13.5	2.7	5.1	67.5	13.5	1.6	5
Lakes and rivers	2	1.3	1.8	0.36	0.7	0.1	0.02	0	0.1
Total emersed lands	100.5	67.4		19.61	37.1		84.41	10.2	4.3
Total continental	149	100		52.87	100		828.04	100	15.7
Open Ocean	332	89.5	0.56	18.59	70.9	0.1	3.32	73.1	0.2
Continental platform	36.6	9.9	1.62	5.93	22.6	0	0.01	0.3	0
Estuaries, swamps excluded	1.4	0.4	6.75	0.95	3.6	4.5	0.63	13.9	0.7
Alga beds and coral barrels	0.6	0.2	11.25	0.68	2.6	9	0.54	11.9	0.8
Emergent lands	0.4	0.1	2.25	0.09	0.3	0.9	0.04	0.8	0.4
Total seas	371	100		26.23	100		4.54	100	0.2
General total	520		79.1			832.58			10.53

Table 2.2 Estimated distribution of World's biomass carbon (adapted from Table 2.1)

	Forests	Savanna and grasslands	Swamp and marsh	Remaining terrestrial	Marine
Area (10^6 km^2)	48.5	24.0	2.0	74.5	361
Percent, %	9.5	4.7	0.4	14.6	70.8
Net C production (Gt/year)	33.26	8.51	2.70	8.40	24.62
Percent, %	42.9	11.0	3.5	10.8	31.8
Standing C (Gt)	744	33.5	14.0	37.5	4.5
Percent, %	89.3	4.0	1.7	4.5	0.5

Each ecosystem on the earth is considered in terms of area, mean net carbon production per year, and standing biomass carbon. Standing biomass carbon is that contained in biomass on the earth's surface and does not include the carbon stored in biomass underground. Condensing these data in Table 2.2, the interpretation is earlier: of the total net carbon fixed on the earth each year, forest biomass, which is produced on only 9.5% of the earth's surface, contributes more than any other source. Marine sources of net fixed carbon are also high, as might be expected because of the large area of the earth occupied by water. But the high turnover rates of carbon in a marine environment result in relatively small steady-state quantities of standing carbon. In contrast, the low turnover rates of forest biomass make it the largest contributor to standing carbon reserves. According to the Table 2.2, the forests produce about 43% of the net carbon fixed each year and contain over 89% of the standing biomass carbon of the earth. Tropical forests are the largest sources of these carbon reserves. Temperate deciduous and evergreen forests are also major sources of biomass carbon. Next in order of biomass carbon supply would probably be the savanna and grasslands.

Terrestrial

Forest biomass

About one-third of the world's land area is forestland. Broad-leaved evergreen trees are a dominant species in tropical rain forests near the equator. In the northern hemisphere, stands of coniferous, softwood trees such as spruce, fir and larch dominate in the boreal forests at the higher latitudes, while both the broad-leaved deciduous hardwoods such as oak, beach and maple and the conifers such as pine and fir are found in the middle latitudes. Although the prime purpose is to produce wood fibre for the manufacture of paper products, the pulp and paper companies have operated large tree plantations that yield energy as a by-product for decades. Heat, steam and electricity are produced from wood wastes and also black liquor which is generated in the paper manufacturing process.

One of the tree species that has been studied in great detail as a renewable energy resource is the eucalyptus, evergreen hardwood trees that belong to the myrtle family. The eucalyptus is a rapidly growing tree native to Australia and New Guinea, and Hawaii for a variety of construction purposes. It appears to be a prime candidate for energy use because it reaches a size suitable for harvesting in about 7 years (Klass 1998).

Grasses

Grasses are very abundant forms of biomass. About 400 genera and 6000 species are distributed all over the world and grow in all land habitats capable of supporting higher forms of plant life (Klass 1998). Grass family (Gramineae) includes the great fruit crops, wheat, rice, corn, sugarcane, sorghum, millet, barley and oats. Grass also includes the many species of sod crops that provide forage or pasturage for all types of farm animals. Grasses are grown as farm crops, for decorative purposes, for preserving the balance of productive capacity of lands by crop rotation, for controlling erosion on sloping lands, for the protection of watersheds and for the stabilization of arid areas. Perennial grasses have been suggested as candidate feedstocks for conversion to synfuels. Most perennial grasses can be grown vegetatively, and they re-establish themselves rapidly after harvesting. Moreover, more than one harvest can usually be obtained per year.

Aquatic

With the exception of phytoplankton, which generally has lower net productivities, aquatic biomass seems to exhibit higher net organic yields than most terrestrial biomass. Aquatic biomass species that are considered to be the most suitable for energy applications include the unicellular and multicellular algae, freshwater plants and marine species.

Algae

Microalgae are prokaryotic or eukaryotic photosynthetic microorganisms that can grow rapidly and live in harsh conditions due to their unicellular or simple multicellular structure. Microalgae are present in all existing earth ecosystems, not just aquatic but also terrestrial, representing a big variety of species living in a wide range of environmental conditions. It is estimated that more than 50,000 species exist, but only a limited number, of around 30,000, have been studied and analysed (Mata et al. 2010). They have been under development as renewable energy resources and other useful products for a long time. For the past 50 years, extensive research has been performed on microalgae and how they can be used in a wide variety of processes or to manufacture many practical and economic important products. The first large-scale culture of microalgae started in the early 1960s in Japan by Nihon Chlorella with the culture of Chlorella. Other common types of microalgae are Dunaliella, Nannochloropsis, Scenedesmus and Spirulina. Microalgae cultivation can be done in open-culture systems such as lakes or ponds and in highly controlled closed-culture systems called photo-bioreactors (PBR). Open-culture systems are normally less expensive to build and operate, more durable than large closed reactors and with a large production capacity when compared with closed systems. On the other hand, PBRs are flexible systems that can be optimized according to the biological and physiological characteristics of the algal species being cultivated, allowing one to cultivate algal species that cannot be grown in open ponds. Depending on their shape or design, PBRs are considered to have several advantages over open ponds: offer better control over culture conditions and growth parameters (pH, temperature, mixing, CO_2 and O_2), prevent evaporation, reduce CO_2 losses, allow to attain higher microalgae densities or cell concentrations, higher volumetric productivities, offer a more safe and protected environment, prevent contamination or minimize invasion by competing microorganisms. Despite their advantages PBRs suffer from several drawbacks that need to be considered and solved. Their main limitations include: overheating, bio-fouling, oxygen accumulation, difficulty in scaling up, the high cost of building, operating and of algal biomass cultivation, and cell damage by shear stress and deterioration of material used for the photo-stage.

Nowadays, microalgae are seen as an alternative feedstock for biodiesel production. Recent research efforts have concentrated on applying metabolic engineering and genetic methods to microalgae in order to develop organisms optimized for high productivity and energy value, in order to achieve their full processing capabilities (Mata et al. 2010). Since microalgae do usually have no cell differentiation, genetic manipulations to increase its content of higher value compounds is very tempting. Nevertheless, progress in the genetic engineering of algae was extremely slow until recently and these promising advances should be viewed with caution because transgenic algae potentially pose a considerable threat to the ecosystem and thus will most likely be banned from outdoor cultivation systems.

Water Plants

The productivity of some salt marshes is similar to that of seaweeds. Their annual organic productivity on optimum sites is about 30–35 t/(ha-year) for Spartina alterniflora and emergent communities in brackish water (Klass 1998) but there are not sufficient information available to judge their value in biomass energy systems. Other species such as *Arundo donax*, *Scirpus lacustris* and Cattail, if can be sustained, should be suitable candidates for biomass energy usage.

2.1.1.2 Waste Biomass

Another large source of renewable carbon supplies is waste biomass. Waste biomass is energy-containing materials that are discarded or disposed of and that are mainly derived from or have their origin in virgin biomass. It is generated by anthropological activities and some natural events. As it is derived from primary biomass like trees, vegetables, meat during the different stages of their production or use, it is defined secondary biomass. It consists of a wide range of materials and includes municipal solid waste (MSW), municipal biosolids (sewage), industrial waste, animal manures, agricultural crops and forestry residues, landscaping and tree clippings and trash, and dead biomass that result from nature's life cycles. They are lower in cots than virgin biomass and often have negative costs.

Municipal Wastes

There are basically two types of municipal wastes that offer opportunities for combined waste disposal and energy recovery.

Municipal solid waste—MSW

As the populations of urban areas grow, the production of MSW increases, sometimes in a disproportional way. The collection and disposal costs increase and proper disposal becomes more difficult to achieve with the passage of time. At the same time, the loss of natural resources in the MSW occurs if no effort is made to recover them. The opportunities for combined waste disposal and energy recovery are thus evident. The major part of MSW comes from renewables like food scraps, lawn clippings, leaves and papers. Non-renewable components of MSW like plastics, glass and metals are not considered biomass. Thus, waste biomass is a potential energy resource in the same manner as virgin biomass.

To assess the **energy potential** from waste biomass on supplying energy demand, it is necessary to consider the amounts of different types of wastes generated, their energy contents and their availabilities. In 2013, in Europe, the total amount of MSW was about 243.2 Million of tonnes (ISPRA 2015). Every person, discards about 1.5 kg of MSW per day (2013) distributed as shown in Table 2.3.

From an energy standpoint, one short tonne of MSW has an as-received energy content of about 9.5 GJ, so about 4.2 EJ/year of energy potential resides in the MSW generated in Europe.

Biosolids

Municipal wastewater treatment plants in industrialized countries receive wastewaters from residential sources, industry, groundwater infiltration and stormwater runoff. The pollutants associated with these sources include a wide range of suspended and dissolved compounds and oxygen-demanding materials, many of which are toxic. Pathogenic components are present, including certain bacteria, viruses, organic compounds, inorganic nutrients and heavy metals. The purpose of most wastewater treatment processes is to remove or reduce these components, other pollutants and biological oxygen demand before discharge to receiving waters. Primary biosolids (settleable and suspended solids) are present at a level of a few percent in the influent wastewater.

The **energy potential** of municipal biosolids is small. At an average higher heating value (HHV) of 19.9 MJ/kg_{dm} (Klass 1998), the energy content of all the primary and treated biosolid produced in 2005 in Europe (Iranpour et al. 2000) can be estimated to be around 0.166 EJ/year, much less than the energy potential of MSW.

Table 2.3 MSW pro capite in Europe, 2013

Country	MW produced	MW treated	MW treated (%)			
	(kg/year) pro capite	(kg/year) pro capite	Recycling	Composting	Incineration	Landfill
UE28	481	470	28	15	26	31
UE15	521	517	29	16	29	26
Belgium	439	440	34	21	44	1
Bulgaria	432	428	25	3	2	70
Czech Republic	307	307	21	3	20	56
Denmark	747	747	28	17	54	2
Germany	617	617	47	17	35	0
Estonia	293	253	14	6	64	16
Ireland	586	531	34	6	18	42
Greece	510	510	16	4	0	81
Spain	449	449	20	10	10	60
France	530	530	21	17	34	28
Croatia	404	393	14	2	0	85
Italy	491	474	26	15	21	38
Cyprus	624	624	12	9	0	79
Leetonia	312	312	11	6	0	83
Lithuania	433	421	21	8	7	64
Luxemburg	653	653	28	20	35	17
Hungary	378	378	21	5	9	65
Malta	570	526	6	5	0	88
The Netherland	526	526	24	26	49	1
Austria	578	559	25	34	36	4
Poland	297	249	16	13	8	63
Portugal	440	440	13	13	24	50
Romania	272	220	3	0	0	97
Slovenia	414	287	55	7	1	38
Slovakia	304	278	4	8	12	77
Finland	493	493	19	14	42	25
Sweden	453	453	34	15	50	1
UK	482	476	28	16	21	35

Agricultural Solid Wastes

Application of animal wastes to land is one of the most economical choices for disposal as well as providing fertilizing benefits. However, the utilization of livestock and poultry manures as waste biomass resources for energy applications could help mitigate pollution and at the same time open new markets. Agricultural crop residues are also examined.

Livestock and poultry manures

Intuitively, high populations of specific animals would be expected to offer the greatest opportunity to serve as sources of waste biomass because waste generation is maximized. Because of the relationship of waste productivity and animal size, this is not always the case as will be shown here. Domestic farm animals and those confined to feedlots are appropriate choices. The animals that produce large, localized quantities of excreta are cattle, hogs and pigs, sheep and lambs and poultry.

The **energy potential** of each category can be derived from the heating values estimated by Stanford Research Institute (Stanford Research Institute 1976) and reported in Table 2.4.

Agricultural crop residues

Agricultural crop residues are those left in the field or accumulated during sorting and cleaning of produce. Being a predominantly agricultural-based economy, large quantities of biomass residues from the crop are generated throughout the country. It is a common practice to burn the residues in cultivated fields as a means of agricultural land preparation. Residues that are also generated from agricultural processing facilities are burdensome to processors because of costs incurred in their disposal. However, their use as an energy source is still very limited in the country. One of the most important steps in developing biomass energy supply from residue is to evaluate their spatial and temporal availability. Such an analysis would provide useful information for decision-makers on the opportunities for using biomass residues for energy application in the country.

Here the focus is on biowastes related to some of the main food crops of Southern Europe, which are known to yearly generate very large amounts of biowastes according to the Food and Agriculture Organization of the United Nations (FAO). FAO reported that the top five primary earth products of South Europe in 2012 were wheat, grapes, tomatoes, sugar beet and olives, which were produced as shown in Table 2.5 together with the heating value.

Because of the discontinuity in growing seasons, many crops that are grown, differences between specific crops, variation in crop yields in different areas,

Table 2.4 Heating values of livestock/poultry (Stanford Research Institute 1976)

Livestock/poultry	Heating value (MJ/dry kg)
Cattle	15.73
Hogs and pigs	16.99
Sheep and lambs	17.82
Chickens	13.53
Commercial broilers	13.53
Turkeys	13.49

Table 2.5 Production of target crops and related wastes in South Europe according to reviewed literature (FAO 2009 and 2012)

Earth product	Mt	Solid wastes	Mt	Heating value (MJ/kg)
Olives	9.31	Olive pomace	4.5–10	14.6–16.7 (Intelligent Energy EUROPE 2010)
Grapes	13.92	Grape pomace	1.3–2.8	16.1–18.9[a] (Burg et al. 2016)
Tomatoes	12.13	Tomato pomace	0.1–0.2	23.77 (Tillman and Jahn 2016)
Sugar beets	9.76	Sugar beet pulp	0.5–1.5	3.20 (Akram et al. 2015)

[a]Lower calorific values were determined in the pomace after separation of the seeds: 14.60–17.75 MJ/kg, whereas the highest calorific values were determined in the seeds themselves, i.e. 19.78–21.13 MJ/kg

difficulty of acquiring reliable data and long-term time effects, an inventory of the annual production of agricultural crop residues and their disposition might seem to be an impossible task. Fairly reliable data can, however, be obtained estimating a residue factor defined as the ratio of a field weight of residue per mass unit of crop yield. The quantity of residue generated will be the product of the residue factor and the country yield total for that crop.

To estimate the **energy potential** of crop residues, several parameters are needed, namely, annual crop production and the residue availability, dry weight and ash factors. In literature, for the selected residues, the energy potential is shown in Table 2.5.

Forestry Residues

Forestry residues consist of slash left on the forest floor following logging operations; stems, stumps, tops, foliage and damaged trees that are not merchantable, wood and bark residues accumulated at primary wood manufacturing plants during production of lumber. Underground tree roots can also be included in the list of forestry residues. The difficulty of accurately assessing the amounts of forestry residues that are and can be realistically collected and utilized as waste biomass for an entire country has been encountered by almost all who have embarked on the task (Klass 1998). Regional forest offices, for example, do not keep and maintain detailed records of residues production and its disposition; surveys that are done periodically and for different countries for a given time period are subject to considerable error.

Table 2.6 lists the composition and heating values of some waste biomass products.

2.1.2 *Chemical–Physical Composition*

A proper understanding of the physical and the chemical properties of biomass feedstock is essential for the design of a reliable biomass conversion system.

Table 2.6 Typical composition of some waste biomass (Basu 2013)

Biomass	Moisture (wt%)	Organic matter (dry wt%)	Ash (dry wt%)	HHV (MJ/dry kg)
Cattle manure	20–70	76.5	23.5	13.4
Sewage	90–98	73.5	26.5	19.9
RDF	15–30	86.1	13.9	12.7
Sawdust	15–60	99.0	1.0	20.5

Table 2.7 Typical composition of different biomass species (Adapted from Vigouroux 2001)

Species	Cellulose	Hemicellulose	Lignin	Extractives	Ash
Scandinavian Birch[a]	40.0	39.0	21.0	n.a.	0.3
Softwood[a] (average)	45.8	24.4	28.0	n.a.	1.7
Hardwood[a] (average)	45.2	31.3	21.7	n.a.	2.7
Scandinavian pine	40.0	28.5	27.7	3.5	n.a.
Wood bark[a]	24.8	29.8	43.8	n.a.	1.6
Bagasse[b]	41.8	22.9	18.5	13.9	2.9
Coconut shell[b]	36.6	25.3	29	8.4	0.7
Corn stalks[b]	42.7	23.6	17.5	9.8	6.8
Olive husk[a]	24.0	23.6	48.4	n.a.	4.0
Wheat straw[b]	34.1	32.3	18.3	2.7	12.5
Rice husk[b]	30.7	23.9	14	8.3	23.1
Rice straw[b]	43.9	27	3.1	2.5	23.5

n.a. not analysed
[a]wt% dry and extractive free basis
[b]wt% dry basis

This chapter discusses some important properties of biomass that are relevant to such processes.

As far as it regards chemical–physical composition, the main materials constituting plant biomass are lignin and carbohydrates, structural support of wood plants. Carbohydrates are also divided in cellulose and hemicellulose. Long polymers of cellulose constitute the fibres which provide mechanical strength for plant structure, while lignin works as fibre glue; hemicellulose is the cementing matter of the wood parts. There may be also a broad variety of chemical compounds (resins, waxes, fats, oils, amid, sugar, tannic substances and pigments, etc.) whose name derives from the capability of being extracted with solvents. There are also small quantities of inorganic matter constituted with alkaline species (Na, K, Mg, Ca), heavy metals (Cd, Zn, As, Pb, Cu, etc.) and S, Cl, N, P, Si, Al, etc.; the remaining matter of combustion is ash. Table 2.7 points out the composition of some important species for food or industrial production, while the Table 2.8 points out variation ranges of the three main components.

Table 2.8 Cellulose/lignin content of selected biomass (wt%) (McKendry 2002)

Biomass	Lignin (%)	Cellulose (%)	Hemicellulose (%)
Softwood	27–30	35–40	25–30
Hardwood	20–25	45–50	20–25
Wheat straw	15–20	33–40	20–25
Switchgrass	5–20	30–50	10–40

2.1.3 Composition of Biomass

Biomass contains a large number of complex organic compounds, moisture, and a small amount of inorganic impurities known as ash. The organic compounds comprise four principal elements: carbon (C), hydrogen (H), oxygen (O) and nitrogen (N). Biomass (e.g. MSW and animal waste) may also have small amounts of chlorine (Cl) and sulphur (S). The latter is rarely present in biomass except for secondary sources like demolition wood, which comes from torn-down buildings and structures.

Thermal design of a biomass utilization system, whether it is a gasifier or a combustor, necessarily needs the composition of the fuel as well as its energy content. In the context of thermal conversion like combustion, the following two types of compositions are mostly used:

1. Proximate composition.
2. Ultimate or elemental composition.

Besides these, there is also the polymeric composition of biomass, which is important for chemical conversions like torrefaction, pyrolysis and gasification.

Important properties for energy conversion processes of woody and herbaceous plants are:

- The moisture content: high moisture is a major characteristic of biomass. The root of a plant biomass absorbs moisture from the ground and pushes it into the sapwood. The moisture travels to the leaves through the capillary passages. Photosynthesis reactions in the leaves use some of it, and the rest is released to the atmosphere through transpiration. For this reason, there is more moisture in the leaves than in the tree trunk.
- Heating value: it represents the heating value when the biomass is combusted. This heating value can be defined in different two ways: HHV and lower heating value (LHV).
- Proportion of fixed carbon and volatiles: the volatile matter of a fuel is the condensable and non-condensable vapour released when the fuel is heated. Its amount depends on the rate of heating and the temperature to which it is heated.
- The ash/residue content: Ash is the inorganic solid residue left after the fuel is completely burned. Its primary ingredients are silica, aluminium, iron and calcium; small amounts of magnesium, titanium, sodium and potassium may also be present. The ash content of biomass is generally very small but may play a

Table 2.9 Average property values for specified biomass groups (Phyllis ECN)

Property	Woody plants	Herbaceous plants	Aquatic plants	Wastes
Example	*Untreated wood*	*Grass/plant*	*Algae*	*Manure*
Moisture content (wt% wet)	18.6	29.8	31.9	74
Lower heating value[a] (kJ kg^{-1})	18,772	18,298	23,147	18,906
Fixed carbon[a] (wt%)	18.1	17.5	14.8	19.1
Volatile matter[a] (wt%)	81.9	82.5	85.2	80.9
Ash content[b] (wt%)	2.2	6.9	6.1	28.5
Cellulose content[b] (wt%)	39.8	43.9	30	23.1
Hemicellulose content[b] (wt%)	23.3	19.7	35	26.7
Lignin content[b] (wt%)	24.8	10.9	–	11.3

[a]The values are on dry ash free (daf) basis
[b]The values are on dry basis

significant role in biomass utilization especially if it contains alkali metals such as potassium or halides such as chlorine. Straw, grasses and demolition wood are particularly susceptible to this problem. These components can lead to serious agglomeration, fouling and corrosion in boilers or gasifiers.

- The alkali metal content: many biomass types among the agricultural residues show high contents of alkali salts, in particular potassium based (straw, perennial grasses). When the temperature exceeds approximately 700 °C, these (eutectic) salts are evaporated into the gas phase (Stevens 2001).
- The cellulose/lignin ratio: the cellulose and lignin are important in biochemical processes, because the biodegradability of cellulose is greater than of lignin. This is important for the selection of plants for biochemical conversion.

In dry biomass conversion processes, the first five properties are important. While for wet biomass conversion processes, the first and the last property are of more importance.

Because of the diverse nature of biomass, properties in these categories can vary widely (Williams et al. 2001; Khan 2007). In Table 2.9, some average values of properties are given for specified biomass groups (Phyllis ECN).

2.1.3.1 "Proximate" Analysis

Proximate analysis gives the composition of the biomass in terms of gross components such as moisture, volatile matter, ash and fixed carbon. It is a relatively simple and inexpensive process.

Two fundamental characteristics which discriminate the technology suitable for the production of energy coming from biomass are the humidity content and volatile matter. *Moisture* can be present in biomass in two ways:

- Intrinsic is the moisture captured within the biomass, e.g. the content of water which does not depend on whether condition since it is deeply linked to the structure of the biomass. It is hard to remove;
- Extrinsic is the moisture which is influenced by the weather condition during harvesting, e.g. the water content determined from the weather condition present during the harvest. This type of moisture is easy to remove.

The extrinsic moisture content is of special interest, since the intrinsic moisture content is hard to change (McKendry 2002). A possible drying (both natural and artificial) of biomass mainly reduces the extrinsic humidity; the intrinsic humidity is the limit up with the drying can go through. Drying of biomass is energy intensive, so if possible it should be avoided. Often the composition of biomass is pointed out without making any distinction between the two types of humidity, reporting only the total value.

The *fixed carbon* is the mass remaining after the release of all the volatiles, excluding the ash and moisture content. According to "EN 15148-2009 solid biofuels—determination of the content of volatile matter", the volatile matter is the portion of released gas by heating of the solid biofuel to 900 ± 10 °C for 7 min. *Volatile matter* is the fraction which volatilizes after heating (without any chemical attack). Being the reactivity of volatile gases much higher than the solid matrix, the quantity of free volatile matter (and the release velocity) is an important factor for the choice and dimensioning of thermos-chemical systems (pyrolysis, combustion, gasification). The fixed carbon together with the volatile matter is measure for the ease if ignition and further gasification or oxidation.

The fraction which still remains after having kept the biomass at a temperature of 900/950 °C for several minutes, is composed of *ashes* and "fixed carbon". In other words, the ash content is the amount of solid residue left over when the biomass is completely combusted. The ash content affects both the handling and the processing costs of the overall biomass energy conversion (McKendry 2002). The analysis used in laboratory to determine humidity, volatile matter, fixed carbon and ashes is called proximate because it consists of the simply furnace heating in an inert atmosphere. Table 2.10 reports some examples of proximate analysis of biomass fuels, biomass residues and coals in terms of percentages of moisture, volatile matter (VM), fixed carbon (FC) and ash.

2.1.3.2 Elementary Composition

In the thermo-chemical processes typically adopted for the energy production from dry biomass (combustion, gasification, pyrolysis, liquefaction), the chemical-structural components described at Sect. 2.1.2, are separated and destroyed. The composition of the combustion (or gasification or pyrolysis) products and the mass and energy balances exclusively depend on the atomic species content of the fuel, that is, on its "ultimate composition". In addition to the humidity, the biomass volatile matter is primarily composed by carbon, hydrogen

Table 2.10 Typical proximate analysis of various biomass fuels, biomass residues and coals (De Jong and Van Ommen 2015)

Fuel	Proximate analysis (wt% as received a.r.)			
	Moisture	VM	FC	Ash
Wood pellets, clean	4.9	80.4	14.5	0.2
Wood pellets, demolition	9.1	69.6	19.7	1.7
Wheat straw	8.5	69.5	15.0	7.0
Sunflower pellets	11.2	65.2	19.5	4.1
Olive cake pellets	11.9	64.2	15.7	8.2
Pepper plant residue	6.5	60.5	19.5	13.5
Greenhouse residue	2.5	61.0	5.50	31.0
Sewage sludge, dried	8.7	47.2	4.7	39.4
MBM[b]	2.5	61.7	12.4	23.4
Microalgae	5.2	77.5	14.9	2.4
Macroalgae (seaweed)	8.0	42.4	19.5	30.1
Peat, (young surface), dry	12.5[a]	63.1	22	2.4
Brown coal, dried	15.6	44.1	36.0	4.3
Bituminous coal	5.5	30.1	56.6	7.8
Anthracite	1.9	7.6	87.7	2.8

[a]Average value
[b]Meat and bone meal

and oxygen with less content of nitrogen, sulphur and possibly, other elements such as sodium, potassium, chlorine, etc. Tables 2.11, 2.12 and 2.13 show some examples of ultimate composition of biomass and carbon and where available, their heating value.

With respect to the fossil fuels, biomass is characterized by:

- Less carbon content;
- Higher hydrogen and especially oxygen content;
- Significant less nitrogen and sulphur content;
- Less ashes content.

Coal contains between 75 and 90% carbon while biomass carbon content is about 50%. This means that the heating value of biomass is lower. Biomass fuels contain more volatile components and are more reactive than coal. These differences are explained by the O/C and H/C ratios of each fuel, shown in the Van Krevelen diagram.

Figure 2.1 shows according to the original Van Krevelen diagram, the variation ranges of oxygen and carbon in the volatile matter of biomass and the more common fossil fuels with their heating values calculated accordingly to Dulong–Berthelot equation (reported below).

Table 2.11 Typical ultimate analysis of various biomass fuels, biomass residues and coals (De Jong and Van Ommen 2015)

Fuel	Ultimate analysis (wt% dry and ash free (d.a.f.))					
	C	H	O	N	S	Cl
Wood pellets, clean	45.6	6.6	47.8	BDL	BDL	BDL
Wood pellets, demolition	51.2	7.1	40.6	1.0	BDL	0.1
Wheat straw	47.3	5.5	45.3	0.9	0.5	0.5
Sunflower pellets	52.1	6.1	41.0	0.6	0.1	0.1
Olive cake pellets	52.7	6.3	38.9	1.6	0.1	0.4
Pepper plant residue	42.3	5.0	48.9	3.1	0.6	0.1
Greenhouse residue	70.8	11.1	16.4	1.5	BDL	0.2
Sewage sludge, dried	54.3	7.7	27.4	8.4	2.2	ND
MBM[b]	56.6	8.0	20.6	12.0	1.7	1.1
Microalgae	54.1	7.4	29.6	8.2	0.5	0.2
Macroalgae (seaweed)	41.1	7.5	46.2	5.2	6.3[a]	21.7[a]
Peat, (young surface), dry	52.6	5.8	40.6	0.9	0.1	0.0
Brown coal, dried	56.3	5.0	37.6	0.6	0.4	0.1
Bituminous coal	82.3	5.1	10.3	1.4	0.8	0.1
Anthracite	91.6	3.5	2.4	1.6	0.8	0.1

ND, not determined; BDL, below the lower detection limit
[a]wt% in dry ash (S and Cl contents can be high)
[b]Meat and bone Meal

Table 2.12 Elementary composition and high heating value (HHV) of some substances (Vigouroux 2001). The HHV value in the right column is specific to kg of dry ash free matter, the other one is specific to dry matter kg, ash inclusive (Consonni 2006)

Biomass	%Moisture and a.f.			%mf			HHV		
	C	H	O	N	S	TOT	Ash	MJ/kg_{dm}	MJ/kg_{daf}
Birch	48.8	6.0	44.2	0.5	0.01	99.5	0.5	20.0	20.1
Poplar	49.8	5.9	44.4	0.6	0.04	100.7	1.9	18.3	18.7
Trembling poplar	48.8	6.0	44.7	0.5	0.01	100.0	0.5	20.0	20.1
Pine	49.3	6.0	44.2	0.5	0.01	100.0	0.5	20.1	20.2
Bark	47.2	5.6	46.9	0.3	0.07	100.1	3.9	20.9	21.7
Straw (Denmark)	49.6	6.2	43.6	0.6	n.a.	100.0	4.7	18.6	19.5
Miscanthus (Germany)	49.7	6.1	43.9	0.3	n.a.	100.0	2.3	18.4	18.8
Miscanthus (Italy)	49.5	6.2	43.7	0.6	n.a.	100.0	3.3	18.5	19.1
Sugar cane (Germany)	49.5	6.2	43.8	0.5	n.a.	100.0	3.7	18.5	19.2
Cane from Canarie	49.4	6.3	42.7	1.6	n.a.	100.0	8.8	18.8	20.6

n.a.: not analysed; dm: dry matter; daf: dry ash free; mf: moisture free

Table 2.13 Ultimate analyses for typical biomass materials (wt%) (McKendry 2002)

Material	C	H	O	N	S	Ash	TOT
Cypress	55.0	6.5	38.1	–	–	0.4	100.0
Ash	49.7	6.9	43.0	–	–	0.3	99.9
Beech	51.6	6.3	41.4	–	–	–	99.3
Miscanthus	48.1	5.4	42.2	0.5	<0.1	2.8	99.0
Wheat straw	48.5	5.5	41.6[a]	0.3	0.1	4.0	100.0
Barley straw	45.7	6.1	38.3	0.4	0.1	6.0	96.6
Rice straw	41.4	5.0	39.9	0.7	0.1	12.9[a]	100.0
Bituminous coal	73.1	5.5	8.7	1.4	1.70	9.0	99.4

[a]100's complement of other elements percentage

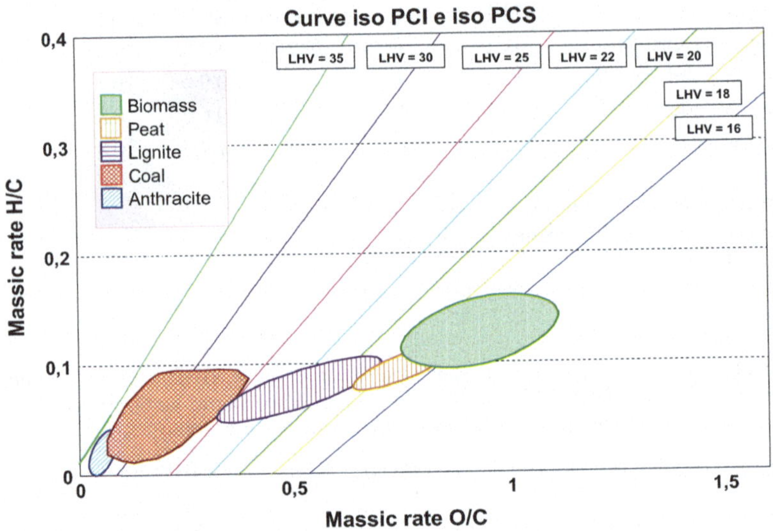

Fig. 2.1 Variation ranges of oxygen and carbon in the volatile matter of biomass and the more common fossil fuels. The LHV (MJ/kg) reported is dry and ash free and is calculated according to Dulong–Berthelot equation assuming nitrogen and sulphur content of 0.5 and 0.1%, respectively

2.1.4 *Heating Value*

The heating value of a fuel stands for the heat that can be released by its complete oxidation under standard conditions. Such a heat depends on the physical state of the water present in the combustion products, hence the distinction between high heating value (HHV) and low heating value (LHV). The first one corresponds to the complete condensation of water vapour in the combustion products; the second one to the case of water present as vapour in the products. Since this one is by far the most common case, in the devices used for energy production from biomass, LHV should be the reference value. Obviously, an exception may be given to the technologies where water vapour is condensed (e.g. boilers running on natural gas) that is a rare case for appliances powered with biomass.

The heating value of dry matter of some biomass is reported in Table 2.12. Compared to the heating value of fossil fuels, the heating value of biomass is noticeably lower, mainly because of the huge presence of oxygenated bonds.

In practical applications, the main energy parameter is the heating value per kg of humid biomass; given the high moisture content typical of biomass, the heating value is generally significant lower than the one of the dry matter reported in Table 2.12. The discrepancy is higher for LHV than for HHV because, in order to have the water as vapour in the combustion products, a huge quantity of heating is necessary for the evaporation process.

Given the variety of biomass in terms of composition and physical state, experimental tests are needed in order to determine its heating value. HHV is experimentally determined using a so-called bomb calorimeter, a constant volume calorimeter, in which in a closed vessel a fuel portion is oxidized using pure oxygen. In this device, the heat transferred to a precisely known amount of water is measured by its temperature increase. The sample is ignited electrically. This is a standard method (e.g. DIN 51,900).

When these data are not available, empirical correlations based on ultimate analysis can give a first-attempt value. One of the most reliable is the Dulong–Berthelot equation:

$$\mathrm{HHV}\left[\frac{\mathrm{kcal}}{\mathrm{kg_{dvm}}}\right] = 8137 \cdot y_{\mathrm{C}} + 34{,}500 \cdot \left(y_{\mathrm{H}} - \frac{y_{\mathrm{O}} + y_{\mathrm{N}}}{8}\right) + 2250 \cdot y_{\mathrm{S}} + 43.125 \tag{2.1}$$

where y_C, y_H, y_O, y_N and y_S are, respectively, the mass fractions of carbon, hydrogen, oxygen, nitrogen and sulphur in the dry ash free matter (daf), expressed in kg of each atomic specie per kg of daf, or per kg of dry biomass ash free. Considering the relation between HHV and LHV, from Eq. (2.1), LHV is derived:

$$\mathrm{LHV} = \mathrm{HHV} - \left(\frac{W_{\mathrm{H2O}}}{2 \cdot W_{\mathrm{H}}} \cdot y_{\mathrm{H}}\right) \cdot \Delta h_{\mathrm{ev}} = HHV - 9 \cdot y_{\mathrm{H}} \cdot \Delta h_{\mathrm{ev}} \tag{2.2}$$

where W_{H2O} and W_H are, respectively, the molecular weight of water and hydrogen, y_H is always the hydrogen mass fraction (kg_H/kg_{daf}) and Δh_{ev} the heat for water evaporation at standard conditions adopted for the LHV definition (at 25 °C, Δh_{ev} = 2.4425 MJ/kg_{H2O}). Replacing in (2.2) and using the I.S. units:

$$\mathrm{LHV}\left[\frac{\mathrm{MJ}}{\mathrm{kg_{dvm}}}\right] = 34.061 \cdot y_{\mathrm{C}} + 122.43 \cdot y_{\mathrm{H}} - 18.052 \cdot (y_{\mathrm{O}} + y_{\mathrm{N}}) + 9.419 \cdot y_{\mathrm{S}} + 180.52 \tag{2.3}$$

Based on this expression, the lines LHV-constant of Fig. 2.1 were plotted, assuming respectively for nitrogen and sulphur a concentration of 0.5 and 0.02%.

The heating value LHV_{ar} of the biomass as received is derived from the LHV_{daf} multiplying per kg_{daf}/kg_{ar}. Expressing the ash and moisture content as y_{ash} (kg ash/kg biomass as received) and y_m (kg H_2O/kg biomass as received):

$$HHV_{ar} = HHV_{dvm} \cdot (1 - y_{ash} - y_m) \tag{2.4}$$

$$LHV_{ar} = LHV_{dvm} \cdot (1 - y_{ash} - y_m) - y_m \cdot \Delta h_{ev} \tag{2.5}$$

Or, for LHV:

$$\frac{LHV_{ar}}{LHV_{dvm}} = (1 - y_{ash}) - y_m \cdot \left(1 + \frac{\Delta h_{ev}}{LHV_{dvm}}\right) \tag{2.6}$$

Being Δh_{ev} of about 2.45 MJ/kgH_2O and the LHV_{daf} of woody biomass typically between 18 and 19 MJ/kg_{daf}, their rate is about 2/15 = 0.133.

In the systems where the combustion products are discharged as gaseous products (that is almost the entire thermochemical processes of any practical interest), the heat needed to evaporate the humidity implies a reduction of energy available in a quantity equal to:

$$\text{Fraction heat lost} = \frac{y_m \cdot \Delta h_{ev}}{LHV_{dvm} \cdot (1 - y_{ash} - y_m)} \tag{2.7}$$

Figure 2.2 shows how the fraction of heat lost varies varying the humidity content. When the moisture is about 60%, the heat needed for the evaporation exceeds the available energy of more than 10%; for even higher values, the fraction of heat lost increases rapidly, reaching 100% when the humidity is about 90–95%.

When the moisture content of biomass as received makes the LHV too low for any thermochemical process, biological processes are promoted.

2.2 Biomass Productivity and Energy Value

The production of biomass for energy purposes is simply a method to use the solar energy. As such, it is comparable to other kinds of use of the same source: photovoltaic, solar thermal, eolic energy and hydroelectric.

The use of solar energy through biomass may take different times, from some months for herbaceous plantations to some decades for plants to high stalk. Being these times comparable with the cycle of human activities, biomass represent a source of renewable energy, which can be indefinitely recollected until solar radiation and suitable climatic conditions will be available on our planet.

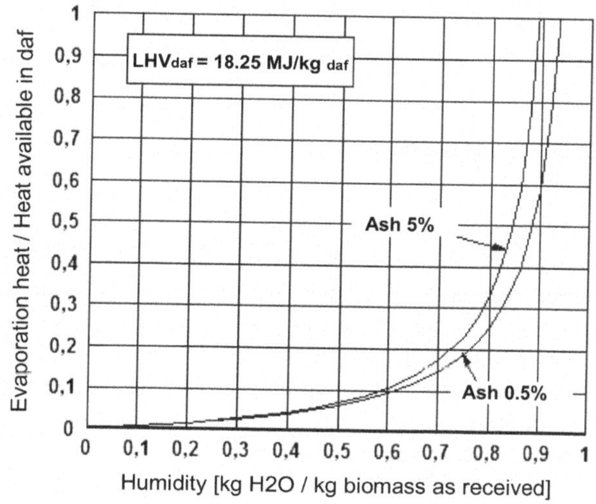

Fig. 2.2 Heat lost for moisture evaporation depending on the moisture content in the biomass as received (Consonni 2006)

2.2.1 Chlorophyll Photosynthesis

The accumulation of chemical energy in biomass is the result of the complex chain of reactions defined as chlorophyll photosynthesis. The prime engine of such a mechanism is solar radiation especially the radiation visible with a wavelength within the range between 0.4 and 0.7 nm[6]. Photosynthesis develops in cellular organelles defined as chloroplasts by chlorophyll, the green pigment responsible for the capture of radiation and for a complex series of enzymes and catalysts like nitrogen, phosphorus, potassium and calcium. Among all compounds, ATP (adenosine-tri-phosphate) and ADP (adenosine-di-phosphate) assume an important role. The process has two steps, a dark phase and a bright one: such a definition does not relate to the presence of light, but simply to the role of solar irradiation, which is fundamental for the bright phase but irrelevant for the dark one. During the bright phase radiation promotes the formation of hydrogen ions and of free electrons, charging with energy the ATP derived from ADP; during the dark phase, ATP recovers the energy accumulated by returning to ADP status and promoting the synthesis of the bound (CH_2O), which is itself the essential building block for the composition of carbohydrates $(CH_2O)_n$ and $C_n(H_2O)_m$. Overall, such a mechanism can be represented by the following equation:

$$n \cdot CO_2(g) + n \cdot H_2O(l) \longrightarrow \text{light} + \text{chlorophyll}(CH_2O)_n + n \cdot O_2(g), \quad (2.8)$$

where (g) and (l) specify the reagent physical state (both gas and liquid). At reference conditions of 25 °C, the energy stored in this process is about 470 MJ per mole of CO_2 converted in CH_2O (Klass 1998).

The accumulation of chemical energy in the carbohydrates formed during chlorophyll photosynthesis resolves two main issues of renewable energies: the intermittence of the source and the storage (Klass 1998). The chlorophyll photosynthesis is the way to convert solar energy in a chemical energy storage using a break-off process.

2.2.2 *Efficiency*

To synthesize CO_2 in CH_2O, eight photons of solar radiation are needed. The maximum efficiency with which photosynthesis can occur has been estimated by several methods. The upper limit is the maximum amount of solar energy trapped as chemical energy in the biomass, that is 8–15% of energy content of the incident solar radiation (Klass 1998). A first method to estimate the photosynthesis efficiency is the rate between the chemical energy stored through (2.8) and the energy the photons need to promote this reaction. The relationship of the energy and frequency of a photon is given by:

$$e_f = h \cdot \frac{c}{\lambda}, \tag{2.9}$$

where

e_f energy content of one photon, J;
h Planck's constant, 6.624×10^{-34} J s;
c velocity of light, 3.00×10^8 m/s;
λ wavelength of light

The eight photons absorbed during the reaction, have different wavelengths, thus, to correctly estimate their energy content, each photochemical elementary reaction should be considered (Twidell and Weir 2006). A first-attempt value for the total energy absorbed from the solar irradiation is given, assuming the same wavelength for all the photons and equal to 575 nm (the average value of the visible wavelength). The energy absorbed in the fixation of 1 mol CO_2, which requires eight photons per molecule, is then given by:

$$8\left[\frac{(6.624 \times 10^{-34}\,\mathrm{J\,s}) \cdot (3 \times 10^8\,\mathrm{m/s})}{575 \times 10^{-9}\mathrm{m}}\right] \cdot 6.023 \times 10^{23} = 1665\,\mathrm{kJ}, \tag{2.10}$$

where 6.023×10^{23} is the Avogadro number. Since 470 kJ of solar energy is trapped as chemical energy in this process, the maximum efficiency for total white light absorption is 28.2%. Further adjustments are usually made to take into account inefficiencies that can occur. Following Klass' path (Klass 1998), there are three main reasons of inefficiencies:

- The fraction of photosynthetically active radiation in solar radiation that reaches the earth is estimated to be about 43%. The energy of the visible radiation in fact, that is the number of photons with a wavelength between 400 and 700 nm, represents only part of the solar radiation energy whose spectrum spread along ultraviolet ($\lambda < 400$ nm) and infrared ($\lambda > 700$ nm) zone;

- The fraction of the incident light absorbed is a function of many factors such as leaf size, canopy shape and reflectance of the plant. It is estimated to have an upper limit of 80%;
- A portion of the stored energy is used by the plant, the amount of which depends on the properties of the particular biomass species and the environment. For purposes of calculation, Klass (1998) assumes that about 25% of the solar energy trapped as chemical energy is used by the plant, thereby resulting in an upper limit for retention of the no-respired energy of 75%.

From these considerations, the upper limit for the efficiency of photosynthetic fixation of biomass is as follows:

$$\text{Max capture efficiency} = \eta_{\max} = 0.282 \cdot 0.43 \cdot 0.80 \cdot 0.75 = 7.2\%$$

From this estimation, it is possible to calculate, depending on isolation, the upper productivity value in ideal conditions as:

$$P = \frac{\eta_{\max} \cdot I}{\text{HHV}_{\text{dvm}}}, \tag{2.11}$$

where

P	biomass produced ($\text{kg}_{\text{daf}}/\text{m}^2$ year)
I	insolation
HHV_{daf}	High heating value of the dry ash free matter produced (J/kg_{daf})

The use of HHV is coherent with the fact that the photosynthesis reaction goes with liquid water and the energy store refers to that physical state.

For example, at 45° Nord latitude (like Milan), the average annual insolation is more or less t 4500 MJ/m^2 year (1250 kWh/m^2 year). Assuming a HHV_{daf} of 20 MJ/kg and ideal conditions in terms of isolation, weather, water and nutriments availability, the productivity will be around 16 $\text{kg}_{\text{daf}}/\text{m}^2$ year, or 160 tonn_{daf} per hectare per year. As explained in the further paragraph, the real productivity is more than ten times less.

2.2.3 Production and Biomass Energy Yield

At real cultivation conditions, the capture yield of solar energy and the productivity are a lot less than the forecasted value. This is due to a lot of reasons:

- The fraction of incident light really absorbed and used for the photosynthesis can be significantly lower than the values used for $\eta_{\max}$ estimation;
- The capacity of completing the photosynthesis and producing storage matter strongly vary with the vegetal specie;

- For caduceus leaves species, the photosynthesis does not take place all the yearlong but only for a part of it;
- The real availability of water and nutriments limits the growing of vegetables organisms below the potential offered by the photosynthesis;
- Illness and parasitic limit the growth below the potential level.

A Dutch study realized by Nonhebel (Nonhebel 1997) for 58 European areas, shows the productivity of three interesting species used for energy production: the willow, the eucalyptus and the wheat. Willow and eucalyptus are interesting for energy production and wheat for food production. In Figs. 2.3, 2.4, and 2.5 the results are showed. We have three series of points, each one for a different growth condition:

- The "real" series is the effective productivity in the European agricultural;
- The "max without irrigation" series shows the productivity in ideal condition of fertilizings supply and illnesses and parasites lack;
- The "max potential" series shows the productivity in ideal conditions of water availability and without illnesses or parasites.

Over a latitude of 45° the last two series coincide. This means that the water availability is a limiting factor only for the Southern European agriculture (like Portuguese, Italian, Greek and Spanish agriculture). We can also notice a huge

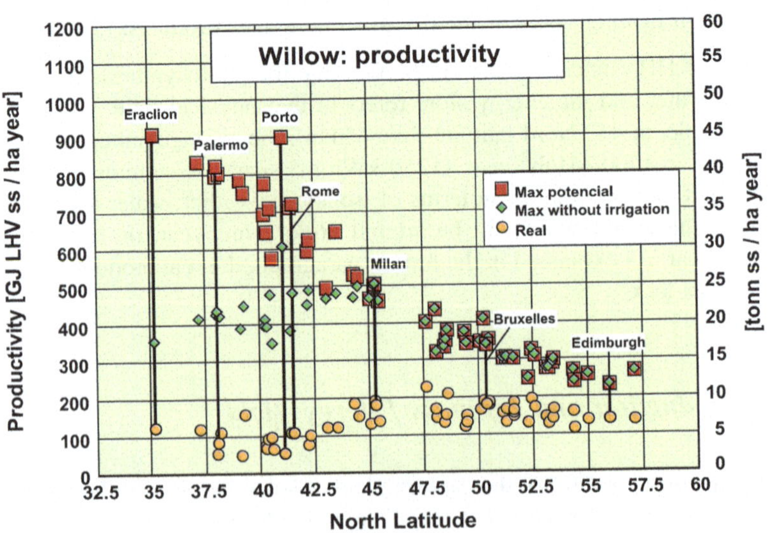

Fig. 2.3 Willow productivity in (GJ of LHV_{ss}/(ha-year)) and in (tonnes/(ha-year)) for different European areas at different latitude studied by Nonhebel (1997). For each 58 fields the figure shows the real productivity, the maximal one without irrigation and the best one. To avoid confusion, the figure represents only the name of some of the considered areas. The conversion between mass and energetic productivity is done assuming a LHV_{dm} = 19.8 MJ/kg_{dm} (Consonni 2006)

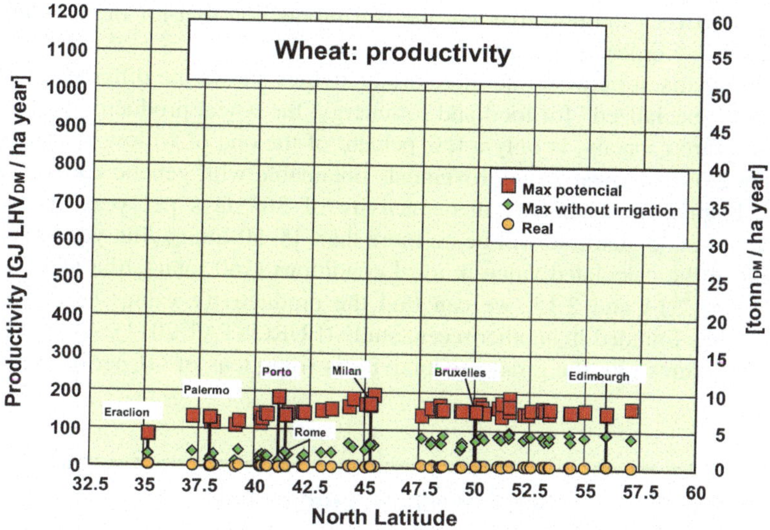

Fig. 2.4 Wheat productivity in the same European areas already considered in Fig. 2.3. The conversion between mass and energetic productivity is done assuming a $LHV_{dm} = 15.1\ MJ/kg_{dm}$ (Consonni 2006)

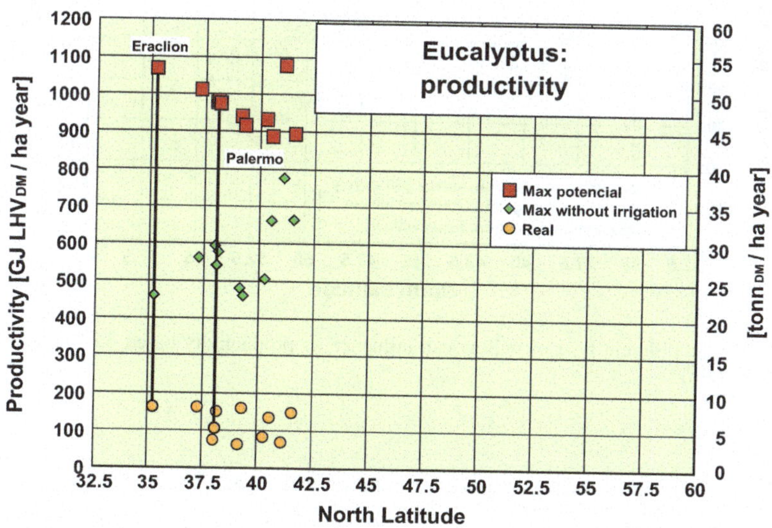

Fig. 2.5 Eucalyptus productivity in the same European areas already considered in Fig. 2.3. The conversion between mass and energetic productivity is done assuming a $LHV_{dm} = 18.1\ MJ/kg_{dm}$ (Consonni 2006)

difference between the potential and the real production that for the low latitudes can exceed one order of magnitude.

The comparison between the three productivities shows the difference between organisms "specialized" for food and for energy: the wheat productivity, that is its energy capture capacity, is only a few percent of the one of willow or eucalyptus.

Figure 2.6 quantifies the improvements obtainable with genetic changes of the willow. Even with a photosynthetic activity of 365 days per year, the willow productivity at 45° latitude will be no more than 45–50 $tonn_{dm}$/(ha year), less than one-third of the calculated value in ideal conditions (160 $tonn_{dm}$/(ha year)).

In Tables 2.14 and 2.15, we can find the productivity values of the oil and starchy plants founded in another recent study (EUROSTAT 2011). As we can see, for these cultures too, the gross productivity is some tens of GJ per ha per year.

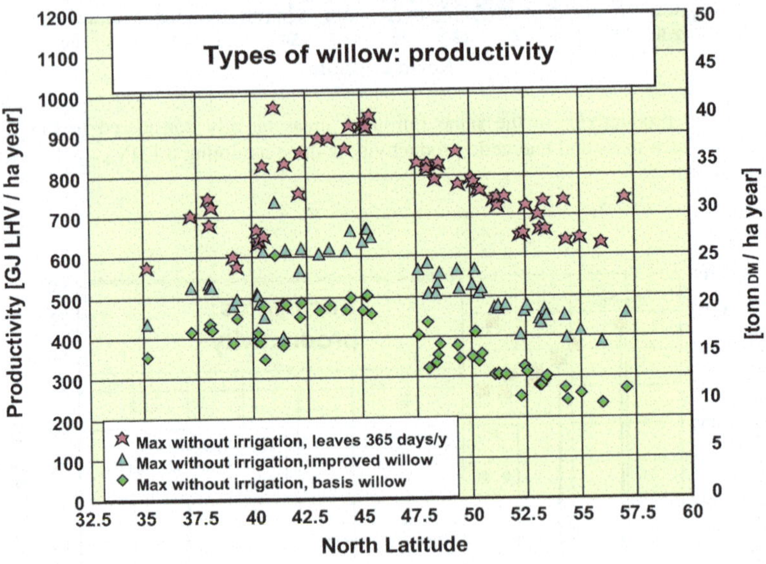

Fig. 2.6 How a different type of willow can influence its productivity based on Nonhebel study (Consonni 2006)

Table 2.14 Oil productivity from rapeseed. Sunflower and soya in Europe (EUROSTAT 2011)

Cultivation	Seeds productivity	Oil productivity		Oil specific energy	Energy oil content	Energetic input
	t/ha	% wt seeds	t/ha	MJ/kg	GJ/ha	GJ/ha
Rape	0.7–3.4	35–40%	0.3–1.4	37.4	11.2–52.3	13–37
Sunflower	0.5–2.5	40–48%	0.2–1.2	38.4	7.7–46.1	20–38
Soya	0.7–3.6	18–20%	0.1–0.7	36.4	3.6–25.5	15–35

Table 2.15 Productivity of some amylaceous species and sugar species in Europe (EUROSTAT 2011)

	Wheat	Corn	Mais	Sorgho	Sugar beet
Charbohydrate (% in wt)	59–61	53–57	62–63	58–60	14–15
Average productivity (t/ha)					
Northern Europe	2.34	2.98	2.18	–	26.61
Central Europe	4.88	3.84	6.63	6.17	40.36
Southern Europe	2.77	2.77	4.88	5.70	41.13
Average productivity (GJ/ha)					
Northern Europe	22.7	26.5	21.5	–	62.3
Central Europe	47.3	34.1	65.3	58.8	94.4
Southern Europe	26.8	24.6	48.0	54.3	96.3

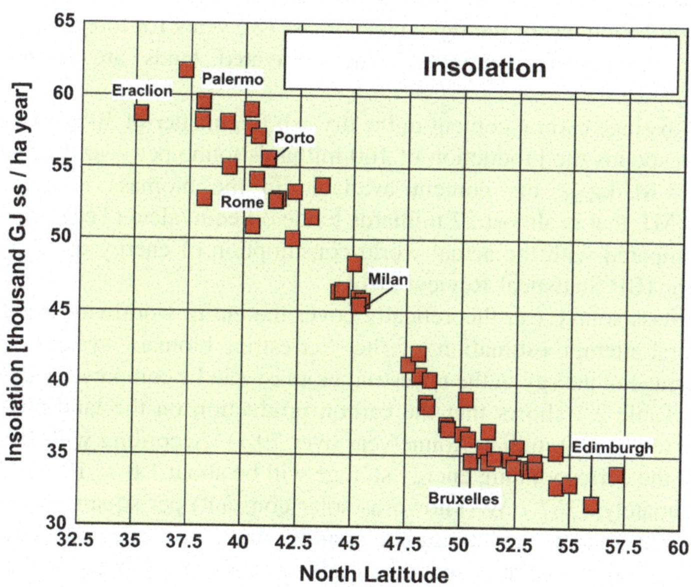

Fig. 2.7 Average insolation in the areas already considered. The calculation was done with the procedure available on the website: http://re.jrc.ec.europa.eu/pvgis/solradframe.php?en&europ (Consonni 2006)

The productivity values shown in Figs. 2.3–2.5 and the capture yields of the solar radiation shown in Figs. 2.8, 2.9 and 2.10 were calculated from the productivity and insolation values showed in Fig. 2.7.

Also for species, like willow and eucalyptus, strongly voted to the Energy production, the energy capture yields are very low, like 0.2–0.6% for willow and 0.2–0.4% for eucalyptus; if there is enough water and feed supply these yields can reach values of 0.8–1.5% and 2%, respectively.

2.2.4 Worldwide Productivity

The low values of yields showed in the previous paragraph, could bring mistrust in the contribution that the biomass source can give to the energy need of our society. Anyway, the diffusion of biomass is so widespread and pervasive that its potential can be more than enough to satisfy the world energy need.

The photosynthesis activity aims at fixing the CO_2-carbon in carbohydrates. This is a key role for the greenhouse gases mitigation. Thus, the biomass production can be measured in terms of carbon fixed in the biomass organic material. Table 2.1 shows the world prospective. Every year, the photosynthetic activity fixes more or less 80 milliards tonne of carbon in organic matter. To highlight the renewability of the biomass source, the last right column shows the rate supply/production. That column means the time necessary to build again the existing biomass reserves. We can say that the biomass reserve can be built again in more or less 10 years; that time is of course longer for the ligneous biomass (22 years for forestry) and shorter for aquatic biomass (2–3 months). The cultivated lands are in the middle (1–2 years).

With an average carbon content in the dry ash free matter of 50%, to 80 Gton of carbon corresponds the production of 160 milliards tonne per year. For an average LHV of 15 MJ/kg_{daf}, the content available in the biomass is more or less 2.4×10^{15} MJ, that is almost 57 milliards tonne oil equivalent (Tep) per year. This value is compared with the actual world consumption of energy of 13.5 milliards Tep per year (BP Statistical Review 2014).

The biomass source can theoretically cover the entire worldwide need.

For a first-attempt estimation of the "terrestrial biomass system" yield, the biomass storage of energy in the terrestrial biomass can be compared with its annual insolation. Table 2.1 shows that the carbon production on the land area is about 67% of the total (52.9 milliard tonne/year over 79.1). According with the previous hypothesis, the corresponding energy storage will be about 1.6×10^{15} MJ/year. Of the approximately 1367.7 W (known as solar constant) per square metre that hits the earth's atmosphere, considering the quantity that it has reflected away from the earth or absorbed by the atmosphere, clouds, etc., about 1 kW/m^2 is the radiation that invests an area perpendicular to solar rays. This power invests the section of our planet that can be considered as a circle of 6378 km radius. Follows:

$$1 \frac{\mathrm{kW}}{\mathrm{m}^2} \cdot \pi \cdot \left(6.378 \times 10^6\,\mathrm{m}\right)^2 \cdot \frac{8760\,\mathrm{h}}{\mathrm{year}} \cdot \frac{3600\,\mathrm{s}}{\mathrm{h}} = 4.03 \times 10^{18} \frac{\mathrm{MJ}}{\mathrm{year}}$$

Assuming that this energy is split between sea and earth proportionally to their surface, the energy that reaches the 149 million km of land area, will be about 1.15×10^8 MJ/year (149/520, see surfaces reported in Table 2.1). Following that the average yield of "biomass system" of land area is about 0.14% that is not so far from the "real" values shown in Figs. 2.8–2.10.

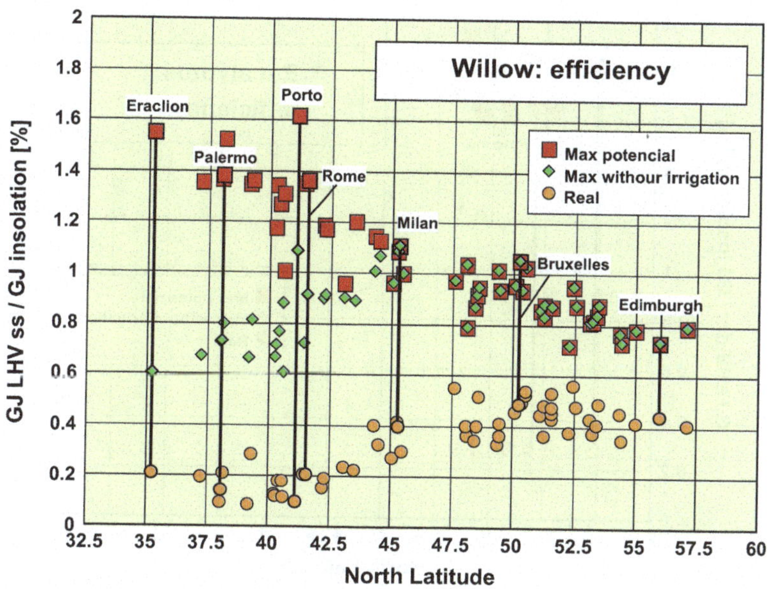

Fig. 2.8 Average yield of the solar insolation capture of willow. The *points* in the diagram are the rate between the productivity and the insolation (Consonni 2006)

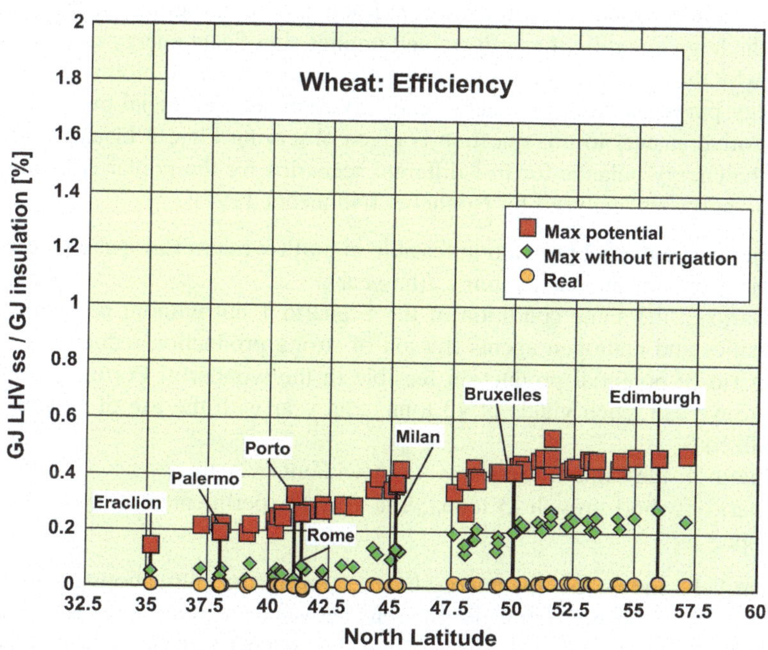

Fig. 2.9 Average yield of the solar insolation capture of wheat. The *points* in the diagram are the rate between the productivity and the insolation (Consonni 2006)

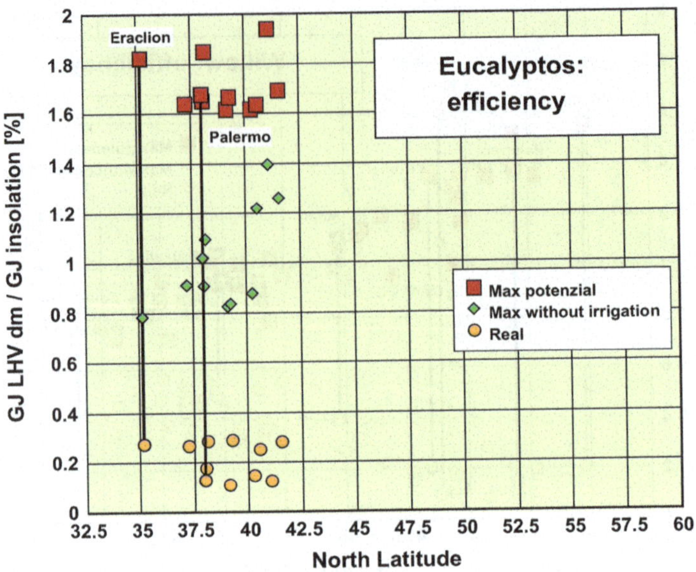

Fig. 2.10 Average yield of the solar insolation capture of eucalyptos. The *points* in the diagram are the rate between the productivity and the insolation (Consonni 2006)

Apart from economic, technical or environmental considerations, the exploitation of the huge quantity of energy valued implies significant energy consumptions. This is why they wonder about the "energetic" convenience of biomass cultivation for energy purposes before the economic, environmental or social one.

A positive answer to this question is given observing Figs. 2.11 and 2.12. That shows the energy balance for five different scenarios for the poplar cultivation for energetic purposes analysed by Nonhebel (Nonhebel 1997):

- Scenario 1: potential production feasible in north-western Europe, where we can obtain yields of almost 12 $tonn_{daf}$/(ha year);
- Scenario 2: the same condition of the Scenario 1 but without protection from parasitics and pathogen agents (reason of strong production reduction);
- Scenario 3: potential production feasible in the wonderful Portuguese climate where we can reach yields of 43 $tonn_{dm}$/ha year with the use of irrigation and fertilizers;
- Scenario 4: the same of Scenario 3 but without irrigation;
- Scenario 5: short growth (5 $tonn_{daf}$/(ha year)) experimented today in the south Europe.

Figure 2.11 shows that the energy balance for all the cultivations considered is positive even if the net output is different between the different cases. The rate between the energy produced (LHV of the dry matter) and the energy consumed varies between a minimum of 8.4 (Scenario 3) and a maximum of 22.1 (Scenario 5). The use of intensive techniques we have in Scenario 3, particularly the irrigation,

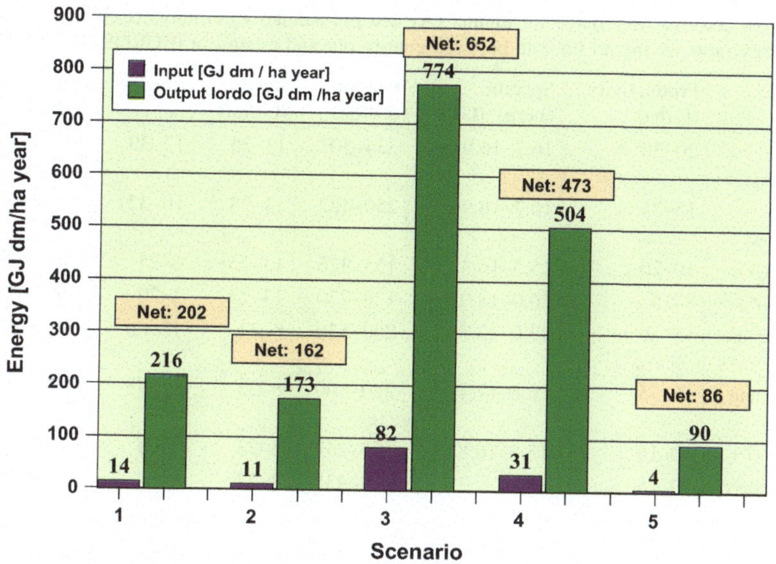

Fig. 2.11 Energetic input and output [GJ_{dm}/(ha year)] for five scenarios of poplar cultivation studied by Nonhebel (1997). The gross output is the product between the dry matter [$tonn_{dm}$/(ha year)] and its LHV_{dm}, (18 MJ/kg_{dm}) (Consonni 2006)

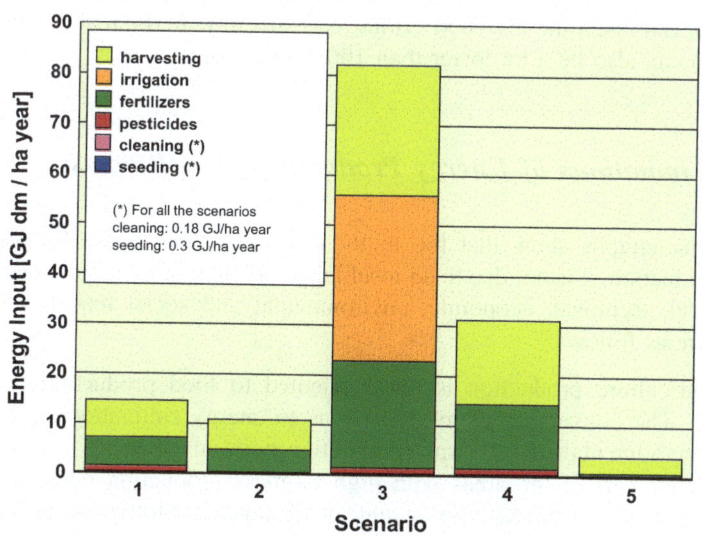

Fig. 2.12 Detailed energy input for the scenarios previous analysed (Consonni 2006)

Table 2.16 Productivity range for annual (A) and poliannual (P) coltures (EUROSTAT 2011). For Energy ratio we intend the rate between energy out and energy in (EUROSTAT 2011)

	Productivity (t_{ss}/ha)	Specific energy (GJ/t)	Output (GJ/ha)	Input (GJ/ha)	Energy ratio	Output–input (GJ/ha)
Sorgho fibre (A)	20–30	16.7–16.9	334–507	13–25	13–39	309–494
Sorgho sugar (A)	15–25	16.7–16.9	250–422	13–25	10–32	225–409
Kenaf (A)	10–20	15.5–16.3	155–326	13–25	6–25	130–313
Hemp (A)	8–15	16.0–18.0	128–270	13–25	5–20	103–257
Mischantus (P)	15–30	17.6–17.7	260–530	8–22	12–66	238–522
Sugar cane (P)	15–35	16.5–17.4	240–600	8–22	11–75	118–592
Thistle (P)	10–15	15.5–16.8	155–252	8–22	7–31	133–244
Panìc grass (P)	10–25	17.4	174–435	8–22	8–54	152–427

increases a lot the productivity but reduces the “efficiency” of the operation that is the rate between the produced energy and the consumed one. The Scenario 5, representative of the warm countries in Southern Europe (Italy, Spain, Greece), has the lowest productivity but the highest “efficiency”.

In Table 2.16, we have the net productivity values for some herbaceous species.

Finally, at our latitudes, the potential for the energy production, in the best conditions, can reach the 500–600 GJ/ha/year. Anyway, in the real conditions, the production can also be a lot lower than 100 GJ/ha/year.

2.2.5 *Limitations of Energy Production from Biomass*

Previous paragraphs show that the minor role played nowadays by biomass in energy production, it is not due to an insufficient potential production but to a series of historical, technical, economic, environmental and social reasons. The most relevant are as follows:

- The agriculture production is fairly oriented to food production rather than energy. The conversion of relevant areas to energy cultivations is a problem from the alimental/energy aims competition point of view;
- The major part of the areas with high biomass production potential (forests, woods, savannas, grasslands) is high environmental-biological and landscape value ecosystems. Their conversion to energy cultivations sets difficult problems related to ecosystem protection and social/political acceptance;
- A consistent part of biomass production reported in Table 2.1 takes place in remote and impassable places, where the intensive exploitation of mineral resources for energy purposes would be difficult and expensive;

- Biomass is an energy source more difficult and expensive than fossil fuels;
- The production efficiencies of electricity, heat and fuels from biomass are typically lower than the ones from fossil fuels;
- Technologies and systems required for biomass collect and transport are typically more complex and expensive than the ones required by fossil fuels;
- Technologies and systems adopted for the generation of energy vectors from biomass are more expensive and difficult to manage than the ones adopted for fossil fuels;
- The biomass exploitation for energy purposes can show serious problems from an environmental point of view, sometimes more difficult to solve than the ones presented by the exploitation of fossil fuels. The only exception is for the greenhouse gases emissions, almost null;
- A wide diffusion of energy cultivations would require a significant change in habits of people that act within the agriculture sector.

All these reasons are not listed to discourage the use of biomass for energy aims but simply to make aware of the problems that have to be solved.

References

Akram M et al (2015) Co-firing of pressed sugar beet pulp with coal in a laboratory-scale fluidised bed combustor. Appl Energy 139(2015):1–8

Basu P (2013) Biomass gasification, pyrolysis and torrefaction: practical design and theory. Academic Press

BP Statistical Review of World Energy 2014—BP Global (2014). http://bp.com/statisticalreview

Burg P et al (2016) Calorific evaluation and energy potential of grape pomace. Int Agrophys 30:261–265. doi:10.1515/intag-2015-0082

Consonni S (2006) Caratteristiche e potenzialità della risorsa energetica biomassa, dispensa del corso di Tecnologie e prospettive della produzione di energia da biomasse. Politecnico di Milano

De Jong W, Van Ommen R (2015) Biomass as a sustainable energy source for the future. Fundamentals of Conversion Processes. Wiley

EUROSTAT (2011) Agriculture statistics

FAO statistics related to 2009 and 2012 production data (http://faostat3.fao.org)

Intelligent Energy EUROPE (2010) Market of olive residues for energy—D5.1: business plan for energy facilities exploiting olive-milling solid residues

Iranpour R et at (2000) Regulations for biosolids Land Application in U-S. and European Union. Rev J Residuals Sci Technol 1(4)

ISPRA (2015) Rapporto rifiuti urbani 2015. 230/2015

Khan AA (2007). Combustion and co-combustion of biomass in a bubbling fluidized bed boiler. Ph.D Thesis, Delft University of Technology

Klass DL (1998) Biomass for renewable energy, fuels and chemicals. Academic Press

Mata TM, Martins AA, Caetano NS (2010) Microalgae for biodiesel production and other applications: A review. Renew Sustain Energy Rev 14:217–232

McKendry P (2002) Energy production from biomass (part 1): overview of biomass. Bioresour Technol 83(1):37–46

Nonhebel S (1997) Harvesting the sun's energy using agro-ecosystems. Quantitative Approaches in System Analysis No.13, Dept. of Theoretical Production Ecology, AB-DLO, Wageningen

Phyllis. Database for biomass and waste. http://www.ecn.nl/phyllis/. ECN

Stanford Research Institute (1976) An evaluation of the use of agricultural residues as an energy feedstock, vol 1. National Science Foundation, Washington

Stevens DJ (2001) Hot gas conditioning: recent progress with large-scale biomass gasification systems. Update and Summary of Recent Progress. NREL, Golden, CO. Report nr NREL/SR-51029952

Tillman DA, Jahn, EC (eds) (2016) Progress in biomass conversion, vol 5. Elsevier

Twidell J, Weir T (2006) Renewable energy resources, 2nd edn. Taylor & Francis, London

Vigouroux RZ (2001) Pyrolisis of biomass. PhD Thesis, Royal Institute of Technology, Dept. of Chemical Engineering and Technology, Stockholm. ISSN 1104-3466

Williams A, Pourkashanian M, Jones JM (2001) Combustion of pulverized and biomass. Prog Energy Combust Sci 27(6):587–610

Author Biography

Martina Fantini, Ph.D. is a researcher in the area of Process and Energy at "Laboratory for Energy and Environment Piacenza" (Consorzio LEAP), a private body linked with Politecnico di Milano. She graduated in Mechanical Engineering (2005) from Politecnico di Milano (IT) and she got her Ph.D. in Energy Technology (2010) from the same institute. She carried out her postdoc activity at Technical University of Delft in The Netherlands (2009–2014). Currently, she supports teaching activity for the course "Energy Systems" at Politecnico di Milano. She has experience in research and technology development in the EU project on sustainable sources of energy, on hydrodynamic transport of biomass particles and on biomass gasification and hot gas filtration.

Chapter 3
Biomass Conversion Technologies: Fast Pyrolysis Liquids from Biomass: Quality and Upgrading

A.V. Bridgwater

Abstract A thorough assessment has been made of the characteristics of bio-oil from fast pyrolysis of biomass. Fast pyrolysis uniquely gives high yields of a homogenous mobile liquid for direct use for heat and power and indirect use for biofuels and green chemicals. An improved understanding of the significance of the different aspects of quality of bio-oil helps to establish standards and key quality requirements which help to define limitations for use. An appreciation of the potential for bio-oil to meet a broad spectrum of applications in renewable energy has led to a significantly increased R&D activity in studying the science and technology of fast pyrolysis with increased emphasis on quality improvement. This increased activity is evident in North America, Europe and Asia with many new entrants as well as expansion of existing activities. The only disappointment is the continued limited industrial development and deployment of fast pyrolysis that are necessary to provide the basic bio-oil raw material for the development and exploitation of applications.

3.1 Introduction to Fast Pyrolysis and Bio-oil

3.1.1 Introduction

Pyrolysis is thermal decomposition occurring in the absence of oxygen. Lower process temperatures and longer vapour residence times favour the production of charcoal; high temperatures and longer residence times increase biomass conversion to gas; and moderate temperatures and short vapour residence time are optimum for producing liquids which is referred to a bio-oil. Three products are always produced, as summarised in Table 3.1, but the proportions can be varied over a

A.V. Bridgwater (✉)
Bioenergy Research Group, European Bioenergy Research Institute, Aston University, Birmingham B4 7ET, UK
e-mail: a.v.bridgwater@aston.ac.uk

M. Rabaçal et al. (eds.), *Biorefineries*, Lecture Notes in Energy 57,
DOI 10.1007/978-3-319-48288-0_3

Table 3.1 Typical product weight yields (dry wood basis) obtained by different modes of pyrolysis of wood

Mode	Conditions	Liquid	Solid	Gas (%)
Fast	~500 °C, short hot vapour residence time ~1 s	75% (bio-oil)	12% char	13
Intermediate	~500 °C, hot vapour residence time ~10–30 s	40% in 2 phases	40% char	20
Slow (carbonisation)	~400 °C, long vapour residence hours → days	30%	35% char	35
Gasification	~700–900 °C	1–5% (tar)	10% char	85
Torrefaction (slow low temperature pyrolysis)	~290 °C, solids residence time ~10–60 min	0% unless condensed, then up to 15%	80% solid	20

wide range by adjustment of the process parameters. Fast pyrolysis for liquids production is of particular interest as high yields of a liquid are obtained which can be stored and transported, and used for energy, chemicals or as an energy carrier (Bridgwater 2011, 2012; Czernik et al. 2004). Fast pyrolysis is the focus of this chapter.

Bio-oil, the main product from fast pyrolysis at temperatures of around 500 °C, is obtained in yields of up to 75 wt% on a dry-feed basis. The by-product char is normally combusted within the process to reheat the fluidising medium, usually sand, so char is not usually a by-product. The gas can also be utilised with in the process, so there are no waste streams other than hot flue gas and ash. Liquid yield and quality depend on many factors, which are described below.

3.2 Fast Pyrolysis Technology

3.2.1 Principles

In fast pyrolysis, biomass decomposes very quickly to generate mostly vapours and aerosols and some charcoal and gas. After cooling and condensation, a dark brown homogenous mobile liquid is formed, which has a heating value about half that of conventional fuel oil. A high yield of liquid is obtained with most biomass feeds low in ash. The essential features of a fast pyrolysis process for producing liquids are:

- Very high heating rates and very high heat transfer rates at the biomass particle reaction interface. This usually requires a finely ground biomass feed of typically less than 3 mm as biomass generally has a low-thermal conductivity,
- Carefully controlled pyrolysis reaction temperature of around 500 °C to maximise the liquid yield for most biomass types,

- Short hot vapour residence times of typically less than 2 s to minimise secondary reactions,
- Rapid removal of product char to minimise cracking of vapours,
- Rapid cooling of the pyrolysis vapours to give the bio-oil product.

As fast pyrolysis for liquids occurs in a few seconds or less, heat and mass transfer processes and phase transition phenomena, as well as chemical reaction kinetics, play important roles. The critical issue is to bring the reacting biomass particles to the optimum process temperature and minimise their exposure to the lower temperatures that favour formation of charcoal. One way this objective can be achieved is by using small particles, for example in the fluidised bed processes that are described later. Another possibility is to transfer heat very fast only to the particle surface that contacts the heat source which is used in ablative processes that are also described later.

The main product, bio-oil, is obtained in yields of up to 75 wt% on a dry-feed basis, together with by-product char and gas which can be used within the process to provide the process heat requirements so there are no waste streams other than flue gas and ash. Liquid yield depends on biomass type, temperature, hot vapour residence time, char separation, and biomass ash content, the last two having a catalytic effect on vapour cracking.

A fast pyrolysis process includes drying the feed to typically less than 10% water in order to minimise the water in the product liquid oil, grinding the feed to give sufficiently small particles to ensure rapid reaction, fast pyrolysis, rapid and efficient separation of solids (char), and rapid quenching and collection of the liquid product (often referred to as bio-oil).

Virtually any form of biomass or biogenic materials can be considered for fast pyrolysis. While most work has been carried out on wood because of its consistency and comparability between tests, well over 100 different biomass types have been tested by many laboratories, ranging from agricultural wastes, such as straw, olive pits and nut shells to energy crops such as miscanthus and sorghum, forestry wastes such as bark and solid wastes such as sewage sludge and leather wastes.

In all cases, a commercial process comprises three main stages from feed reception to delivery of one or more useful products:

- Feed reception, storage, handling, preparation and pretreatment such as comminution and drying;
- Fast pyrolysis of the prepared biomass to a more usable form of energy in liquid form known as bio-oil;
- Conversion of this primary liquid product by processing, refining or upgrading into a marketable end-product for production of electricity, heat, biofuels and/or chemicals.

3.2.2 *Fast Pyrolysis Reactors*

At the heart of a fast pyrolysis process is the reactor. Although it probably represents only about 10–15% of the total capital cost of an integrated system, most research and development has focused on developing and testing different reactor configurations on a variety of feedstocks, although increasing attention is now being paid to control and improvement of liquid quality and improvement of liquid collection systems. Several comprehensive reviews of fast pyrolysis processes for liquids production are available such as (Mohan et al. 2006; Kersten et al. 2005; Bridgwater 2003, 2009; Bridgwater et al. 2002a):

3.2.2.1 Bubbling Fluid Beds

All the early work on fluid beds was carried out at the University of Waterloo in Canada, which pioneering the science of fast pyrolysis and established a clear lead in this area for many years (Scott et al. 1985; Scott and Piskorz 1982; Scott 1997). Bubbling fluid beds were selected for further development by several companies, including Dynamotive, who have built a 100 t/d and a 200 t/d plant in Canada; Wellman, who built a 250 kg/h unit (McLellan 2000) in the UK which has not operated; and Fortum who built and extensively tested a 500 kg/h plant in Finland which has now been dismantled (Gust 2002). More recent activities include Ikerlan who are developing a spouted fluid bed in Spain (Fernandez 2010), Metso who are working with UPM and VTT in Finland who have constructed and are operating a 4 MWth unit in Tampere Finland (Lehto et al. 2010) and Anhui University of Science and Technology in China who have overseen the construction of a 10,000 t/y demonstration plants in China (Zhu 2009; Ming-qiang 2006). Many research units have also been built at universities and research institutions around the world, as fluid beds are relatively easy to construct and operate and give good results.

Bubbling fluid beds have the advantages of a well-understood technology that is simple in construction and operation, provides good temperature control and very efficient heat transfer to biomass particles arising from the high solids density. This makes it ideal for laboratory scale research. Figure 3.1 shows a typical configuration with quench to cool and collect condensable vapours and an electrostatic precipitator to collect aerosols. These are incompletely depolymerised lignin fragments which seem to exist as a liquid with a substantial molecular weight. Evidence of their liquid basis is found in the accumulation of liquid in the ESP which runs down the plates to accumulate in the bio-oil product. Demisters for agglomeration or coalescence of the aerosols have been used but published experience suggest that this is less effective. Fluid-bed pyrolysers give good and consistent performance with high liquid yields of typically 70–75 wt% from wood on a dry-feed basis. Small biomass particle sizes of less than 2–3 mm are needed to achieve high biomass heating rates, and the rate of particle heating is usually the rate-limiting step.

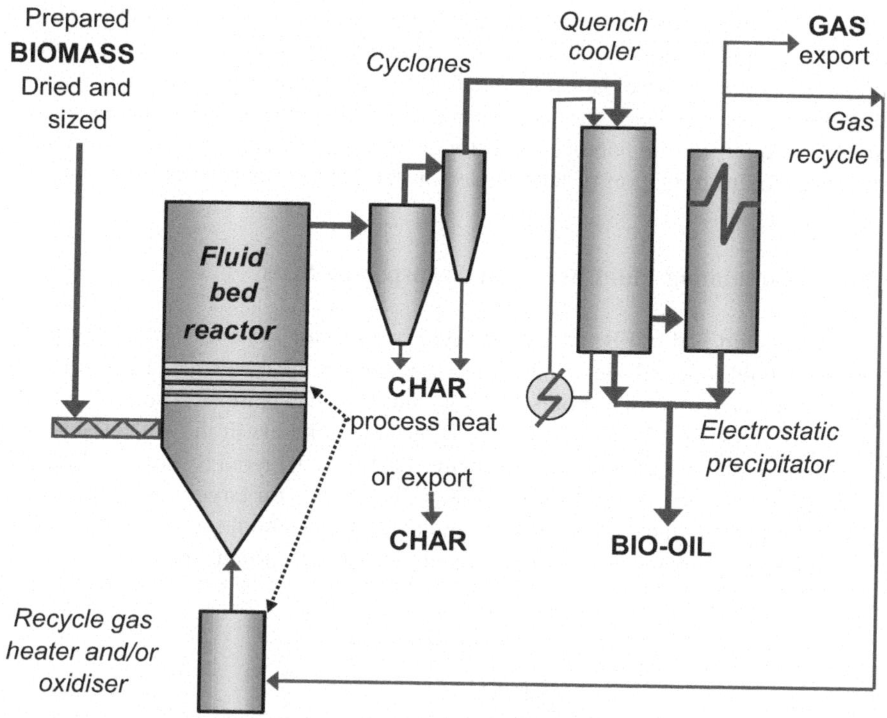

Fig. 3.1 Bubbling fluid bed reactor with electrostatic precipitator

Heating can be achieved in a variety of ways and scaling is well understood. However, heat transfer to fluid beds at large scales of operation has to be considered carefully because of the scale-up limitations of different methods of heat transfer. Dynamotive focused on indirectly heated fluid beds at all scales of processing using natural gas as fuel, although the byproduct char could in principle be used as fuel.

Vapour and solid residence time is controlled by the fluidising gas flow rate and is higher for char than for vapours. As char acts as an effective vapour cracking catalyst at fast pyrolysis reaction temperatures, rapid and effective char separation is important. This is usually achieved by ejection and entrainment followed by separation in one or more cyclones so careful design of sand and biomass/char hydrodynamics is important. The high level of inert gases arising from the high permanent gas flows required for fluidisation result in very low partial pressures for the condensable vapours and thus care is needed to design and operate efficient heat exchange and liquid collection systems. In addition the large inert gas flowrates result in relatively large equipment thus increasing cost.

The byproduct char is typically about 15 wt% of the products but about 25% of the energy of the biomass feed. It can be used within the process to provide the process heat requirements by combustion or it can be separated and exported, in which case an alternative fuel is required. Depending on the reactor configuration and gas velocities, a large part of the char will be of a comparable size and shape as the biomass fed. The fresh char is pyrophoric, i.e. it spontaneously combusts when exposed to air so careful handling and storage is required. This property deteriorates with time due to oxidation of active sites on the char surface.

3.2.2.2 Circulating Fluid Beds and Transported Beds

Circulating fluid bed (CFB) and transported bed reactor systems have many of the features of bubbling beds described above, except that the residence time of the char is almost the same as for vapours and gas, and the char is more attrited due to the higher gas velocities. This can lead to higher char contents in the collected bio-oil unless more extensive char removal is included. A typical layout is shown in Fig. 3.2. An added advantage is that CFBs are potentially suitable for larger throughputs even though the hydrodynamics are more complex as this technology is widely used at very high throughputs in the petroleum and petrochemical industry.

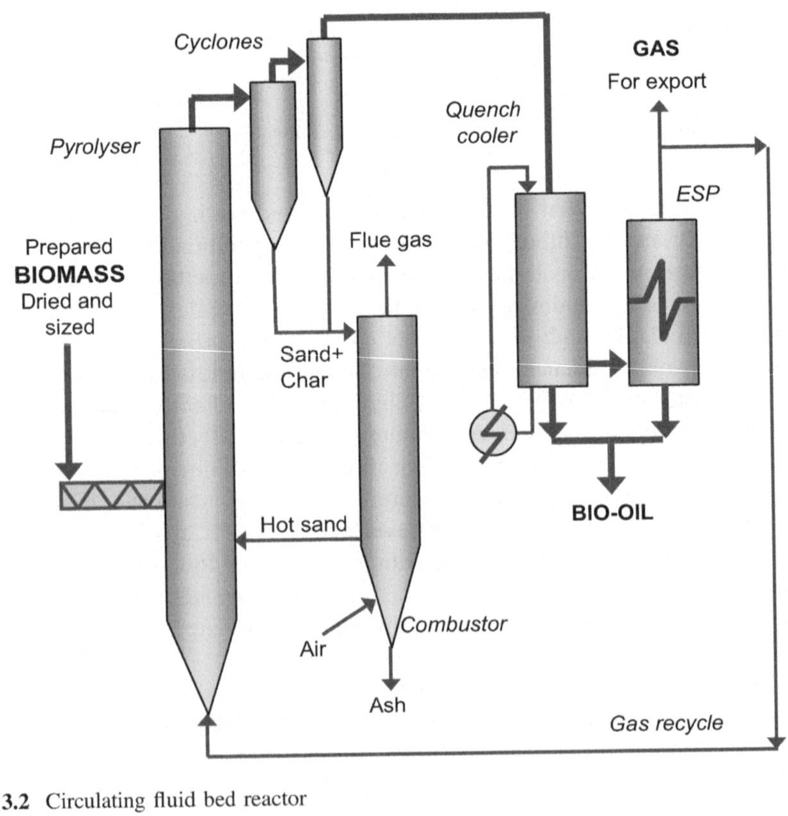

Fig. 3.2 Circulating fluid bed reactor

Heat supply is usually from recirculation of heated sand from a secondary char combustor, which can be either a bubbling or CFB. In this respect the process is similar to a twin fluid-bed gasifier except that the reactor (pyrolyser) temperature is much lower and the closely integrated char combustion in a second reactor requires careful control to ensure that the temperature, heat flux and solids flow match the process and feed requirements. Heat transfer is a mixture of conduction and convection in the riser. One of the unproven areas is scale up and heat transfer at high throughputs. All the char is burned in the secondary reactor to re-heat the circulating sand, so there is no char available for export unless an alternative heating source is used. If separated the char would be a fine powder.

Ensyn in Canada has built several units, mostly for production of liquid smoke for the food industry. The company has a significant R&D operation including a 2 t/h unit which also currently produces bio-oil for utilisation in eastern Canada and the USA (Muller 2010). BTG has recently started up their 125 t/d rotating cone technology known as Empyro which is a version of a CFB. This is described below.

The rotating cone reactor, invented at the University of Twente (Prins and Wagenaar 1997) and developed by BTG (Wagenaar et al. 2001), is a relatively recent commercial development and effectively operates as a transported bed reactor, but with transport effected by centrifugal forces in a rotating cone rather than gas. A 250 kg/h unit is now operational, and a scaled up version of 50 t/d was commissioned in Malaysia in mid 2005. A 125 t/d plant was commissioned in 2015 (Muggen 2010). As with CFB and transported beds all the char is burned so is not a by-product.

3.2.2.3 Ablative Pyrolysis

Ablative pyrolysis is substantially different in concept compared with other methods of fast pyrolysis. In all the other methods, the rate of reaction is limited by the rate of heat transfer through the biomass particles, which is why small particles are required. The mode of reaction in ablative pyrolysis is like melting butter in a frying pan—the rate of melting can be significantly enhanced by pressing the butter down and moving it over the heated pan surface. In ablative pyrolysis, heat is transferred from the hot reactor wall to “melt” wood that is in contact with it under pressure. As the wood is moved away, the molten layer then vapourises to a product very similar to that derived from fluid bed systems.

The pyrolysis front thus moves unidirectionally through the biomass particle. As the wood is mechanically moved away, the residual oil film both provides lubrication for successive biomass particles and also rapidly evaporates to give pyrolysis vapours for collection in the same way as other processes. There is an element of cracking on the hot surface from the char that is also deposited. The rate of reaction is strongly influenced by pressure of the wood onto the heated surface; the relative velocity of the wood and the heat exchange surface; and the reactor surface temperature.

As reaction rates are not limited by heat transfer through the biomass particles, larger particles can be used and in principle there is no upper limit to the size that can be processed. The process, in fact, is limited by the rate of heat supply to the reactor rather than the rate of heat absorption by the pyrolysing biomass, as in other reactors. There is no requirement for inert gas, so the processing equipment is smaller and the reaction system is thus more intensive. In addition the absence of fluidising gas substantially increases the partial pressure of the condensable vapours leading to more efficient collection and smaller equipment. However, the process is surface-area-controlled so scaling is less effective and the reactor is mechanically driven, and is thus more complex.

Much of the pioneering fundamental work on ablative pyrolysis reactors was performed by the CNRS laboratories in Nancy, France, where extensive basic research has been carried out onto the relationships between pressure, motion and temperature (Lédé et al. 1985). The National Renewable Energy Laboratory (NREL) in Boulder, Colorado developed the ablative vortex reactor, in which the biomass was accelerated to supersonic velocities to derive high tangential pressures inside a heated cylinder (Diebold and Power 1988). Unreacted particles were recycled and the vapours and char fines left the reactor axially for collection. Liquid yields of 60–65 wt% on dry-feed basis were typically obtained. Aston University has developed an ablative plate reactor (Peacocke and Bridgwater 1995) in which pressure and motion is derived mechanically, obviating the need for a carrier gas. Liquid yields of 70–75 wt% on dry-feed basis are typically obtained. A second-generation reactor has recently been built and commissioned and has been patented (Bridgwater et al.). Another configuration is the mechanically driven PyTec process in Germany (Meier 2005). The company has built and tested a laboratory unit based on hydraulically feeding wood rods onto a rotating electrically heated cone. The liquid collection system is analogous to other systems described above (Meier 2005). A 6 t/d unit has been built in north Germany in 2006 with the bio-oil fuelling an engine for power generation.

Ablative pyrolysis is an attractive concept that has too many challenging technical issues currently to be a serious commercial prospect.

3.2.2.4 Other Reaction Systems

Entrained flow fast pyrolysis is, in principle, a simple technology, but most developments have not been so successful because of the poor heat transfer between a hot gas and a solid particle. High gas flows are required to affect sufficient heat transfer, which requires large plant sizes and entails difficult liquid collection from the low vapour partial pressure. Liquid yields have usually been lower than fluid bed and CFB systems at 50–55 wt% as in Georgia Tech Research Institute (Kovac and O'Neil 1989) and Egemin (Maniatis et al. 1993) but neither is now operational. There is some basic research in this area in China.

Vacuum pyrolysis, as developed in Canada by the University of Laval and Pyrovac, is arguably not a true fast pyrolysis as the heat transfer rate to and through the solid biomass is much slower than in the previously described reactors although the vapour residence time is comparable. The basic technology was developed at the University of Laval using a multiple hearth furnace but was upscaled to a purpose-designed heated horizontal moving bed (Yang et al. 2001). The process operated at 450 °C and 100 kPa. Liquid yields of 35–50% on dry feed were typically obtained with higher char yields than fast pyrolysis systems. The process was complex and costly because the high vacuum necessitates the use of very large vessels and piping. The advantages of the process are that it can process larger particles than most fast pyrolysis reactors, there is less char in the liquid product because of the lower gas velocities, and no carrier gas is needed. The process has not operated for some years and no activities are currently known using vacuum pyrolysis.

There have been a number of developments that mechanically move biomass through a hot reactor rather than using fluids. These include screw and augur kilns. Heating can by external heating or with recycled hot sand as at the Bioliq plant at KIT (FZK until 2009) (Raffelt et al. 2006a), or with heat carriers such as steel or ceramic balls. The nature of mechanically driven reactors is that very short residence times comparable to fluid and CFBs are difficult to achieve, and hot vapour residence times can range from 5 to 30 s depending on the design and size of reactor. Examples include the Lurgi LR reactor at Karlsruhe Institute of Technology (KIT) (Ingram et al. 2008) and the Bio-oil International reactors which have been studied at Mississippi State University (Ingram et al. 2008). Screw and augur reactors have also been developed as intermediate pyrolysis systems such as Haloclean also at KIT (Hornung et al. 2007) and the twin screw Pyroformer technology at Aston (Yang et al. 2014). Screw reactors are particularly suitable for feed materials that are difficult to handle or feed, or are heterogeneous. The liquid product yield is lower than fluid beds and is often phase separated due to the longer residence times and contact with byproduct char. Also, the char yields are higher. KIT has promoted and tested the concept of producing a slurry of the char with the liquid to maximise liquid yield in terms of energy efficiency (Raffelt 2006a), but this would require an alternative energy source to provide heat for the process.

There have been claims of fast pyrolysis in fixed beds but it is difficult to envisage a fixed bed pyrolysis process that satisfies the basic requirements of fast pyrolysis which can be constructed at anything above laboratory or bench scale.

Some basic research has been carried out on microwave driven pyrolysis. Microwave heating is fundamentally difference from all other pyrolysis techniques as the biomass particles are heated from within and not by external heat transfer from a high temperature heat source. Microwave heating requires a material with a high dielectric constant or loss factor, of which water is a good example. So, in microwave pyrolysis, water is rapidly driven off then the particle heats up to start forming char. It is not clear that this can be considered fast pyrolysis. This is electrically conductive and eddy currents are created that provide very rapid heating. Therefore, control of a microwave system is quite challenging. A further problem to be considered is that penetration of microwaves is limited to typically 1–2 cm, so the design of a microwave reactor presents interesting scale up challenges.

One of the potentially valuable aspects of microwave pyrolysis is that due to the absence of thermal gradients, an environment is created for studying some of the fundamentals of fast pyrolysis. This offers possibilities to examine the effect of the thermal gradient in a pyrolysing particle and the secondary reactions that occur both within and without the biomass particle.

In an effort to reduce the oxygen content of the bio-oil product within a single step process, some attention has returned to the concept of integrating pyrolysis and hydrocracking in which hydrogen is added to the pyrolysis reactor. GTI is starting a new hydropyrolysis and hydroconversion programme to make gasoline and diesel in early 2010 (Marker et al. 2009) and a new patent has been applied for that includes hydrogen in the pyrolysis reactor with claims of producing hydrocarbons, alcohols and other oxygenates. The concept has some contradictory requirements—high pressure in pyrolysis increases char yields, e.g. Antal et al. (1996), and reduces liquid yields, while high pressures are required to provide effective hydrogenation.

3.3 Liquid Characteristics and Quality

3.3.1 Bio-oil General Characteristics

Crude fast pyrolysis liquid or bio-oil is dark brown and approximates to biomass in elemental composition. It is composed of a very complex mixture of oxygenated hydrocarbons with an appreciable proportion of water from both the original moisture and reaction product. Solid char may also be present as fine particles. The liquid is formed by rapid quenching of the vapours and aerosols which are thus prevented from secondary reactions. The product, therefore, is not as stable as many liquid fuels and has a tendency to slowly change some physical and chemical characteristics over time. This is referred to as ageing. Fast pyrolysis liquid has a higher heating value of about 17 MJ/kg as produced with about 25 wt% water that cannot readily be separated. There are some important characteristics of this liquid that are summarised in Table 3.2 and discussed below.

The liquid is formed by rapidly quenching, and thus 'freezing' of the intermediate products of flash degradation of hemicellulose, cellulose and lignin. The liquid thus contains many reactive species, which contribute to its unusual attributes. Bio-oil can be considered a micro-emulsion in which the continuous phase is an aqueous solution of holocellulose decomposition products, which stabilises the discontinuous phase of pyrolytic lignin macromolecules through mechanisms such as hydrogen bonding. The dispersed phase is micelles of pyrolytic lignin of around 500 Å in diameter and the micro-emulsion is maintained by naturally derived surfactants. Ageing or instability is believed to result from a breakdown in this emulsion. In some ways, it is analogous to asphaltenes found in petroleum.

Table 3.2 Typical properties of wood-derived crude bio-oil

Physical property	Typical value
Moisture content (%)	25
pH	2.5
Specific gravity	1.2
Elemental analysis	
C (%)	56
H (%)	6
O (%)	38
N (%)	0–0.1
HHV[a] as produced (MJ/kg)	17
Miscibility with hydrocarbons	Very low
Viscosity (40 °C and 25% water) (cP)	40–100
Solids (char) (%)	0.05
Stability	Relatively poor
Vacuum distillation residue (%)	Up to 50

[a]*HHV* higher heating value

The typical product distribution all the products from fast pyrolysis of aspen wood is shown in Fig. 3.3. Bio-oil is a homogenous mixture of the organics, the reaction water as shown in the figure and the feed water.

Bio-oil is a homogenous mixture of organics + reaction water + feed water. The liquid has a distinctive odour, an acrid smoky smell due to the low molecular weight aldehydes and acids, which can irritate the eyes on prolonged exposure. The liquid contains several hundred different chemicals in widely varying proportions,

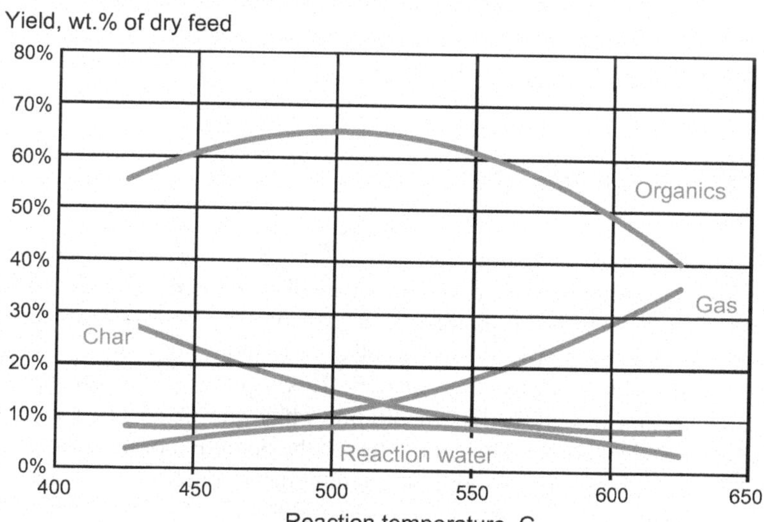

Fig. 3.3 Products from fast pyrolysis of Aspen

ranging from formaldehyde and acetic acid to complex high molecular weight phenols, anhydrosugars and other oligosaccharides.

The liquid contains varying quantities of water, which forms a stable single-phase mixture, ranging from about 15 wt% to an upper limit of about 35 wt % water, depending on the feed material and how it was produced and subsequently collected. A typical feed material specification is a maximum of 10% moisture in the dried feed material, as both this feed moisture and the water of reaction from pyrolysis, typically about 12% based on dry feed, both report to the liquid product.

The density of the liquid is very high at around 1.2 kg/L, compared with light fuel oil at around 0.85 kg/L. This means that the liquid has about 42% of the energy content of fuel oil on a weight basis, but 61% on a volumetric basis. This has implications for the design and specification of equipment such as pumps and atomisers in boilers and engines.

Viscosity is important in many fuel applications (Diebold et al. 1997). The viscosity of the bio-oil as produced can vary from as low as 25 cSt (1 cSt = 1 mm^2/s) to as high as 1000 cSt (measured at 40 °C) or more depending on the feedstock, the water content of the oil, the amount of light ends collected, and the extent to which the oil has aged.

Pyrolysis liquids cannot be completely vaporised once they have been recovered from the vapour phase. If the liquid is heated to 100 °C or more to try to remove water or distill off lighter fractions, it rapidly reacts and eventually produces a solid residue of around 50 wt% of the original liquid and some distillate containing volatile organic compounds, including cracked compounds and water. While bio-oil has been successfully stored for several years in normal storage conditions in steel and plastic drums without any deterioration that would prevent its use in any of the applications tested to date, it does change slowly with time; most noticeably there is a gradual increase in viscosity. More recent samples that have been distributed for testing have shown substantial improvements in consistency and stability, demonstrating the improvement in process design and control as the technology develops.

3.3.2 Upgrading Bio-oil

As the unusual nature of bio-oil has become more widely known, considerably greater effort has been undertaken to improve quality, or upgrade bio-oil, often with little appreciation of what properties are important. The objective or purpose of upgrading bio-oil is to improve its quality; that is, to reduce or remove one or more of its undesirable characteristics or properties. This requires a thorough appreciation of the characteristics or properties of bio-oil, which are listed in Table 3.3 with causes, effects and solutions. More important is the definition of the term "quality," since different applications have different requirements in terms of characteristics, some of which have been reviewed (Czernik et al. 2004). Each of these characteristics is described in more detail later in this chapter, with an emphasis on those aspects that have attracted most interest and attention in recent years and which are of potentially

Table 3.3 Characteristics of bio-oil

Characteristic	Cause	Effect	Solution	Comments
Acidity or low pH	Organic acids from biopolymer degradation	Corrosion of vessels and pipework	Careful materials selection, such as polyolefins or stainless steel	Important in all applications
Ageing	Continuation of secondary reactions including polymerisation	Slow increase in viscosity from secondary reactions such as condensation Potential phase separation	Do not store for long periods Avoid exposure to air Add water Add co-solvents	Important
Alkali metals	Nearly all alkali metals report to char so not a big problem High ash feed Incomplete solids separation	Catalyst poisoning Deposition of solids in combustion Erosion and corrosion Slag formation Damage to turbines	Pretreat feed to remove ash Hot-vapour filtration Process oil Modify application	Almost all alkali metals report to char, so good char separation minimises alkali metals in oil
Char	Incomplete char separation in process	Ageing of oil Sedimentation Filter blockage Catalyst blockage Engine injector blockage Alkali metal poisoning	Improved cyclones Multiple cyclones Hot-vapour filtration	
Chlorine	Contaminants in biomass feed	Catalyst poisoning in upgrading	Include suitable cleaning processes either upstream or downstream	Important for biofuel production
Colour	Cracking of biopolymers and char	Discolouration of some products such as resins	Efficient char filtration Deoxygenation	Only important where visible, such as resins or blended products
Contamination of feed	Poor harvesting practice	Contaminants, notably soil act, as catalysts and can increase particulate carryover	Improve harvesting practice Wash biomass	

(continued)

Table 3.3 (continued)

Characteristic	Cause	Effect	Solution	Comments
Distillability is poor	Reactive mixture of degradation products	Bio-oil cannot be distilled—maximum 50% typically. Liquid begins to react at below 100 °C and substantially decomposes above 100 °C	None known	Important for biofuel production and refinery integration
High viscosity		Gives high pressure drop, increasing equipment cost High pumping cost Poor atomisation	Careful heating up to 50 °C; rapid in-line heating to 80 °C is also possible Add water Add co-solvents	Important in heart and power applications
In-homogeneity	See Phase separation			
Low H:C ratio	Biomass has low H:C ratio	Upgrading to hydrocarbons is more difficult	Add hydrogen and/or remove oxygen	See upgrading to biofuels section
Low pH	See acidity			
Materials incompatibility	Phenolics and aromatics	Destruction of seals and gaskets	Careful materials selection	
Miscibility with hydrocarbons is very low	Highly oxygenated nature of bio-oil	Will not mix with any hydrocarbons, so integration into a refinery is more difficult	Upgrading by hydrotreating or cracking with zeolites	See upgrading to biofuels section
Nitrogen	Contaminants in biomass feed High-nitrogen feed, such as proteins in wastes	Unpleasant smell Catalyst poisoning in upgrading NOx in combustion	Careful feed selection Feed blending Include suitable cleaning processes Add NOx removal in combustion applications	Important for biofuel production
Odour	See smell			

(continued)

Table 3.3 (continued)

Characteristic	Cause	Effect	Solution	Comments
Oxygen content is very high	Biomass composition	Poor stability Nonmiscibility with hydrocarbons	Reduce oxygen thermally and/or catalytically	Important for biofuel production
Phase separation or inhomogeneity	High feed water High ash in feed Poor char separation	Phase separation Partial phase separation Layering Poor mixing Inconsistency in handling, storage and processing	Modify or change process Modify or change feedstock Add co-solvents Control water content	May be important for biofuel production
Smell	Aldehydes and other volatile organics, many from hemicellulose	While not toxic, the smell is often objectionable	Better process design and management Reduction in hemicelluloses content of feed Containment and/or venting to flare	
Solids	See also char Particulates from reactor such as sand Particulates from feed contamination	Sedimentation Erosion and corrosion Blockage	Filtration of vapour or liquid	
Structure	The unique structure is caused by the rapid de-polymerisation and rapid quenching of the vapours and aerosols	Susceptibility to ageing, such as viscosity increase and phase separation	None known	See ageing
Sulfur	Contaminants in biomass feed	Catalyst poisoning in upgrading	Include suitable cleaning processes	Important for biofuel production

(continued)

Table 3.3 (continued)

Characteristic	Cause	Effect	Solution	Comments
Temperature sensitivity	Incomplete reactions	Irreversible decomposition of liquid into two phases >100 °C Potential phase separation >60 °C	Store liquids at room temperature, preferably in absence of air	Important for all applications
Toxicity	Biopolymer degradation products	Human toxicity is positive but small Eco-toxicity is negligible	Health and safety precautions	
Viscosity	Nature of bio-oil	Fairly high and variable with time Greater temperature influence than hydrocarbons	Water and/or solvent addition reduces viscosity	Important for heat and power
Water content	Pyrolysis reactions Feed water	Complex effect on viscosity and stability: increased water lowers heating value, density, stability, and increase pH Affects catalysts	Control water in feed Optimise at 25% for consistency and miscibility Optimise for application	Important in biofuel production

greater significance. This includes biofuels by hydro-treatment, biofuels by zeolite cracking, biofuels by gasification and synthesis, hydrogen production by steam reforming, chemicals recovery and stability improvement.

3.4 Significant Factors Affecting Bio-oil Characteristics and Quality

3.4.1 Feed Material

The composition of the biomass feed has a significant effect on both the yield and quality of the resulting bio-oil. The main parameters are ash, water content biomass, composition of biomass and contamination of biomass. Each of these is discussed below.

3.4.1.1 Ash Content and Composition

Biomass contains a variety of metals that are necessary for the movement of nutrients within the plant, of which the most significant are potassium and sodium. Both are catalytically active in fast pyrolysis through cracking to water and CO_2 in the vapour phase, and ash contents above around 2.5 wt% often lead to a phase-separated product in significantly lower yields (Chiaramonti et al. 2007). Ash and inorganics can also arise from contamination during harvesting, as discussed below.

3.4.1.2 Water Content of Prepared Biomass

As bio-oil is oleophobic, all feed water reports to the bio-oil. Water is also formed from fast pyrolysis reactions; so, to maintain a reasonable water level in bio-oil, feed water is usually limited to 10 wt%. This typically gives water content of 25 wt% in the bio-oil. This level minimises the potential for phase separation and gives a manageable viscosity.

3.4.1.3 Composition of Biomass

In addition to hemicellulose, cellulose, and lignin, biomass can contain other components, including ash and water (as discussed above), contaminants (discussed below), and minor organic components (such as extractives, oils and proteins). The extractives and oils can lead to a separate phase separating at the top of the bio-oil. Proteins have high nitrogen content and lead to a distinctive and unpleasant odour.

3.4.1.4 Contamination of Biomass

Biomass can be contaminated in many ways, such as: chlorine from seaside locations and biocide applications; sulphur from fertiliser applications; metals and inorganic compounds from soil during harvesting, such as mud splashing during rain and accumulation from dragging over soil. All contaminants will have an influence on the yield and quality of the bio-oil produced, as discussed below.

3.4.2 Reactors

At the heart of a fast pyrolysis process is the reactor. Although it represents only one stage in the overall fast pyrolysis system, most research and development has focused on the reactor, although increasing attention is now being paid to control and improve liquid quality and improvement of collection systems. The rest of the process consists of biomass reception, storage and handling, biomass drying and grinding, product collection, storage, and, when relevant, upgrading. Several comprehensive reviews of fast pyrolysis processes for liquids production are available, for example, see Refs Czernik et al. (2004), Mohan et al. (2006), Bridgwater et al. (2002b).

Both the time-temperature profile of the gases and vapours, and the char and the liquid collection system influence the quality of bio-oil. Temperatures below 400 °C in the vapour handling section can result in fractional condensation of high molecular weight oligomers derived from lignin. While the removal of these oligomers reduces the viscosity and stability of the remaining bio-oil, control of this fractionation is difficult and usually results in blockage. Temperatures above around 550 °C promote secondary cracking reactions, resulting in lower liquid yields and higher water and gas yields, especially CO_2. This increases the propensity of the oil to age and phase separate. Systems with sand recycle thus need to balance the flow rate of hot sand with sand temperature to minimise these effects. Entrained-flow reactors require such a high temperature gradient between gas and pyrolysis to affect the necessary heating rates and reaction temperature that the liquid yield and quality usually suffer.

3.5 Norms and Standards

For bio-oil to successfully become a traded commodity, norms and standards are required. The first attempts at defining quality was carried out in 1995 by an international group from North America and Europe (Diebold et al. 1997). This was developed subsequently with exploration and development of standard tests for bio-oil (Oasmaa et al. 1997; Oasmaa and Czernik 1999; Oasmaa and Peacocke 2001; Gust et al. 2003) up to certification by CEN in Europe and ASTM in North America (ASTM D7544—09 2009; Oasmaa et al. 2009). The evaluation and development of test methods is very important in defining quality and setting standards for definition and marketing.

3.6 Bio-oil Upgrading

Bio-oil can be upgraded in a number of ways: physically, chemically and catalytically. This has been extensively reviewed (Diebold 2002; Czernik and Bridgwater 2004; Bridgwater 1966, 1994; Chheda et al. 2007). Some aspects are particularly significant and have attracted more extensive attention, including stability improvement, which is reviewed in Section Ageing.

3.6.1 Acidity or Low pH

Bio-oil has a typical pH of around 2.5 from the organic acids formed by degradation or cracking of the biopolymers that make up biomass, particularly the cellulose and hemicellulose. Hemicellulose can be preferentially reduced by washing in hot water or dilute acid and by torrefaction (see Section Smell below for references to torrefaction). Neither method is very effective, since cellulose is also affected in both methods. There has been little work on corrosion of metals in bio-oil (Darmstadt et al. 2004). The general view is that polyolefins and stainless steel are acceptable materials of construction. High acidity can be managed in several ways, including esterification and addition of magnesium powder, an alloy or a magnesium compound (Nemoto et al. 2013).

3.6.2 Ageing

Ageing or instability is a known problem that affects most bio-oils. It is caused by continued reaction of the decomposition products from fast pyrolysis (Diebold 2002). Polar solvents have been used for many years to homogenise and reduce the viscosity of biomass oils. The addition of solvents, especially methanol, showed a significant effect on the oil stability. Diebold and Czernik (1997) found that the rate of viscosity increase ("ageing") for the oil with 10 wt% of methanol was almost 20 times less than for the oil without additives. A stability test based on changes in viscosity is described in Section Viscosity.

3.6.3 Alkali Metals

All biomass contains ash in which alkali metals, notably potassium and sodium, dominate. Potassium, in particular, and other alkali metals are catalytically active and enhance secondary cracking reactions. This results in loss of liquid yield, higher water (and carbon dioxide) production, and potential phase separation from higher water content and loss of naturally derived surfactants that maintain the micro-emulsion of bio-oil.

Woody feeds typically contain up to 1 wt% ash, while grasses can range up to 8 wt% ash. The amount of ash in perennial crops depends upon harvesting time, since a significant proportion of the nutrients in above-ground plant matter returns to the rhizome at the end of the growing season. In addition, ash will be leached from the standing crop during winter from rainfall to potentially give ash contents below 2.5 wt%. The limiting value of ash content to reduce or avoid this effect is believed to be around 2.5 wt%, although this depends on other process parameters and the composition of the ash.

Washing biomass with water or dilute acid is feasible to remove ash, but is costly in financial and energy terms both for washing and subsequent drying. However, a further advantage of acid washing is the potential removal of hemicelluloses, from which are derived aldehydes and related degradation products that contribute an unpleasant odour to bio-oil and are partially responsible for the ageing effect. So, as with many other characteristics of bio-oil, effects are complex and can be difficult to evaluate comprehensively.

3.6.4 *Char*

Char acts as a vapour cracking catalyst, so rapid and effective separation from the pyrolysis product vapours is essential. Cyclones are the usual method of char removal; however, some fines always pass through the cyclones and collect in the liquid product, where they accelerate ageing and exacerbate the instability problem, which is described below. A more effective, but more difficult, method is hot-vapour filtration, which is also described below.

3.6.4.1 Cyclones

It is important to separate char as quickly as possible to minimise vapour cracking reactions. This is conventionally carried out in two or more cyclones, with the first, higher capacity cyclone removing coarse particles and the second removing fines. However, cyclones are not efficient with very small char particles, so these are usually carried over to the liquid collection system.

3.6.4.2 Filtration

Hot-vapour filtration can reduce the ash content of the oil to less than 0.01% and the alkali content to less than 10 ppm, much lower than reported for biomass oils produced in systems using only cyclones. There is limited information available on the performance or operation of hot-vapour filters, but they can be specified and perform similar to hot-gas filters in gasification processes (Diebold et al. 1994a). An alternative approach is insertion of the filter medium in the fluid bed (Hoekstra et al. 2009). Hot-vapour filtration gives a higher quality product with lower char however, the liquid yield is reduced by about 10–20% due to the char accumulating on the filter surface that cracks the vapours which is countered to some extent by preferential cracking of pyrolytic lignin in aerosol form on the char cake. More recent work using state-of-the-art filter elements reported similar results, with lower viscosities and reduced yields (Sitzmann and Bridgwater 2007).

Diesel engine tests performed on crude and on hot-filtered oil showed a substantial increase in burning rate and a lower ignition delay for the latter, due to the lower average molecular weight for the filtered oil (Shihadeh 1998). Hot-gas filtration has not yet been demonstrated over a long-term process operation. Despite its promise, only a few institutions have investigated hot-vapour filtration, including NREL (Diebold et al. 1994a), VTT, University of Twente (Hoekstra et al. 2009) and Aston University (Baker and Elliott 1988a).

Pressure filtration of the liquid for substantial removal of particulates (down to <1 μm) is very difficult because of the complex interaction of the char and pyrolytic lignin, which appears to form a gel-like phase that rapidly blocks the filter. Filtration down to 10 μm is not so difficult, but it increases process complexity and cost, as well as potentially leading to lower liquids yields due to losses. Modification of the liquid microstructure by addition of solvents, such as methanol or ethanol that solubilise the less-soluble constituents can improve this problem and contribute to improvements in liquid stability, as described below.

3.6.4.3 Slurries

An alternative approach to separation of char is to deliberately leave all of it in the bio-oil to create a bio-oil—char slurry. KIT (which used to be known as FZK) in Karlsruhe, Germany has developed and promoted the production and processing of slurries made from bio-oil and char (Raffelt et al. 2006a, b; Henrich et al. 2007). The liquid from straw pyrolysis in a twin-screw reactor is phase separated, but a homogeneous slurry from this separated liquid and the char is claimed. The slurry has been gasified by what was Future Energy (now Siemens) in Freiberg, Germany in a pressurised oxygen-blown gasifier (Volkmann 2004). There are unresolved questions over the source of energy for the pyrolysis reaction if all the char is used in the slurry; over the separation of char and sand from the reactor; and over the long-term stability of the slurry and its suitability for oxygen pressure gasification. Dynamotive also produced a bio-oil-char slurry known as Intermediate Bio-oil (Dynamotive 2007) that has also been tested at the Future Energy gasifier unit in Germany (now Siemens) (Dynamotive 2009).

3.6.5 Chlorine

Use of chlorine biomass, such as straw, will result in a high chlorine bio-oil from the hydrochloric acid formed in pyrolysis which is dissolved in the bio-oil. There is little work reported on measurement and control, but the levels rarely affect the low pH of the bio-oil, which is mostly due to organic acids. Combustion of a higher chlorine bio-oil will result in hydrogen chloride formation, which can be removed in conventional emissions control. The absence of alkali metals in a relatively char-free bio-oil would not be expected to lead to ash deposition and chlorine mobility.

3.6.6 Colour

The dark brown colour of bio-oil is only noticeable and potentially problematic in applications where it is visible. For example, use of whole bio-oil or fractionated bio-oil as a substitute for phenol in wood panel resins gives a dark brown-coloured resin. In many applications this is not a real problem but may cause customer concern from an aesthetic viewpoint. A similar problem of adverse consumer or customer reaction could arise in any application, where the conventional "white" material is replaced by a dark brown material; for example, diesel-bio-oil emulsions for engine fuel. Even hydrotreated bio-oil with greater than 99% deoxygenation that is water white when produced has been observed to blacken and thicken with time.

3.6.7 Contamination of Feed

Biomass can be contaminated by soil, etc. from the harvesting process. Forest residues that are removed by dragging over ground and energy grasses that are cut and recovered later by ground collection will accrete soil. Soil often contains a variety of metals, such as iron that can be catalytically active in fast pyrolysis (Salter 1999). The solution is either improved harvesting techniques and/or washing to remove contamination. An example is hog fuel washing experiments in Canada (Smith 2005). However, there is an energy and financial cost of washing which is difficult to justify in most cases unless more valuable chemical products are produced.

3.6.8 Distillability

Pyrolysis liquids cannot be completely vapourised once they have been recovered from the vapour phase. If the liquid is heated to 100 °C or more to try to remove water or distil off lighter fractions, it rapidly reacts and eventually produces a solid residue of around 50 wt% of the original liquid and some distillate containing volatile organic compounds and water. The distillate contains those compounds that are volatile, together with thermally cracked products from higher temperatures.

3.6.9 High Viscosity

While bio-oil has been successfully stored for several years in normal storage conditions in steel and plastic drums without any deterioration that would prevent its use in any of the applications tested to date, it does change slowly with time; most noticeably, there is a gradual increase in viscosity. More recent samples that have been distributed for testing in round-robin exercises have shown substantial improvements in consistency and stability, demonstrating the improvement in process design and control as the technology develops. Bio-oil viscosity is important particularly for direct combustion applications, where it needs to be atomised, such as in burners, engines and turbines. There is quite extensive testing of bio-oil in engines reviewed in Czernik et al. (2004), Leech (1997), Ormrod and Webster (2000) and in burners in Czernik et al. (2004), Oasmaa and Czernik (1999), Oasmaa et al. (2001). For engines, the preferred maximum viscosity is 17 cSt, above which the pressure requirements become excessive. Conventional fuels can be preheated to reduce viscosity, but above around 55 °C there is an irreversible change in the bio-oil properties; thus, preheating can only be used on a once through, very short residence time basis, as used in combustion tests by Canmet and MIT (reviewed in Bridgwater 2012). Viscosity is most affected by water content and temperature and is thoroughly covered in Diebold's review (2002).

3.6.10 Inhomogeneity

See section *Phase separation or Inhomogeneity* below.

3.6.11 Low H:C Ratio

The poor C:H ratio for bio-oil conversion to hydrocarbons means that either hydrogen needs to be added and/or carbon lost. This is considered more fully in section *Acidity or low pH above.*

3.6.12 Low pH

See section *Acidity or low pH* above.

3.6.13 Materials Incompatibility

The complex nature of bio-oil and the many different oxygenated compounds present mean that materials selection needs to be carefully considered. Polyolefins and stainless steel are acceptable materials for vessels and pipelines. Seals and gaskets can be sensitive to some of the organics, such as phenolics, and careful compatibility testing is needed for many polymers used as seals and gaskets. For example, synthetic rubber has been known to swell to three times its thickness when in contact with bio-oil.

3.6.14 *Miscibility with Hydrocarbons*

Pyrolysis oils are not miscible with hydrocarbon fuels so co-utilisation applications that require mixing are unsuitable without further processing, such as upgrading, blending, and emulsification.

3.6.14.1 Blending

As bio-oil is not miscible in any proportions with hydrocarbons, blending is only possible with polar compounds or mixtures, such as alcohols. This has been used to improve viscosity and reduce ageing, as reviewed by Diebold (2002). Blends have successfully been produced with biodiesel and butanol as a co-solvent (Alcala and Bridgwater 2013) and more recently also with diesel. Some blending with slow pyrolysis oil, diesel, and alcohols has been reported (Weerachanchai et al. 2009), although the bio-oil used is quoted as having a heating value of nearly 40 MJ/kg, which suggests that it is quite dissimilar to usual fast pyrolysis bio-oil.

3.6.14.2 Emulsions

Bio-oil can be emulsified with diesel oil with the aid of surfactants. A process for producing stable micro-emulsions with 5–30% of bio-oil in diesel has been developed at CANMET (Ikura et al. 1998). The University of Florence, Italy, has been working on emulsions of 5–95% bio-oil in diesel (Baglioni et al. 2001, 2003) to make either a transport fuel or a fuel for power generation in engines that does not require engine modification to dual-fuel operation. There is limited experience of using such fuels in engines or burners, but significantly higher levels of corrosion/erosion were observed in engine applications compared to bio-oil or diesel alone. A further drawback of this approach is the cost of surfactants and the energy required for emulsification.

3.6.15 *Nitrogen*

Biomass contains relatively small quantities of nitrogen which report to bio-oil. In most applications this is not a problem. However, some wastes with a high protein content, such as vegetable oil cake (e.g. rape or colza-meal) can have nitrogen contents above 5 wt% from the proteins in the waste. Two problems result from use of such feedstocks. The first is the very noxious smell of high-nitrogen bio-oils from the extensive nitrogen-containing degradation products. This is substantially worse than wood- or grass-derived bio-oil. Second, much of the nitrogen reports to the liquid, meaning that most applications will need to consider nitrogen pretreatment for its removal (hydro-denitrogenation) or post-use emissions control if the bio-oil is combusted or gasified. An interesting solution is recovery of the high-value proteins prior to pyrolysis, which becomes an embryonic biorefinery.

3.6.16 Other Solid Particulates, Excluding Char

The inert solids in fluid beds and CFBs, usually sand, will suffer from slow attrition and the fines are likely to substantially bypass cyclones. While not significant in terms of concentration (Strenziok et al. 2001), the solids could accumulate over time and create handling, pumping, and erosion problems. Solids are likely to be easier to filter than char, which seems to create a complex with bio-oil, but char would usually be the dominant solid contaminant compared with sand.

A potentially more significant problem is catalyst fines from fluid-bed and circulating-fluid-bed catalytic reactors. There is insufficient evidence that attrition is a serious problem, but the mechanical properties of catalysts need to be considered carefully.

3.6.17 Oxygen Content

Bio-oil approximates biomass in elemental composition with typically 40–45 wt% oxygen from the diverse oxygenated organic compounds. This means that it is not miscible with hydrocarbons, but miscible with polar solvents like methanol, ethanol, acetone, and so on. Upgrading to hydrocarbons for transport fuels or biofuels is reviewed in section *Chemical and catalytic upgrading of bio-oil* below. Hydrocarbon biofuel production thus requires the reduction or removal of oxygen, and there are many methods reviewed below.

3.6.18 Phase Separation or Inhomogeneity

The microstructure of bio-oil is briefly discussed in Sect. 5.20 below. Diebold has provided a thorough review of storage instability and methods for upgrading bio-oil (Lehto et al. 2010).

3.6.19 Smell

The liquid has a distinctive odour, an acrid smoky smell due to the low molecular weight aldehydes and acids, which can irritate the eyes on prolonged exposure. This characteristic smoky smell is exploited in the production of liquid smoke by several companies around the world. Pretreatment of biomass to reduce hemicelluloses, the source of the noxious components, will reduce the problem, but at the expense of lower yields, a more viscous product, and a significant energetic and/or financial cost. Pretreatment includes acid washing to preferentially hydrolyse hemicelluloses (Oasmaa et al. 2009) and torrefaction (Boerrigter et al. 2006; Zanzi et al. 2005; Prins et al. 2006) to pyrolyse the hemicelluloses preferentially. Neither is very efficient, as some cellulose is also converted by both methods, resulting in loss of yield and significant emissions control problems.

3.6.20 Structure of Bio-oil

Bio-oil is a complex mixture of water-soluble derivatives of cellulose and hemicellulose degradation and water-insoluble pyrolytic lignin. It is believed to be a micro-emulsion of pyrolytic lignin micelles around 500 Å in diameter maintained as an emulsion by a surfactant derived from the cracking reactions that create the liquid. This is a poorly researched area with few publications (Fratini et al. 2006). Dilution of the aqueous phase by adding water reduces the surfactant concentration to a level when it is no longer effective, resulting in phase separation. The complex aqueous-oil-surfactant relationship can be destroyed by a number of other circumstances, including cooling, heating, addition of emulsion-breaking additives, shear, etc. Phase separation is one of the consequences of ageing (see sections *Ageing* and *Water content*).

3.6.21 Sulphur

Biomass usually contains little sulphur. Wood is typically 0.05–0.1% S and some crops and wastes or residues can be higher. Sulfur levels in bio-oil have invariably been quoted in trace quantities (for example 0.03 mg/kg (ASTM D7544—09 2009) and have attracted little attention in utilisation activities. Sulfur in biomass feed would mostly report to the gas product. For synfuel applications, parts per million or even parts per billion levels would typically be required, which would normally be met by guard beds, such as zinc oxide.

3.6.22 Temperature Sensitivity

Bio-oil is formed by rapid freezing of a complex degradation process and the resultant liquid is therefore inherently unstable and wanting to continue reacting. This is the cause of the instability discussed previously. Raising the temperature, therefore, will increase the tendency for chemical reactions to continue. Up to around 55 °C the changes are reversible, so preheating to 50 °C or less will have no adverse effects on oil quality or behaviour. Above around 55 °C the changes are increasingly less reversible, and prolonged exposure to higher temperatures causes increased viscosity and an increased propensity for phase separation. Around 100 °C, bio-oil separates into a light bitumen-type phase mostly from the pyrolytic lignin and a low-viscosity aqueous fraction, but both are different to phase separated bio-oil at ambient conditions. The heavier material will hinder heat transfer and as temperatures increase will eventually carbonise to form a coke layer. This is what happens in attempts at distillation.

Temperature is widely used to control viscosity in combustion applications, but for bio-oil this needs to be carefully considered. In-line preheating immediately prior to combustion works well, but recirculation of a heated bio-oil, as for example, in some engine designs, needs to be managed carefully due to the temperature sensitivity of bio-oil.

3.6.23 *Toxicity*

As bio-oil becomes more widely available, attention will be increasingly placed on environment, health and safety aspects. A study was completed in 2005 to assess the ecotoxicity and toxicity of 21 bio-oils from most commercial producers of bio-oil around the world in a screening study, with a complete assessment of a representative bio-oil. The study included a comprehensive evaluation of transportation requirements as an update of an earlier study (Peacocke 2002) and an assessment of the biodegradability (Blin et al. 2007). The results are complex and require more comprehensive analysis, but the overall conclusion is that bio-oil offers no significant health, environment or safety risks.

3.6.24 *Viscosity*

Viscosity is important in many fuel applications (Antal et al. 1996). The viscosity of the bio-oil as produced can vary from as low as 25 cSt to as high as 1000 cSt (measured at 40 °C) or more, depending on the feedstock, the water content of the oil, the amount of light ends collected, and the extent to which the oil has aged.

An accelerated ageing or stability test was introduced to provide an index of stability that represents the effect of long-term storage at ambient conditions. This is known as a stability or ageing index (Oasmaa et al. 1997) and is measured by heating the sample to 80 °C for 24 h and comparing the increase in viscosity with the original viscosity. This is believed to approximate to 12 months' storage and is an indication of the propensity for viscosity increase and possible phase separation. A similar test can be carried out on water content which increases with ageing but this appears to be a less reliable method and does not correlate well with the viscosity method.

3.6.25 Water Content

Pyrolysis liquids can tolerate the addition of some water, but there is a limit to the amount of water which can be added to the liquid before phase separation occurs; in other words, the liquid cannot be dissolved in water. Water addition reduces viscosity, which is useful; it reduces heating value, which means that more liquid is required to meet a given duty; and it can improve stability. The effect of water, therefore is complex and important. Bio-oil is miscible with polar solvents, such as methanol, ethanol, acetone, etc., but totally immiscible with petroleum-derived fuels.

3.7 Chemical and Catalytic Upgrading of Bio-oil

Bio-oil can be upgraded in a number of ways: physically, chemically and catalytically. These have been extensively reviewed (Czernik et al. 2004; Maggi and Elliott 1997; Zhang et al. 2007) and only the more significant features are reported here. A summary of the main methods for upgrading fast pyrolysis products and the products is shown in Fig. 3.4.

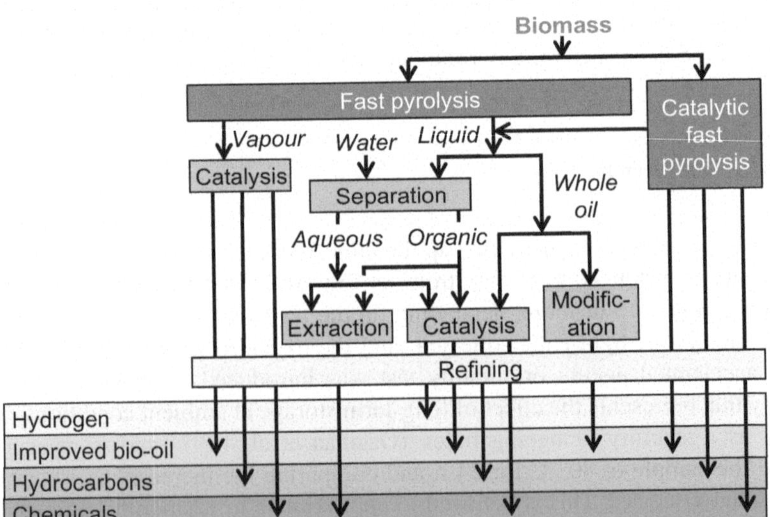

Fig. 3.4 Overview of fast pyrolysis upgrading methods

3.7.1 Physical Upgrading of Bio-oil

The most important properties that may adversely affect bio-oil fuel quality are incompatibility with conventional fuels from the high oxygen content of the bio-oil, high solids content, high viscosity and chemical instability.

3.7.1.1 Filtration

Hot-vapour filtration can reduce the ash content of the oil to less than 0.01% and the alkali content to less than 10 ppm, much lower than reported for biomass oils produced in systems using only cyclones. This gives a higher quality product with lower char; however, the liquid yield is reduced by about 10–20% due to the char accumulating on the filter surface that cracks the vapours. There is limited information available on the performance or operation of hot-vapour filters, but they can be specified and perform similarly to hot-gas filters in gasification processes.

Diesel engine tests performed on crude and on hot-filtered oil showed a substantial increase in burning rate and a lower ignition delay for the latter, due to the lower average molecular weight for the filtered oil (Shihadeh 1998). Hot-gas filtration has not yet been demonstrated over a long-term process operation. A consequence of hot-vapour filtration to remove char is the catalytic effect of the accumulated char on the filter surface, which potentially cracks the vapours, reduces yield by up to 20%, reduces viscosity, and lowers the average molecular weight of the liquid product. A little work has been done in this area by NREL (Diebold et al. 1994a) and VTT and Aston University (Sitzmann and Bridgwater 2007), but very little has been published.

Liquid filtration to very low particle sizes of below around 5 μm is very difficult due to the physico-chemical nature of the liquid and usually requires very high pressure drops and self-cleaning filters.

3.7.1.2 Solvent Addition

Polar solvents have been used for many years to homogenise and reduce the viscosity of biomass oils. The addition of solvents, especially methanol, showed a significant effect on the oil stability. Diebold and Czernik (1997) found that the rate of viscosity increase ("ageing") for the oil with 10 wt% of methanol was almost 20 times less than for the oil without additives.

3.7.1.3 Emulsions

Pyrolysis oils are not miscible with hydrocarbon fuels, but they can be emulsified with diesel oil with the aid of surfactants. A process for producing stable microemulsions with 5–30% of bio-oil in diesel has been developed at CANMET (Ikura et al. 1998). The University of Florence, Italy, has been working on emulsions of 5–95% bio-oil in diesel (Baglioni et al. 2001, 2003) to make either a transport fuel or a fuel for power generation in engines that does not require engine modification to dual-fuel operation. There is limited experience of using such fuels in engines or burners, but significantly higher levels of corrosion/erosion were observed in engine applications compared with bio-oil or diesel alone. A further drawback of this approach is the cost of surfactants and the high energy required for emulsification.

3.7.2 *Catalytic Upgrading of Bio-oil*

3.7.2.1 Natural Ash in Biomass

Before considering catalytic upgrading of bio-oil, it is important to appreciate first that biomass contains very active catalysts within its structure. These are the alkali metals that form ash and which are essential for nutrient transfer and growth of the biomass. The most active is potassium followed by sodium. These act by causing secondary cracking of vapours and reducing liquid yield and liquid quality.

Ash can be managed to some extent by selection of crops and harvesting time, but it cannot be eliminated from growing biomass. Ash can be reduced by washing in water or dilute acid, and the more extreme the conditions in both temperature and concentration, the more complete the ash removal. However, as washing conditions become more extreme, firstly hemicellulose and then cellulose are lost through hydrolysis. This reduces liquid yield and quality. In addition, washed biomass needs to have any acid removed as completely as possible and recovered or disposed of and the wet biomass has to be dried.

So washing is not often considered a viable possibility, unless there are some unusual circumstances, such as removal of contaminants. Another consequence of high ash removal is the increased production of levoglucosan, which can reach levels in bio-oil where recovery becomes an interesting proposition, although commercially, markets need to be identified and/or developed.

3.7.2.2 Upgrading to Biofuels

Upgrading bio-oil to a conventional transport fuel such as diesel, gasoline, kerosene, methane and LPG requires full deoxygenation and conventional refining, which can be accomplished either by integrated catalytic pyrolysis, as discussed

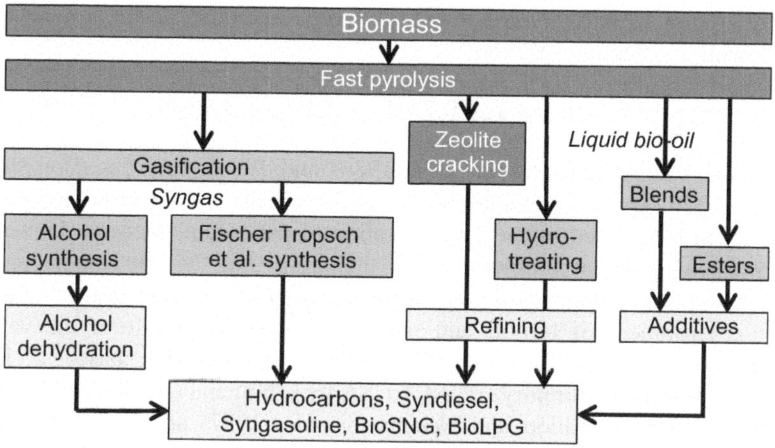

Fig. 3.5 Upgrading of bio-oil to biofuels and chemicals

above, or by decoupled operation, as summarised below and depicted in Fig. 3.5. There is also interest in partial upgrading to a product that is compatible with refinery streams in order to take advantage of the economy of scale and experience in a conventional refinery. Integration into refineries by upgrading through cracking and/or hydrotreating has been reviewed by Huber and Corma (2007):

- Hydrotreating;
- Catalytic vapour cracking;
- Esterification and related processes;
- Gasification to syngas followed by synthesis to hydrocarbons or alcohols.

3.7.2.3 Hydrotreating

Hydroprocessing rejects oxygen as water by catalytic reaction with hydrogen. This is usually considered as a separate and distinct process to fast pyrolysis that can, therefore, be carried out remotely. The process is typically high pressure (up to 200 bar) and moderate temperature (up to 400 °C) and requires a hydrogen supply or source (Elliott and Baker 1987). Full hydrotreating gives a naphtha-like product that requires orthodox refining to derive conventional transport fuels, however complete de-oxygenation is difficult to achieve and requires relatively extreme conditions. This refining would be expected to take place in a conventional refinery to take advantage of know-how, existing processes and economies of scale. A projected typical yield of naphtha equivalent from biomass is about 25% by weight or 55% in energy terms, excluding provision of hydrogen (Bridgwater 1966). Inclusion of hydrogen production by gasification of biomass reduces the yields to around 15 wt% or 33% in energy terms. The process can be depicted by the following conceptual reaction:

$$C_1H_{1.33}O_{0.43} + 0.77\,H_2 \rightarrow CH_2 + 0.43H_2O$$

The catalysts originally tested in the 1980s and 1990s were based on sulfided CoMo or NiMo supported on alumina or aluminosilicate and the process conditions are similar to those used in the desulfurisation of petroleum fractions. However, a number of fundamental problems arose including that the catalyst supports of typically alumina or aluminosilicates were found to be unstable in the high-water content environment of bio-oil and the sulfur was stripped from the catalysts requiring constant re-sulfurisation. The main activities were based at Pacific Northwest National Laboratory (PNNL), USA by Elliott and co-workers (e.g. Baker and Elliott 1988a, b; Elliott and Neuenschwander 1997) at UCL in Louvain la Neuve in Belgium by Maggi and co-workers (e.g. Maggi and Delmon 1994; Grange et al. 1996). This area has been thoroughly reviewed (Oasmaa et al. 1997). A recent design study of this technology for a biomass input of 2000 t/d (dry) for production of gasoline and diesel has been carried out by PNNL (Jones et al. 2009). A comprehensive review of unsupported metal sulphide hydrotreating catalysts was published in 2007 providing an up-to-date view (Eijsbouts et al. 2007).

More recently, attention turned to precious metal catalysts on less susceptible supports, and considerable academic and industrial research has been initiated in the last few years. Of note is the work by UOP in Chicago with (PNNL) in the USA to address the scientific and technical challenges and develop a cost-effective process (UOP 2008). Model compounds were used initially to understand the basic processes (Elliott and Hart 2009) and both whole oil and fractions have been evaluated. Tests have been carried out on both batch and continuous-flow processes and work to date has been based on relatively low-temperature (up to 380 °C) catalytic hydrogenation of bio-oil using different metal catalysts and processing conditions to give a range of products, including petroleum refinery feedstock.

Groningen University in the Netherlands is also active in fundamental research on hydrotreating bio-oils and model compounds using ruthenium on carbon (Wildschut et al. 2008, 2009). Different levels of upgrading are being studied from stabilisation with low levels of oxygen removal through mild hydrotreating to two-stage hydrotreatment with substantial oxygen removal (Ardiyanti et al. 2009).

At the Technical University of Munich, Lercher proposed a "one-pot" approach which is based on aqueous-phase hydro-deoxygenation of phenolic monomers using bifunctional catalysis that couples precious-metal-catalysed hydrogenation and acid-catalysed hydrolysis and dehydration (Zhao et al. 2009). There is still a significant hydrogen requirement which could, in principle, be derived from the aqueous phase after hydrotreatment. A more complex process involving hydrotreatment, esterification and cracking in supercritical ethanol using a palladium on zirconium with a SBA15 catalyst, has been researched in China (Tang et al. 2009). A significant improvement in many properties was reported. Many other organisations are active in hydrotreating, which were listed in Bridgwater (2012) although there has been a rapid increase in activity in this area around the world.

There is a substantial hydrogen requirement in all hydrotreating processes to hydrogenate the organic constituents of bio-oil and remove the oxygen as water. The hydrogen requirement can be represented by processing an additional amount of biomass to provide the hydrogen, by gasification for example. This is about 70–80% of that required to produce the bio-oil. The process is thus less efficient than the simple performance figures often presented. If only the organic fraction of bio-oil after phase separation is hydrotreated, then the hydrogen required can be produced by steam reforming the aqueous phase. There has been extensive research on reforming the aqueous fraction of bio-oil as discussed below. There is also a high cost from the high-pressure requirement (Jones et al. 2009; Cottam and Bridgwater 1994). Catalyst deactivation remains a concern from coking due to the poor C/H ratio.

In all cases the upgraded product will almost certainly require conventional refining to produce marketable products, and this would be expected to take place in a conventional refinery.

3.7.2.4 Zeolite Cracking

Zeolite cracking rejects oxygen as CO_2, as summarised in the conceptual overall reaction below:

$$C_1H_{1.33}O_{0.43} + 0.26\,O_2 \rightarrow 0.65\,CH_{1.2} + 0.34\,CO_2 + 0.27\,H_2O$$

Catalytic vapour cracking over acidic zeolite catalysts provides deoxygenation by simultaneous dehydration-decarboxylation producing mostly aromatics (Chang and Silvestri 1977) at 450 °C and atmospheric pressure. Oxygen is ultimately rejected as CO_2 and CO from a secondary oxidising reactor to burn off the coke deposited on the catalyst. This would operate much as an FCC in a refinery. The low H/C ratio in the bio-oils imposes a relatively low limit on the hydrocarbon yield. A projected typical yield of aromatics suitable for gasoline blending from biomass is about 20% by weight or 45% in energy terms (Diebold et al. 1994b). This publication also reported on technical and economic modelling by an IEA Bioenergy Task (Diebold et al. 1994b). The crude aromatic product would be sent for refining in a conventional refinery.

Catalyst deactivation remains a concern for both routes, although the coking problem with zeolites can in principle be overcome by a conventional FCC arrangement with continuous catalyst regeneration by oxidation of the coke. Some concern has been expressed over the poor control of molecular size and shape with orthodox zeolites and the propensity for formation of more noxious hydrocarbons

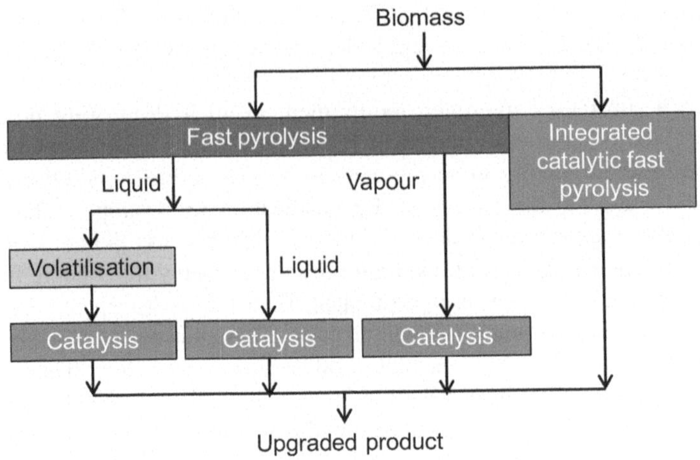

Fig. 3.6 Methods of upgrading fast pyrolysis products with cracking catalysts

(Williams and Horne 1994). The processing costs are high and the products are not competitive with fossil fuels (Bridgwater and Cottam 1992). The approach has only been studied at a relatively basic research level and considerably more development is necessary. Recently, some research is known to have taken place commercially in the USA and Brazil with limited success.

There are several ways in which this can be carried out as summarised in Fig. 3.6. The zeolite upgrading can operate on the liquid or vapours within or close-coupled to the pyrolysis process, or they can be decoupled to upgrade either the liquids or re-vapourised liquids.

A report by hydrocarbon processing for the future of fluid catalytic cracking (FCC) and hydroprocessing in modern refineries states that.

Biomass-derived oils are generally best upgraded by HZSM-5 or ZSM-5, as these zeolitic catalysts promote high yields of liquid products and propylene. Unfortunately, these feeds tend to coke easily, and high TANs [total acid numbers] and undesirable byproducts, such as water and CO_2 are additional challenges. Interestingly, the most recent attempts to use orthodox technology in conventional refineries for deoxygenating bio-oil is co-feeding bio-oil into an FCC unit which has been addressed by several organisations with limited success so far. This approach offers considerable technical and economic potential.

Many other organisations are active in deoxygenation by zeolites, which were listed in (Bridgwater 2012) although there has subsequently been a rapid increase in activity in this area around the world due to the relative lack of success in hydro-deoxygenation.

Integrated Catalytic Pyrolysis

There is increasing interest in improving the quality of bio-oil, often without defining what quality means, and there have been a number of developments in the last few years that integrate or combine catalysis with pyrolysis. These combined pyrolysis—catalysis reaction systems have been studied by a number of organisations including several commercial developments. There is a rapidly expanding research community on catalytic (fast) pyrolysis but with little evidence of the necessary breakthroughs. Two notable commercial developments are included for special mention:

Huber from the University of Massachusetts Amhurst (now at the University of Wisconsin Madison) has also worked extensively in this area and developed a process that pyrolyses biomass in the presence of ZSM-5 into gasoline, and also diesel fuel, heating oil and renewable chemicals including benzene, toluene and xylenes in a one-step process. The product is referred to as Grassoline, and a spin out company has been formed—Anellotech (Huber and Bale 2009).

BioECon from the Netherlands formed a joint venture with KIOR (2010) to exploit their respective technologies. Little information is available other than modified clays are some of the materials studied and that one approach is impregnation of the biomass with nano-catalysts prior to reaction (Jonietz 2007). Success was claimed at temperatures as low as 230 °C (O'Connor et al. 2007). Unfortunately, in spite of considerable investment KIOR was unable to create a commercially viable venture and went into liquidation at the end of 2014 after investments totalling over USD 600 million.

Integration of catalysis and pyrolysis requires operation at a single temperature and sufficiently robust catalyst to withstand the temperature and mechanical environment and with sufficient residence for the catalyst to be effective, which is a challenging requirement. There is, therefore, less flexibility in operating conditions, suggesting that catalytic systems will need to be quite sophisticated and/or sequential with carefully chosen processes and conditions that do not rely on single reactors and catalysts. Since coking is a recognised problem and is the mechanism by which oxygen is rejected from the bio-oil and CO_2, catalyst regeneration is an essential aspect of a reactor design.

Decoupled Vapour Upgrading from Volatilisation of Bio-oil

This approach was extensively investigated at the University of Saskatchewan in the 1990s and widely reported and reviewed (Bridgwater 1966; Valle et al. 2007). Devolatilisation of bio-oil suffers from the temperature sensitivity of bio-oil and the highest yield of vapours would not be expected to be greater than 50 wt%. Data on analyses of the re-volatised bio-oil prior to catalytic conversion are not known to be available. The University of the Basque Country has investigated a close-coupled liquid bio-oil preheated fluid bed zeolite cracking reactor (Cortright et al. 2002). Separation of thermal pretreatment from catalytic upgrading was found to reduce coking, but then the proposal for secondary upgrading of thermally degraded products in the pretreatment section suggest potential for blockage. This is analogous to the work at Saskatchewan.

3.7.3 *Other Methods for Chemical Upgrading of Bio-oil*

This section includes non-physical methods and those catalytic processes not covered in hydrotreating and zeolite-related processes. A list of known research activities was published in 2011 (Bridgwater 2012), but subsequently there has been considerable growth in this area.

Aqueous-phase processing is a relatively new approach that was first proposed by Dumesic and co-workers, who produced hydrogen and alkanes from aqueous solutions of oxygenated hydrocarbons through aqueous-phase reforming and dehydration/hydrogenation (Cortright et al. 2002; Huber and Dumesic 2006). A large fraction of bio-oil is water soluble and the compounds present in its aqueous fraction are mainly oxygenated hydrocarbons. This shows that the concept of aqueous-phase processing can be used to produce hydrogen and alkanes from the aqueous fraction of bio-oil.

Mild cracking is an alternative to orthodox zeolite-based cracking is mild cracking over base catalysts that address only the cellulose-and hemicellulose-derived products and aims to minimise coke and gas formation. Crofcheck at the University of Kentucky (Fisk et al. 2006) has explored ZnO and freshly calcined Zn/Al and Mg/Al layered double hydroxides to upgrade a synthetic bio-oil based on the earlier work in Finland (Nokkosmaki et al. 2000).

Esterification with alcohols up to butanol is being investigated and developed for improving the quality of bio-oil without substantial deoxygenation. Properties that are mostly addressed are water content, acidity, stability and reactivity.

Steam reforming provides a route to convert the water-soluble (carbohydrate-derived) fraction of bio-oil to hydrogen (Wang et al. 1997; Czernik et al. 2002) by steam reforming. This has been accomplished in a fluidised-bed process by several researchers using commercial, nickel-based catalysts under conditions similar to those for reforming natural gas. The process depends on a viable use for the organic lignin-derived fraction of bio-oil; for example use as a phenol replacement in phenol-formaldehyde resins (Mann et al. 1996) or for upgrading this organic fraction.

Model compounds and model bio-oil are often used to better understand the underlying chemical processes. While scientifically credible and providing a consistent and well-characterised feed material, there are dangers in relying too much on synthetic bio-oil, since no mixture can adequately represent the complex composition and behaviour of fast pyrolysis liquid.

3.7.4 Hydrogen

Hydrogen is produced in the syngas from gasification of bio-oil and bio-oil/char slurries as described above. There are also activities in both non-catalytic partial oxidation and catalytic partial oxidation and catalytic steam reforming of the whole bio-oil and the aqueous fraction after phase separation, particularly to meet the hydrogen demands of a hydrotreating process as described above. Catalysts are usually based on nickel or precious metals. A comprehensive list of activities in 2011 has been published (Bridgwater 2012) but as with other catalytic methods, there has been rapid expansion of research in this area.

3.7.5 Chemicals

Since the empirical chemical composition of biomass, approximately $(CH_2O)_n$, is quite different from that of petroleum $(CH_2)_n$, the range of primary chemicals that can be easily derived from biomass and oil are quite different. Hence, any biomass-based chemical industry will necessarily be constructed on quite a different selection of simple 'platform' chemicals than those currently used in the petrochemical industry. Given the chemical complexity of biomass, there is some choice of which platform chemicals to produce, since, within limits, different processing strategies of the same biomass can lead to different breakdown products. However, once a set of platform chemicals has been chosen for bio-based chemical production and the appropriate network of production plants is established, it will become be increasingly difficult to change that choice without disrupting the whole manufacturing infrastructure. Since the available biomass will inevitably show major regional differences, it is quite possible that the choice of platform chemicals derived from biomass will show much more geographical variation than in petrochemical production. Examples of chemicals derived from bio-oil are listed in Table 3.4 (Bridgwater 2012).

Table 3.4 Some chemicals recovered from bio-oil

Acetic acid	Hydrogen	Organic acids
Alcohols	Hydroxyacetaldehyde	Pharmaceuticals
Aldehydes	Ketones	Phenol and phenolics
Anhydrosugars	Levoglucosan	Plastics and polymers
Asphalt	Levoglucosenone	Preservative
Biolime	Liquid smoke and related products	Resins and adhesives
Formic acid	Non-aromatic alcohols	Slow release fertilisers
Furfural		Synthesis gas
		Wood preservative

3.7.5.1 Chemical Composition of Bio-oil

The chemicals in bio-oil are derived from random thermal decomposition of hemicellulose, cellulose and lignin. Over 400 individual chemicals have been identified in bio-oil, and this area has been reviewed by Diebold (2002). There are many papers that provide details of bio-oil analyses, as analytical techniques have rapidly developed.

Strategies for separation or recovery of any of these chemicals need to consider market, value, costs and process. The natural first step is to evaluate components with the highest concentration, since the processing will be easier and the costs lower. However, this may not prove the best strategy, and methodologies need to carefully consider capital and operating costs, product values and residue or waste utilisation or disposal. The opportunities for optimisation are considerable and challenging, and are likely to involve a range of process generation, evaluation and optimisation tools including process synthesis and linear programming.

3.7.5.2 Production of Chemicals

For many centuries wood pyrolysis liquids were a major source of chemicals, such as methanol, acetic acid, turpentine, tars etc. At present, most of these compounds can be produced at a lower cost from fossil fuel feedstocks. Although over 300 compounds have been identified in wood fast pyrolysis oil, their concentrations are usually too low to consider separation and recovery. Up to now, therefore, the development of technologies for producing products from the whole bio-oil or from its major, relatively easy separable fractions is the most developed. A more detailed review on this subject, including consideration of higher value products, was published by Radlein (1999) and a thorough review of the literature on production of chemicals utilising whole oil, fractions of bio-oil, and specific chemicals was published in 2004 (Diebold 2002). A comprehensive review of phenolics recovery and utilisation has been published by Amen Chen et al. (2001).

Chemicals are always attractive commercial possibilities owing to their much higher added value compared to fuels and energy products. This suggests a bio-refinery concept in which the optimum combinations of fuels and chemicals are produced.

3.8 Conclusions

There has been a very considerable expansion of activity in fast pyrolysis in the last 15 years exploring novel processes for production of more useful and valuable products from bio-oil. This is due to the recognition of the value of a crude liquid that can be more easily handled, stored and transported than solid or gas with the potential for enhanced bioenergy and biofuel chains.

Attention has focused on two main areas:

- improving the quality of bio-oil to reduce or avoid problems in direct use, which requires identification and specification of the qualities concerned;
- upgrading to a more conventional product that can be more readily assimilated into existing fuel infrastructures. This includes transport fuels, synthesis gas for transport fuels and chemicals, hydrogen and commodity and speciality chemicals.

Quality can be defined as any one or combination of over 25 characteristics of bio-oil that affect its usage, so it is important to identify which characteristic or characteristics require modification and then address those properties.

There is increasing interest in higher value and more orthodox products, such as transport fuels and hydrogen, which has seen considerable growth of R&D activity. The latter can be partly explained by the requirement for significant amounts of hydrogen for some upgrading processes for production of hydrocarbon fuels and also for decentralised production of hydrogen for fuels cells, as hydrogen is costly to store and transport.

The relatively low hydrogen content of bio-oil invariably results in coking of catalysts in catalytic upgrading processes. Some solutions have been sought in more sophisticated catalyst systems that require less severe conditions and also in multi-stage upgrading, where bio-oil is processed in a series of steps to give a progressively upgraded product. Liquid processing is generally preferred to avoid problems of vapourising bio-oil with consequent loss of carbon as coke unless oxidative processing is included to oxidise any carbon that is formed.

The use of model compounds makes for easier fundamental science, but it is doubtful if any single chemical or even small number of chemicals can adequately reproduce the complexity of whole bio-oil with interactions between the constituent chemicals.

The considerable growth in activity around the world demonstrates the exciting opportunities and future for fast pyrolysis which will move from laboratory to commercial reality in a surprisingly short time.

References

Alcala A, Bridgwater AV (2013) Upgrading fast pyrolysis liquids: blends of bio-oil and biodiesel. Fuel 109:417–426

Amen-Chen C, Pakdel H, Roy C (2001) Production of monomeric phenols by thermochemical conversion of biomass—a review. Bioresource Technol 79:277–299

Antal MJ Jr et al (1996) High yield biomass charcoal. Energy Fuels 10(3):652–658

Ardiyanti AR, Venderbosch RH, Heeres HJ (2009) Process-product studies on pyrolysis oil upgrading by hydrotreatment with Ru/C catalysts. In: AICHE 2009 spring meeting, University of Groningen

ASTM D7544—09 (2009) Standard specification for pyrolysis liquid biofuel, ASTM International

Baglioni P et al (2001) BCO/diesel oil emulsification: main achievements of the emulsification process and preliminary results of tests on diesel engine. In: Bridgwater AV (ed) Progress in thermochemical biomass conversion, Oxford, UK, pp 1525–1539
Baglioni P et al (2003) Development of bio crude oil/diesel oil emulsions and use in diesel engines —Part 1: emulsion production. Biomass Bioenergy 25:85–99
Baker EG, Elliott DC (1988a) Catalytic upgrading of biomass pyrolysis oils. In: Bridgwater AV, Kuester JL (eds) Research in thermochemical biomass conversion. Phoenix, Arizona, USA, pp 883–895
Baker EG, Ellior DC (1988b) Catalytic hydrotreating of biomass derived-oils. In: Soltes J, Milne TA (eds) Pyrolysis oils from biomass—producing, analysing and upgrading. American Chemical Society, Washington DC, pp 228–240
Blin J et al (2007) Biodegradability of biomass pyrolysis oils: comparison to conventional petroleum fuels and alternatives fuels in current use. Fuel 86:2679–2686
Boerrigter H, Kiel J, Bergman P (2006) Biomass pre-treatment by torrefaction. In: Third ThermalNet meeting. Lille, France
Bridgwater AV (1966) Production of high-grade fuels and chemicals from catalytic pyrolysis of biomass. Catal Today 29:285–295
Bridgwater AV (1994) Catalysis in thermal biomass conversion. Appl Catal A 116:5–47
Bridgwater AV (2003) Renewable fuels and chemicals by thermal processing of biomass. Chem Eng J Article 91:87–102
Bridgwater AV (2009) Fast pyrolysis of biomass. In: Hofbauer H, Bridgwater AV, van Loo S (ed) Thermal biomass conversion. CPL Press, England
Bridgwater AV (2011) Review of fast pyrolysis of biomass and product upgrading. Biomass Bioenergy 1–27
Bridgwater AV (2012) Upgrading biomass fast pyrolysis liquids. In: Brown RC (ed) Thermochemical processing of biomass: conversion into fuels, chemicals and power. Wiley, New York
Bridgwater AV, Cottam M-L (1992) Opportunities for biomass pyrolysis liquids production and upgrading. Energy Fuel 6(2):113–120
Bridgwater AV, Czernik S, Piskorz J (2002a) The status of biomass fast pyrolysis. In: Bridgwater AV (ed) Fast pyrolysis of biomass: a handbook. CPL Press Liberty House, Newbury, pp 1–22
Bridgwater AV, Czernik S, Piskorz J (2002b) The status of biomass fast pyrolysis. In: Bridgwater AV (ed) Fast pyrolysis of biomass a handbook, vol 2. CPL Press Liberty House, Newbury, UK, pp 1–22
Bridgwater AV, Peacocke GVC, and Robinson NM
Chang C, Silvestri AJ (1977) The conversion of methanol and other O-compounds to hydrocarbons over zeolite catalysts. J Article Catal 47:249–259
Chen M (2006) Fast pyrolysis of biomass in a spout-fluidized bed reactor—analysis of composition and combustion characteristics of liquid product from biomass. Chin J Article Process Eng 6(2):192–196
Chheda JN, Huber GW, Dumesic JA (2007) Review—liquid-phase catalytic processing of biomass-derived oxygenated hydrocarbons to fuels and chemicals. Angew Chem Int Ed 46 (8):7164–7183
Chiaramonti D, Oasmaa A, Solantausta Y (2007) Power generation using fast pyrolysis liquids from biomass. Renew Sustain Energy Rev 11:1056–1086
Cortright RD, Davda RR, Dumesic JA (2002) Hydrogen from catalytic reforming of biomass-derived hydrocarbons in liquid water. Nature 418:964–967
Cottam M-L, Bridgwater AV (1994) Techno-economic modelling of biomass flash pyrolysis and upgrading systems. Biomass Bioenergy 7:267–273
Czernik S, Bridgwater AV (2004) Applications of biomass fast pyrolysis oil. Energy Fuel 18:590–598
Czernik S et al (2002) Hydrogen by catalytic steam reforming of liquid byproducts from biomass thermoconversion processes. Ind Eng Chem Res 41:4209–4215

Czernik et al (2004) Overview of application of biomass fast pyrolysis oil. Energy Fuels 18:590–598

Darmstadt H et al (2004) Corrosion of metals by bio-oil obtained by vacuum pyrolysis of softwood bark residues. An X-ray photoelectron spectroscopy and auger electron spectroscopy study. Energy Fuels 18(5):1291–1301

Diebold JP (2002) A review of the chemical and physical mechanisms of the storage stability of fast pyrolysis bio-oils. In: Bridgwater AV (ed) Fast pyrolysis of biomass: a handbook. National Renewable Energy Laboratory, Golden, pp 243–292

Diebold JP, Czernik S (1997) Additives to lower and stabilize the viscosity of pyrolysis oils during storage. Energy Fuels 11:1081–1091

Diebold JP, Power A (1988) Engineering aspects of the vortex pyrolysis reactor to produce primary pyrolysis oil vapours for use in resins and adhesives. In: Bridgwater AV, Kuester JL (eds) Research in thermochemical biomass conversion. Elsevier Applied Science, pp 609–628

Diebold JP et al (1994a) Hot-gas filtration to remove char from pyrolysis vapours produced in the vortex reactor at NREL. In: Biomass pyrolysis oil properties and combustion meeting. National Renewable Energy Laboratory, Boulder, CO

Diebold JP et al (1994b) IEA techno-economic analysis of the thermochemical conversion of biomass to gasoline by the NREL process. In: Bridgwater AV (ed) Advances in thermochemical biomass conversion, pp 1325–1342

Diebold JP et al (1997) Proposed specifications for various grades of pyrolysis oils. In: Bridgwater AV, Boocock DGG (eds) Developments in thermochemical biomass conversion. Springer, Netherlands, pp 433–447

Dynamotive (2007) Dynamotive introduces higher energy content BioOil. In: Green car congress

Dynamotive (2009) Dynamotive bio-oil information booklet

Eijsbouts S, Mayo SW, Fujita K (2007) Unsupported transition metal sulfide catalysts: from fundamentals to industrial application. Appl Catal A 322:58–66

Elliott DC, Baker EG (1987) Hydrotreating biomass liquids to produce hydrocarbon fuels. In: Klass D (ed) Energy from biomass and wastes X, pp 765–784

Elliott DC, Neuenschwander GN (1997) Liquid fuels by low-severity hydrotreating of biocrude. In: Bridgwater AV, Boocock DGB (eds) Developments in thermochemical biomass conversion, London, pp 611–621

Elliott DC, Hart TR (2009) Catalytic hydroprocessing of chemical models for bio-oil. Energy Fuels 23(2):631–637

Fernandez AR (2010) Ikerlan, IK4 fast pyrolysis pilot plant: bio-oil and char production from biomass. In: PyNe newsletter. Aston University Bioenergy Research Group, pp 8–10

Fisk C et al (2006) Novel approaches to catalytic upgrading of bio-oil

Fratini E et al (2006) SANS analysis of the microstructural evolution during the aging of pyrolysis oils from biomass. Langmuir 22(1):306–312

Grange P et al (1996) Hydrotreatment of pyrolysis oils from biomass—reactivity of the various categories of oxygenated compounds and preliminary technoeconomic study. Catal Today 29 (1–4):297–301

Gust S, Nieminen J-P (2002) Liquefied wood fuel could soon replace heavy oil! Wood Energy 6:24–34

Gust S et al (2003) Determination of basic fuel quality standards for biomass-derived pyrolysis liquids. In: Pyrolysis and gasification of biomass and waste, p 706

Henrich E et al (2007) The Karlsruhe "bioliq" process for biomass gasification. In: Summer school. University of Warsaw, Poland

Hoekstra E et al (2009) Fast pyrolysis of biomass in a fluidized bed reactor: in situ filtering of the vapors. Ind Eng Chem Res 48(10):4744–4756

Hornung A et al (2007) Thermo-chemical conversion of energy crops—haloclean—intermediate pyrolysis. In: 6th international congress on valorisation and recycling of industrial waste. L'Aquila, Italy

Huber GW, Bale B (2009) Grassoline at the pump. Sci Am 301:52–59

Huber GW, Corma A (2007) Synergies between bio- and oil refineries for the production of fuels from biomass. Angewandte Chemie Int Ed 46(38):7184–7201
Huber GW, Dumesic JA (2006) An overview of aqueous-phase catalytic processes for production of hydrogen and alkanes in a biorefinery. Catal Today 111:119–132
Ikura M, Slamak M, Sawatzky H (1998) Pyrolysis liquid-in-diesel oil microemulsions
Ingram L et al (2008) Pyrolysis of wood and bark in an auger reactor: physical properties and chemical analysis of the produced bio-oils. Energy Fuels 22(1):614–625
Jones SB et al (2009) Production of gasoline and diesel from biomass via fast pyrolysis, hydrotreating and hydrocracking: a design case. In: Pacific Northwest National Laboratory. Richland, Washington
Jonietz E (2007) Oil from wood
Kersten SRA et al (2005) Biomass pyrolysis in a fluidized bed reactor. Part 1: literature review and model simulations. Ind Eng Chem Res 44:8773–8785
KIOR (2010) Welcome to KIOR.com
Kovac RJ, O'Neil DJ (1989) The Georgia Tech entrained flow pyrolysis process. In: Maniatis K, Ferrero GL, Buekens A, Bridgwater AV (eds) Pyrolysis and gasification, pp 169–179
Leech J (1997) Running a dual fuel engine on pyrolysis oil. In: Kaltschmitt M, Bridgwater AV (eds) Biomass gasification and pyrolysis, state of the art and future prospects. Newbury, UK, pp 495–497
Lehto J et al (2010) Integrated heat, electricity and bio-oil. In: PyNe newsletter, Aston University Bioenergy Research Group, pp 2–10
Lédé J et al (1985) Fast pyrolysis of wood: direct measurement and study of ablation rate. Fuel 64:1514–1520
Maggi R, Delmon B (1994) Characterisation of bio-oils produced by pyrolysis. In: Bridgwater AV (ed) Advances in thermochemical biomass conversion, London, pp 1086–1094
Maggi R, Elliott D (1997) Upgrading overview. In: Bridgwater AV, Boocock DGG (eds) Developments in thermochemical biomass conversion, pp 575–588
Maniatis K et al (1993) The Egemin flash pyrolysis process: commissioning and results. In: Bridgwater AV (ed) Developments in thermochemical biomass conversion. Boocock DGG, pp 1257–1264
Mann MK, Spath PL, Kadam K (1996) Technical and economic analysis of renewables-based hydrogen production. Stuttgart, Germany
Marker T, Felix L, Linck M (2009) Integrated hydropyrolysis and hydroconversion process for production of gasoline and diesel fuel from biomass, AICHE
McLellan R (2000) In: PyNe newsletter. Aston University, Birmingham, UK, p 12
Meier D, *New ablative pyrolyser in operation in Germany*, in *PyNe newsletter*, Aston University Boenergy Research Group, Editor. 2005. p. 1–3
Mohan D, Pittman CU, Steele PH (2006) Pyrolysis of wood/biomass for bio-oil: a critical review. Energy Fuels 20:848–889
Muggen G (2010) Empyro project summary. In: PyNe newsletter, Aston University Bioenergy Research Group, pp 3–5
Muller S (2010) Ensyn technologies, In: PyNe newsletter. Aston University Bioenergy Research Group. pp 11–12
Nemoto I et al (2013) Method for treating bio-oil
Nokkosmaki MI et al (2000) Catalytic conversion of biomass pyrolysis vapours with zinc oxide. J Article Anal Appl Pyrol 55:119–131
Oasmaa A, Czernik S (1999) Fuel oil quality of biomass pyrolysis oils state of the art for the end users. Energy Fuels 13(4):914–921
Oasmaa A, Peacocke C (2001) A guide to physical property characterisation of biomass-derived pyrolysis liquids. Technical Research Centre of Finland, Espoo, p 102
Oasmaa A, Leppämäki E, Koponen P et al (1997) Physical characterisation of biomass-based pyrolysis liquids: application of standard fuel oil analyses. VTT Publication, p 87
Oasmaa A et al (1997) Physical characterisation of biomass-based pyrolysis liquids: application of standard fuel oil analyses, VTT Publications, Espoo

Oasmaa A, Elliott DC, Müller S (2009) Quality control in fast pyrolysis bio-oil production and use. Environ Progress Sustain Energy 28(3):404–409
Oasmaa A, Kytö M, Sipilä K (2001) Pyrolysis oil combustion tests in an industrial boiler. In: Bridgwater AV (ed) Progress in thermochemical biomass conversion. Oxford, UK, pp 1468–1481
O'Connor P, van der Meij R (2007) Biomass conversion: a sustainable path to clean renewable energy fuels and chemicals, Vienna
Ormrod D, Webster A (2000) Progress in utilization of bio-oil in diesel engines. In: PyNe newsletter, Birmingham, UK
Peacocke GVC (2002) Transport handling and storage of fast pyrolysis liquids. In: Fast pyrolysis of biomass: a handbook, pp 293–338
Peacocke GVC, Bridgwater AV (1995) Ablative plate pyrolysis of biomass for liquids. Biomass Bioenergy 7:147–154
Prins W, Wagenaar BM (1997) Review of rotating cone technology for flash pyrolysis of biomass. In: Bridgwater AV, Kaltschmitt MK (eds) Biomass gasification and pyrolysis, UK, pp 316–326
Prins MJ, Ptasinski KJ, Janssen FJJG (2006) Torrefaction of wood: Part 1. Weight loss kinetics. J Article Anal Appl Pyrol 77(1):28–34
Radlein D (1999) The production of chemicals from fast pyrolysis bio-oils. In: Bridgwater AV et al (eds) Fast pyrolysis of biomass: a handbook. Newbury, UK, pp 164–188
Raffelt K et al (2006a) Preparation and characterisation of biomass slurries: a new feed for entrained flow gasification. In: Boocock DGB, Bridgwater AV (eds) Science in thermal and chemical biomass conversion
Raffelt K et al (2006b) The BTL2 process of biomass utilization entrained-flow gasification of pyrolyzed biomass slurries. Appl Biochem Biotechnol 129:153–164
Salter EH (1999) Catalytic pyrolysis of biomass for improved liquid fuel quality. Aston University, Birmingham
Scott DS, Piskorz J (1982) The flash pyrolysis of aspen poplar wood. Can J Chem Eng 60:666–684
Scott DS, Piskorz J, Radlein D (1985) Liquid products from the continuous flash pyrolysis of biomass. Ind Eng Chem Process Des Dev 24:581–588
Scott DS et al (1997) Fast pyrolysis of biomass for recovery of speciality chemicals. In: Bridgwater AV, Boocock DGG (eds) Developments in thermochemical biomass conversion. Springer, Dordrecht, pp 523–535
Shihadeh AL (1998) Rural electrification from local resources: biomass pyrolysis oil combustion in a direct injection diesel engine. Massachusetts Institute of Technology
Sitzmann J, Bridgwater AV (2007) Upgrading fast pyrolysis oils by hot vapour filtration. In: 15th European energy from biomass conference, Berlin
Smith PW (2005) CCE-MultiContactor technology—a new process intensification tool with major applications in the cleaning of biomass with energy recovery. In: World congress of chemical engineering, Glasgow
Strenziok R, Hansen U, Künster H (2001) Combustion of bio-oil in a gas turbine. In: Bridgwater AV (ed) Progress in thermochemical biomass conversion, Oxford, UK
Tang Z et al (2009) One step bio-oil upgrading through hydrotreatment, esterification, and cracking. Ind Eng Chem Res 48(15):6923–6929
UOP (2008) Proven pyrolysis oil technology for high quality fuels with a reduced carbon footprint
Valle B et al (2007) Integration of thermal treatment and catalytic transformation for upgrading biomass pyrolysis oil. Int J Article Chem Reactor Eng 5:86
Volkmann D (2004) Update on technology and projects (at future energy). In: Gasification technologies conference, Washington DC
Wagenaar BM et al (2001) Rotating cone bio-oil production and applications. In: Bridgwater AV (ed) Progress in thermochemical biomasss conversion. pp 1268–1280
Wang D et al (1997) Biomass to hydrogen via fast pyrolysis and catalytic steam reforming of the pyrolysis oil or its fractions. Ind Eng Chem Res 36:1507–1518
Weerachanchai P, Tangsathitkulchai C, Tangsathitkulchai M (2009) Phase behaviors and fuel properties of bio-oil-diesel-alcohol blends. World Acad Sci Eng Technol 56:387–393

Wildschut et al (2008) Experimental studies on the upgrading of fast pyrolysis oil to liquid transportation fuels. In: 235th ACS meeting, New Orleans (United States of America)
Wildschut J et al (2009) Catalytic hydrotreatment of fast pyrolysis oil: model studies on reaction pathways for the carbohydrate fraction. Environ Progress Sustain Energy 3
Williams PT, Horne PA (1994) Characterisation of oils from the fluidised bed pyrolysis of biomass with zeolite catalyst upgrading. Biomass Bioenergy 7:223–226
Yang J et al (2001) Modelling, scale-up and demonstration of a vacuum pyrolysis reactor. In: Bridgwater AV (ed) Progress in thermochemical biomasss conversion, pp 1296–1311
Yang Y et al (2014) Intermediate pyrolysis of biomass energy pellets for producing sustainable liquid, gaseous and solid fuels. Bioresour Technol 169:794–799
Zanzi R et al (2005) Biomass torrefaction. In: 14th European biomass conference & exhibition, Paris, France
Zhang Q et al (2007) Review of biomass pyrolysis oil properties and upgrading research. Energy Convers Manag 48:87–92
Zhao C et al (2009) Highly selective catalytic conversion of phenolic bio-oil to alkanes. Angew Chem Int Ed 48(22):3987–3990
Zhu X (2009) Biomass fast pyrolysis for bio-oil

Author Biography

Anthony Bridgwater, Ph.D. is a Professor of Chemical Engineering at Aston University in Birmingham, specialising in thermal conversion of biomass for production of fuels and chemicals. He obtained his first degree in Chemical Engineering from UMIST followed by several years working for BP in Sunbury. After returning to Aston, he gained his Ph.D. and D.Sc. Most of his professional career has been spent at Aston University, where he leads the European Bioenergy Research Institute (EBRI) and Bioenergy Research Group within EBRI. His current interests are focused on the development of fast pyrolysis of biomass and the fuel and chemical products that can be derived from the liquids. He plays a key role in several major EC funded projects focusing on bioenergy, biofuels and biorefineries. He has been actively involved in bioenergy for over 25 years and has published extensively. He was a Technical Director of the UK Flagship SUPERGEN Bioenergy programmes for over 8 years. He has been involved in winning over £25 million in research grants, which established the foundation for the European Bioenergy Research Institute (EBRI) at Aston University.

Chapter 4
Biomass Conversion Technologies: Biological/Biochemical Conversion of Biomass

Luisa Gouveia and Paula C. Passarinho

Abstract The biorefinery concept harbors a wide range of conversion technologies in order to take advantage of all the components of a feedstock. Among those technologies, a wide range of biological/biochemical processes can be envisaged to generate products such as biofuels, value-added products, and other chemical building blocks. These processes are generally defined as fermentative although each of them requires special operational conditions (e.g., anaerobic environment, illumination) and/or different microorganisms (bacteria, yeasts, cyanobacteria, algae). Furthermore, when processing a complex raw material such as a lignocellulosic material or microbial biomass, there is often a need for a previous step to break down the matrix providing suitable substrates for their subsequent conversion. This biomass hydrolysis may be achieved through the action of specialized enzymes which are chosen according to the task to be accomplished, usually cellulases or hemicellulases. This chapter gives short descriptions of all the biological/biochemical processes used today in biorefineries taking into consideration the final product of interest.

4.1 Introduction

Biological and biochemical processes can produce a huge variety of compounds as well as clean energy. Within the biorefinery concept, biological processes can generate several types of molecules: biofuels, either liquid (e.g., alcohols and alkanes) or gaseous (e.g., hydrogen and methane), value-added products (e.g., carotenoids, omega-3 and omega-6 fatty acids and antioxidants) and other chemical building blocks (e.g., acetic acid and lactic acid). These compounds can result from processing feedstocks as lignocellulosic or microbial biomass as well as a wide range of wastes/residues (e.g., forestry and agriculture, industrial processing, municipal solids, sewage, yard, and animal manure).

L. Gouveia (✉) · P.C. Passarinho
Bioenergy Unit, Laboratório Nacional de Energia e Geologia,
Estrada do Paço do Lumiar 22, Ed. G r/c, 1649-038 Lisboa, Portugal
e-mail: luisa.gouveia@lneg.pt

M. Rabaçal et al. (eds.), *Biorefineries*, Lecture Notes in Energy 57,
DOI 10.1007/978-3-319-48288-0_4

Nevertheless, regarding microbial biorefineries the cell itself "works" by excellence as a biorefinery. These organisms can synthesize almost everything through the use of the intrinsic machinery. Algae, for instance, can offer a diverse spectrum of valuable products and pollution solutions, such as food and feed, nutraceutical compounds, energy sources (including jet fuel, biodiesel, gasoline, bioethanol and biohydrogen), organic fertilizers, biodegradable plastics, recombinant proteins, pigments, medicines, pharmaceuticals, and vaccines (Pienkos and Darzins 2009).

4.2 Alcohols

Yeasts and bacteria, as well as microalgae, have the capability to directly produce alcohols by fermentative processes. Although most of these microorganisms rely on processing simple sugars, some microalgae and bacteria species convert carbon oxides.

Ethanol is the alcoholic fermentation product with higher economical impact. Nevertheless, other alcohols such as methanol or butanol can also be produced by fermentation.

4.2.1 Lignocellulosic Feedstock

So far, the largest potential feedstock for alcoholic fermentation is lignocellulosic biomass, which includes materials such as agricultural residues (corn stover, crop straws and bagasse), herbaceous crops (alfalfa, switchgrass), short rotation wood crops, forestry, municipal and agro-industrial residues, and wastes from the paper industry.

Lignocellulosic biomass is a complex matrix consisting mainly of lignin, cellulose, and hemicellulose. The latter are polysaccharides that must be broken down into simple sugars to be prone for microbial action. Consequently, three major steps are involved in cellulosic bioethanol production: (i) biomass pretreatment, (ii) enzymatic hydrolysis, and (iii) fermentation. Among those, (ii) and (iii) are biochemical processes as biomass pre-treatment involves physical, chemical, or physicochemical methods with the final purpose of improving the accessibility of enzymes. After pretreatment, biomass then undergoes enzymatic hydrolysis for conversion of the polysaccharides into monomer sugars such as pentoses (C5 sugars like xylose) and hexoses (C6 sugars like glucose). Subsequently, these simple sugars are fermented to bioethanol by the action of different microorganisms.

Nevertheless, innovative integrated biochemical conversion processes, involving simultaneous enzymatic hydrolysis (saccharification) and fermentation (SSF), started to be increasingly used. The advantages of this methodology are related to the use of a single reactor, lower contamination levels, faster processing time, and reduction of inhibition phenomena due to sugars being released from the biomass concomitantly to sugar conversion into bio-products (Jouzani and Taherzadeh 2015). Although SSF generates higher bioethanol yields with lower enzyme loads, the major drawback of the process are the different optimal pH and temperature conditions of hydrolysis and fermentation. Therefore, in the search for alternatives, SSF technology has evolved to consolidated bioprocessing (CBP) which integrates microbial cellulase and hemicellulase production and hydrolysis with sugar fermentation (Jouzani and Taherzadeh 2015).

4.2.1.1 Enzymatic Hydrolysis

To make sugars available to microbial degradation, lignocellulosic matrixes have to be decomposed. Enzymatic hydrolysis usually targets the split of hemicellulose and cellulose into simple sugars such as glucose. From hemicellulose can also be obtained xylose, mannose, galactose, arabinose, and ramnose.

The efficiency of hemicellulose decomposition relies on the definition of the adequate cocktail of enzymes, generally designated by hemicellulases, taking into account the different constituent polysaccharides. The mixture may include xylanases, mannanases, arabinofuranosidases, glucuronidases, and phenolic acid esterases. These enzymes are produced by a great number of microbes although filamentous fungi are particularly interesting because they excrete the enzymes into the medium at higher levels. *Aspergillus niger, Humicola insolens, Trichoderma reesei, Trichoderma longibrachiatum,* and *Trichoderma koningii* have been used as sources of commercial xylanases while *Bacillus* sp., *Streptomyces* sp., *Caldibacillus cellulovorans*, and *Caldocellum saccharolyticum* have been used to obtain mannanases. One important effect resulting from the enzymatic hydrolysis of hemicellulose is that it also exposes cellulose fibers (Gírio et al. 2010).

The enzymatic hydrolysis of cellulose resumes the joint action of different enzymes, generally designated by cellulases, being the most important endoglucanases, exoglucanases, and celobiases. These microbial cellulolytic enzymes are produced by fungi and bacteria although filamentous fungi of the genera *Trichoderma* are considered to be the most suitable source (Abd Hamid et al. 2015). The major drawback of enzymatic hydrolysis of cellulose is cellulase inhibition by the released sugars (glucose and cellobiose).

One important feature to consider while facing the need for enzymatic hydrolysis is the cost of the enzymes which may be reduced by improving existing enzyme efficiency, identifying new and more efficient ones, and reducing its production cost.

4.2.1.2 Alcohol Fermentation

Alcohol fermentation is the general designation of the microbial process converting sugars into alcohols. The metabolic pathway for glucose conversion starts in the glycolysis, where glucose is converted into pyruvate, and then the latter is converted to ethanol. Hexoses, other than glucose, are primarily converted into glycolysis intermediary products prior to enter the pathway. For instance, galactose is first converted to glucose-6-phosphate while mannose is transformed into fructose-6-phosphate. The overall process yields two ethanol molecules per hexose.

Yeasts such as *Saccharomyces cerevisiae* and bacteria like *Zymomonas mobilis* are traditionally used in alcohol fermentation due to their good fermentative capacity, high ethanol tolerance, and capability to grow under anaerobic conditions. Nevertheless, those are not able to efficiently metabolize pentoses, namely xylose, decreasing the lignocellulosic hydrolysates conversion yields.

The metabolism for pentoses is much more complex and may occur through several different pathways. The overall process yields 5 ethanol molecules per 3 of xylose. Among wild strains, *Scheffersomyces stipitis* (actual designation of *Pichia stipitis*) and *Clebsiella shehatae* yeasts are believed to be the best for C5 conversion into ethanol although improved DNA recombinant strains may also be used (Kuhad et al. 2011).

Besides being a mixture of C5 and C6 sugars, lignocellulosic hydrolysates may include high levels of inhibitory compounds formed during the pretreatment stage (Palmqvist and Hahn-Hägerdal 2000). Therefore, several biotechnological approaches have been used to overcome such problems, including the development of new and more robust strains able to cope with lignocellulosic fermentative harsh conditions and also, in the case of CBP, to produce and release enzymes suitable for breaking down cellulose and hemicellulose. Microorganism improvement may be accomplished choosing one of two ways: good ethanologenic strains, as *Saccharomyces cerevisiae*, may be engineered to be cellulolytics or lignocellulose degraders, as *Clostridium thermocellum*, may be engineered to become efficient ethanol producers (Hasunuma and Kondo 2012).

Besides ethanol, also butanol may be concomitantly produced by fermentation in a process usually referred to as ABE (acetone/butanol/ethanol) which is mainly carried out by *Clostridia* strains, such as *Clostridium acetobutylicum* and *Clostridium beijerinckii* (Jouzani and Taherzadeh 2015; Lee et al. 2008).

4.2.1.3 Syngas Fermentation

Other biological path to produce alcohols in a biorefinery platform using lignocellulosic feedstock is syngas fermentation. This is an anaerobic microbial process where the synthesis gas (a mixture of carbon monoxide, carbon dioxide and hydrogen) from the thermochemical conversion of lignin is transformed into products such as ethanol, butanol, acetic acid, butyric acid and methane.

Several studies have already reported syngas utilization by microorganisms via the Wood–Ljungdahl pathway. For ethanol production, the most used strains of acetogenic bacteria belong to the genus *Clostridium* and include *Clostridium ljungdahlii* (Aghbashlo et al. 2016) and *Clostridium autoethanogenum* (Martin et al. 2016). However, syngas fermentation has limitations related to gas–liquid mass transfer and low ethanol yields.

4.2.2 Microbial Feedstock

Ethanol may also be produced by some algae species which transform atmospheric carbon dioxide into chemical energy using photons from the sunlight in its photosynthetic machinery. Algae metabolism is able to accumulate high carbohydrate contents which include starch, glucose, cellulose, and hemicellulose, and various types of polysaccharides. Algae biomass has advantages over lignocellulosic biomass as it does not usually contain lignin. This characteristic reduces biomass pretreatment harshness, or even avoids it, being only needed a single enzymatic hydrolytic step.

The ethanol production from algae can be achieved by two different ways, both involving autonomous processes for sugar production and fermentation

1. Photosynthetic production and storage of carbohydrates and subsequent dark-anaerobic fermentation of the carbon source, all within the algae itself.
2. Photosynthetic production and storage of carbohydrates within the algae itself and biomass treatment for the release of sugars to be used in yeast fermentation.

The algae strains *Chlamydomonas perigranulata* (Hon-Nami 2006) and *Chlamydomonas reinhardtii* (Costa et al. 2015) have already been used for producing ethanol through in situ production and fermentation of starch, while *Scenedesmus obliquus* (Miranda et al. 2012), *Chlorella vulgaris* (Moncada et al. 2013; Ho et al. 2013) and *Tribonema* sp. (Wang et al. 2014) have been studied for increasing carbohydrates' production to use as carbon source in *Saccharomyces cerevisiae* fermentation. In addition to ethanol, it is possible to use algae to produce other alcohols, such as methanol and butanol, using similar technological processes, although the recovery of heavier alcohols from the culture medium and headspace has proven to be problematic.

Although algae do not have the capability of producing ethanol under illumination, the biofuels quest has led researchers to engineer some strains in the last few years. One example is *Synechocystis* sp. PCC 6803 a well-studied cyanobacteria that has already been improved with a *Zymomonas mobilis* DNA cassette encoding the heterologous expression of pyruvate decarboxylase and alcohol dehydrogenase enzymes to promote ethanol production (Borirak et al. 2015). Targeting butanol production has also been tested an engineered *Synechococcus elongatus* integrating five new genes (Machado and Atsumi 2012).

4.3 Acetic Acid

Acetic acid is a compound present in natural processes which can be used as a bactericide, a solvent as well as a building block for the production of vinyl acetate and other esters. Industrial acetic acid is mainly produced by chemical transformation of methanol.

Although it may be produced biologically from sugars and ethanol, within a biorefinery concept the production derives from bacterial dark fermentation of syngas (Sect. 2.1.3). The fermentative process relies on the capability of acetogenic bacteria of the genus *Clostridium* to reduce one-carbon compounds via the Wood–Ljungdahl pathway being ethanol and butanol by-products of the process (Verma et al. 2016). The overall reaction from carbon dioxide may be resumed as

$$2CO_2 + 4H_2 \rightarrow CH_3COOH + 2H_2O$$

4.4 Lactic Acid

Lactic acid (2-hydroxypropanoic) is a biobased high value-added chemical with a wide range of industrial applications, including a fast expanding market as building block for polylactic acid (PLA), a biodegradable and biocompatible polymer used as substitute of petrochemical-based plastics. Lactic acid can be produced from acetaldehyde by chemical synthesis but it is mainly produced by fermentation. Lactic acid bacteria, from the genera *Lactobacillus, Leuconostoc, Pediococcus* and *Streptococcus*, mediate the conversion of carbohydrates into lactic acid (Marques et al. 2008). These bacteria are characterized by a high tolerance to acidic conditions. Lactic fermentation may follow one of two main routes, according to the microorganism chosen:

- Homolactic fermentation where one molecule of glucose is transformed into pyruvate in the glycolytic pathway. Then, pyruvate is split into two lactic acid moieties through the action of the enzyme lactate dehydrogenase.
- Heterolactic fermentation that uses the pentose phosphate pathway. The yield regarding lactic acid is equimolar because it occurs simultaneously with the formation of carbon dioxide and ethanol or acetic acid.

As in the case of alcohol fermentation, microorganisms that are able to process hexoses do not have the capacity to use pentoses. However, there are some facultative heterofermentative lactic acid bacteria (e.g., *Lactobacillus casei, L. rhamnosus, L. pentosus*) that possess this capability being more adequate to deal with lignocellulosic feedstocks such as recycled paper sludge (Marques et al. 2008). In a biorefinery context, lactic acid bacteria are particularly suited for SSF processes due to its thermotolerance.

Genetically engineered/modified microorganisms, other than lactic acid bacteria, may also produce lactic acid as in the case of a mutant of cyanobacteria *Synechocystis* sp. PCC 6803 which harbors lactate dehydrogenase from *Lactococcus lactis* and pyruvate kinase from *Enterococcus feacalis* (Borirak et al. 2015).

4.5 Biohydrogen

Hydrogen can be produced in a number of ways mainly from petroleum reforming, refining, and gasification. Nevertheless, hydrogen can also be obtained through biological processes such as dark or photo-fermentation of organic materials and through the photolysis of water performed by certain microalgae and cyanobacteria. Biohydrogen production usually involves nitrogenases and bidirectional hydrogenases. The interplay of these enzymes is the key in hydrogen production by fermentation.

In a biorefinery context, the biohydrogen fermentation may convert almost any substrate available (Kapdan and Kargi 2006), including microbial biomass from other biological processes such as leftovers of yeasts or algae (Sargsyan et al. 2016; Batista et al. 2014). The dark-fermentation biological process is restricted due to incomplete degradation of the organic material. Therefore, to improve biohydrogen production yield, a synergistic fermentative system must be implemented, combining dark and photo-fermentations where the organic acids resulting from dark fermentation are further processed as substrate for hydrogen production by photo-fermentation (Sargsyan et al. 2016; Singh and Wahid 2015).

4.5.1 Dark Fermentation

Biohydrogen can be produced by dark fermentation using mainly *Enterobacter* and *Clostridium* bacteria strains (Kapdan and Kargi 2006; Elsharnouby et al. 2013). The latter display natural high hydrogen production rates and they are fast growing and strict anaerobes. *Clostridia* produce biohydrogen essentially by the action of a reversible [Fe–Fe]hydrogenase enzyme but this reaction is product inhibited and the enzyme is inactivated by molecular oxygen (Mathews and Wang 2009).

Unlike *Clostridia*, enteric bacteria such as *Escherichia coli* and *Enterobacter aerogenes* are facultative anaerobes, meaning that they can be grown fast in the presence of oxygen and then use energy to produce hydrogen when the oxygen supply stops. In these microorganisms hydrogen is produced by the action of several hydrogenase isoenzymes.

Dark fermentation processes generate also organic acids, which are toxic to the bacteria at high levels, and carbon dioxide, which has to be subsequently separated from biohydrogen. In *Clostridium*, a carbon source such as glucose is broken down into pyruvate that is further converted to acetyl-CoA and hydrogen. Acetyl-CoA is then converted to organic acids (acetate, butyrate, and/or propionate). The yield of the conversion into acetate and bio-H_2 is 4 molecules of hydrogen per molecule of glucose (Mathews and Wang 2009).

4.5.2 Photo-fermentation

Photo-fermentation is a fermentation process where light is required as a source of energy relying on photosynthesis to maintain the cell energy levels. Photosynthetic microorganisms capable of producing hydrogen by photo-fermentation include green algae such as *Chlamydomonas reinhardtii*, cyanobacteria such as *Anabaena* sp. and bacteria from the genus *Rhodobacter* as *Rhodobacter sphaeroides* (Kapdan and Kargi 2006; Marques et al. 2011).

To direct photobiological production of hydrogen, cyanobacteria possess nitrogenases that catalyze the production of hydrogen concomitantly with the reduction of nitrogen to ammonia, an uptake hydrogenase catalyzing the consumption of hydrogen produced by the nitrogenase, and a bidirectional hydrogenase which has the capacity to both take up and produce hydrogen (Tamagnini et al. 2002). In green algae, hydrogen is produced photosynthetically by the ability to harness the solar energy resource to drive hydrogen production from water with the participation of [Fe–Fe]hydrogenases and [Fe–Ni]hydrogenases.

Besides, biohydrogen can also be indirectly produced by cyanobacteria or algae in light/dark cycles. First, carbon compounds are formed photosynthetically and then are converted to hydrogen in dark-anaerobic conditions by the final action of an hydrogenase (Mathews and Wang 2009).

Regarding *Rhodobacter* metabolism, it requires organic acids and is unable to perform water-splitting. The substrate is oxidized through the citric acid cycle and the produced electrons are transferred to the nitrogenase catalyzed reduction of protons to hydrogen. However, the action of the nitrogenases has to exceed the hydrogen uptake by the hydrogenases (Tamagnini et al. 2002).

There is a high potential for improving hydrogen yield by metabolic engineering (Mathews and Wang 2009; Wang et al. 2016; Sinha et al. 2016). For dark fermentation, the bacteria *Clostridium* could be improved towards hydrogen production enhancement by disabling the oxygen system and the uptake hydrogenase. Regarding photo-fermentation, the option could be to disable the uptake-hydrogenase or to decrease the expression of pigments, which shields the photo-system.

4.6 Biogas

Organic material such as crop biomass, residues including microbial residues, or liquid manure can be used to produce biogas via anaerobic digestion and fermentation. Mixtures of bacteria are used to hydrolyze and break down the organic biopolymers (i.e., carbohydrates, lipids and proteins) into monomers, which are then converted into a methane-rich gas via fermentation.

Anaerobic digestion refers then to the decomposition of the organic material through the metabolic pathways of naturally occurring microorganisms' consortia in an oxygen depleted environment. There are two key processes: mesophilic (30–38 °C) and thermophilic (50–57 °C) digestion.

The digestion process involves four main biological processes (Shah et al. 2014):

- Hydrolysis, where major molecules are broken down to simpler units mainly by extracellular enzymes produced within the consortium;
- Acidogenesis, where simple molecules are converted into volatile fatty acids (2–5 carbons) by acidogenic bacteria as *Propionibacterium* sp. and *Lactobacillus* sp;
- Acetogenesis, where acetate is generated from the fermentation products by the joint action of several reducing bacteria (*Acetobacter* sp.). Some produce acetate, carbon dioxide and hydrogen and others convert carbon dioxide and hydrogen also to acetate;
- Methanogenesis, where acetate mainly but also carbon dioxide and hydrogen are converted into methane. The microorganisms involved are acetoclastic (*Methanosarcina* sp., *Methanosaeta* sp.) and hydrogenotrophic (*Methanococcus* sp. *Methanospirillum* sp.) methanogens.

An anaerobic digestion plant produces biogas and digested sludge. Biogas is a gaseous mixture mainly composed by methane (50–75%) and carbon dioxide (20–35%) that also includes small amounts of hydrogen sulfide, hydrogen, moisture, and even siloxanes (which derive from the decomposition of materials commonly found in soaps and detergents). Biogas may be combusted in the presence of oxygen and used as fuel. Also, it can be upgraded to bio-methane to be used and disseminated in the natural gas grid or converted to power and heat using cogeneration equipments but, in this case, CO_2 has to be previously removed from the mixture. The digested sludge can be further processed to be finally used as a soil conditioner or fertilizer (Weiland 2010).

4.7 Other Compounds

Although in an early commercial stage, microbial biorefineries (such as algae or yeast based) present strong potential in converting carbon from residues or polluting streams, either gaseous or liquid, into valuable compounds. Some of those were already referred to in previous sections (alcohols, lactic acid, and biohydrogen) as they are already being produced in the few operating microbial biorefineries. However, there is still a huge potential to take advantage of. Among the products that can be obtained hydrocarbons, pigments, aminoacids, and polyunsaturated fatty acids (PUFAs) are high value-added compounds.

Hydrocarbons may be produced in either heterotrophic or autotrophic metabolic pathways. Almost all cyanobacteria are able to produce mixtures of hydrocarbons of different types from fatty acids. The compounds—linear, branched, saturated and with double bonds, and bearing chains of 9–34 carbons—may be part of a wide range of fuels (Coates et al. 2014). Besides almost all cyanobacteria, also *Scenedesmus obliquus* produces alkanes (Jamil et al. 2016). The hydrocarbons can be secreted and recovered directly from the fermentation broth without the need for dewatering and extraction.

Long-chain fatty acids, especially PUFAs bearing double bonds in the omega-6 and omega-3 positions, are other value-added compounds of microbial origin that are classed as essential fatty acids due to a broad range of health and medical applications. Certain strains of algae as well as fungi possess the capability to generate PUFAs. Algae commercial sources of PUFAs are species of *Arthrospira, Porphyridium, Nannochloropsis,* and *Phaeodactylum* (Harwood and Guschina 2009) while the most studied fungi nowadays are *Yarrowia lipolytica* and *Rhodosporidium toruloides*. Microorganisms able to produce PUFAs are oleaginous and produce concomitantly high amounts of lipids that can be used for biodiesel or jet fuel (Gouveia and Oliveira 2009; Sitepu et al. 2014; Fu et al. 2015).

Other valuable compounds obtained from microbial biomass are pigments. These products have an important and widespread utilization mainly in food industry, textile dying, pharmaceuticals and cosmetics and are produced by all algae (D'Alessandro and Antoniosi Filho 2016) as well as by some yeast (Mannazzu et al. 2015) and bacteria species (Venil et al. 2013). Pigments are intrinsic to the metabolism of each microorganism and so are produced in line with other compounds such as lipids, ethanol or biohydrogen.

Natural pigments are more stable to light, heat, and pH and their color range comprises all the visible light spectra from blue to red. Pigments can be classified based on structural similarities and carotenoids and chlorophylls are the most found in microorganisms.

- Carotenoids are yellow to red pigments that have important antioxidant roles namely to photoprotect the photosynthetic systems inside the algae cells. They are also found in red-yeasts and bacteria being the most common pigments and include beta-carotene, cantaxanthin, astaxanthin, lycopene, lutein, zeaxanthin, violaxanthin, neoxanthin, torulene, and torularhodin.
- Chlorophylls are greenish pigments existing in three forms (a, b, and c). They are responsible for converting solar energy into chemical energy in photosynthetic microorganisms.

Algae can also produce phycobilins (phycoerythrin and phycocyanin) that enhance the light capture especially in cyanobacteria. Some strains of bacteria also produce other pigments such as violacein (purple) and prodigiosin (red).

As for other biorefinery products, pigment biosynthetic pathways can also be manipulated to change pigment's molecular structure and consequently its colour. Carotenogenesis can be induced in many microalgae species by changing environmental conditions (e.g., nitrogen depletion, salt and light increases) (Gouveia et al. 1996; Campenni' et al. 2013). *Streptomyces coelicolor*, which produces the blue pigment actinorhodin can be genetically modified to produce orange or yellow-red anthraquinones (Venil et al. 2013).

References

Abd Hamid SB, Islam MM, Das R (2015) Cellulase biocatalysis: key influencing factors and mode of action. Cellulose 22(4):2157–2182

Aghbashlo M, Tabatabaei M, Hosseini SS, Younesi H, Najafpour G (2016) Performance analysis of a continuous bioreactor for ethanol and acetate synthesis from syngas via *Clostridium ljungdahlii* using exergy concept. Clean Technol Environ Policy 18(3):853–865

Batista AP, Moura P, Marques PASS, Ortigueira J, Alves L, Gouveia L (2014) *Scenedesmus obliquus* as feedstock for biohydrogen production by *Enterobacter aerogenes* and *Clostridium butyricum*. Fuel 117:537–543

Borirak LJ, de Koning AD, van der Woude HC, Hoefsloot HL, Dekker W, Roseboom CG, de Koster Hellingwerf KJ (2015) Quantitative proteomics analysis of an ethanol—and a lactate-producing mutant strain of *Synechocystis* sp. PCC6803. Biotechnol Biofuels 8:111–124

Campenni' L, Nobre BP, Santos CA, Oliveira AC, Aires-Barros MR, Palavra AF, Gouveia L (2013) Carotenoid and lipid production by the autotrophic microalga *Chlorella protothecoides* under nutritional, salinity, and luminosity stress conditions. Appl Microbiol Biotechnol 97:1383–1393

Coates RC, Podell S, Korobeynikov A, Lapidus A, Pevzner P, Sherman DH, Allen EE, Gerwick L, Gerwick WH (2014) Characterization of cyanobacterial hydrocarbon composition and distribution of biosynthetic pathways. PLoS ONE 9(1):e85140

Costa RL, Oliveira TV, Ferreira JD, Cardoso VL, Batista FRX (2015) Prospective technology on bioethanol production from photofermentation. Bioresour Technol 181:330–337

D'Alessandro EB, Antoniosi Filho NR (2016) Concepts and studies on lipid and pigments of microalgae: a review. Renew Sustain Energy Rev 58:832–841

Elsharnouby O, Hafez H, Nakhla G, El Naggar MH (2013) A critical literature review on biohydrogen production by pure cultures. Int J Hydrog Energy 38(12):4945–4966

Fu J, Yang CY, Wu JH, Zhuang JL, Hou ZY, Lu XY (2015) Direct production of aviation fuels from microalgae lipids in water. Fuel 139:678–683

Gírio FM, Fonseca C, Carvalheiro F, Duarte LC, Marques S, Bogel-Łukasik R (2010) Hemicelluloses for fuel ethanol: a review. Bioresour Technol 110:4775–4800

Gouveia L, Oliveira AC (2009) Microalgae as a raw material for biofuels production. J Indus Microbiol Biotechnol 36:269–274

Gouveia L, Veloso V, Reis A, Fernandes HL, Empis J, Novais JM (1996) Evolution of pigment composition in *Chlorella vulgaris*. Bioresour Technol 57:157–163

Harwood JL, Guschina IA (2009) The versatility of algae and their lipid metabolism. Biochimie 91:679–684

Hasunuma T, Kondo A (2012) Development of yeast cell factories for consolidated bioprocessing of lignocellulose to bioethanol through cell surface engineering. Biotechnol Adv 30:1207–1218

Ho SH, Huang SW, Chen CY, Hasunuma T, Kondo A, Chang JS (2013) Bioethanol production using carbohydrate-rich microalgae biomass as feedstock. Bioresour Technol 135:191–198

Hon-Nami K (2006) A unique feature of hydrogen recovery in endogenous starch-to-alcohol fermentation of marine microalga, *Chlamydomonas perigranulata*. In: McMillan JD (ed) Twenty-Seventh symposium on biotechnology for fuels and chemicals (session 5), pp 808–828

Jamil TS, Abdel Aty AM, Ghafar HHA, Abdo SM (2016) Separation and identification of hydrocarbons and other organic compounds from *Scenedesmus obliquus* and three cyanobacterial species. Desalination Water Treat 57(2):908–915

Jouzani GS, Taherzadeh MJ (2015) Advances in consolidated bioprocessing systems for bioethanol and butanol production from biomass: a comprehensive review. Biofuel Res J 5:152–195

Kapdan IK, Kargi F (2006) Bio-hydrogen production from waste materials. Enzyme Microbial Technol 38(5):569–582

Kuhad RC, Gupta R, Khasa YP, Singh A, Zhang YHP (2011) Bioethanol production from pentose sugars: current status and future prospects. Renew Sustain Energy Rev 15(9):4950–4962

Lee SY, Park JH, Jang SH, Nielsen LK, Kim J, Jung KS (2008) Fermentative butanol production by Clostridia. Biotecnhol Bioeng 101(2):209–228

Machado IMP, Atsumi S (2012) Cyanobacterial biofuel production. J Biotechnol 162:50–56

Mannazzu I, Landolfo S, Lopes da Silva T, Buzzini P (2015) Red yeasts and carotenoid production: outlining a future for non-conventional yeasts of biotechnological interest. World J Microbiol Biotechnol 31(11):1665–1673

Marques S, Santos JAL, Gírio FM, Roseiro JC (2008) Lactic acid production from recycled paper sludge by simultaneous saccharification and fermentation. Biochem Eng J 41:210–216

Marques AE, Barbosa AT, Jotta J, Coelho MC, Tamagnini P, Gouveia L (2011) Biohydrogen production by *Anabaena* sp. PCC 7120 wild-type and mutants under different conditions: Light, nickel, propane, carbon dioxide and nitrogen. Biomass Bioenergy 35:4426–4434

Martin ME, Richter H, Saha S, Angenent LT (2016) Traits of selected clostridium strains for syngas fermentation to ethanol. Biotechnol Bioeng 113(3):531–539

Mathews J, Wang G (2009) Metabolic pathway engineering for enhanced biohydrogen production. Int J Hydrog Energy 34(17):7404–7416

Miranda JR, Passarinho PC, Gouveia L (2012) Bioethanol production from *Scenedesmus obliquus* sugars: the influence of photobioreactors and culture conditions on biomass production. Appl Microbiol Biotechnol 96:555–564

Moncada J, Jaramillo JJ, Higuita JC, Younes C, Cardona CA (2013) Production of bioethanol using *Chlorella vulgaris* cake: a technoeconomic and environmental assessment in the colombian context. Indus Eng Chem Res 52(47):16786–16794

Palmqvist E, Hahn-Hägerdal B (2000) Fermentation of lignocellulosic hydrolysates. II: inhibitors and mechanisms of inhibition. Bioresour Technol 74(1):25–33

Pienkos PT, Darzins A (2009) The promise and challenges of microalgal-derived biofuels. Biofuels Bioprod Biorefin 3(4):431–440

Sargsyan H, Trchounian K, Gabrielyan L, Trchounian A (2016) Novel approach of ethanol waste utilization: biohydrogen production by mixed cultures of darkand photo-fermentative bacteria using distillers grains. Int J Hydrog Energy 41(4):2377–2382

Shah FA, Mahmood Q, Shah MM, Pervez A, Asad SA (2014) Microbial ecology of anaerobic digesters: the key players of anaerobiosis. Scientific World J, p 21 (Article ID 183752)

Singh L, Wahid ZA (2015) Methods for enhancing bio-hydrogen production from biological process: a review. J Ind Eng Chem 21:70–80

Sinha P, Roy S, Das D (2016) Genomic and proteomic approaches for darkfermentative biohydrogen production. Renew Sustain Energy Rev 56:1308–1321

Sitepu IR, Garay LA, Sestric R, Levin D, Block DE, German JB, Boundy-Mills KL (2014) Oleaginous yeasts for biodiesel: current and future trends in biology and production. Biotechnol Adv 32:1336–1360

Tamagnini P, Axelsson R, Lindberg P, Oxelfelt F, Wünschiers R, Lindblad P (2002) Hydrogenases and hydrogen metabolism of cyanobacteria. Microbiol Mol Biol Rev 66(1):1–20

Venil CK, Zakaria ZA, Ahmad WA (2013) Bacterial pigments and their applications. Process Biochem 48:1065–1079

Verma D, Singla A, Lal B, Sarma PM (2016) Conversion of biomass-generated syngas into next-generation liquid transport fuels through microbial intervention: potential and current status. Curr Sci 110(3):329–336

Wang H, Ji CL, Bi SL, Zhou P, Chen L, Liu TZ (2014) Joint production of biodiesel and bioethanol from filamentous oleaginous microalgae *Tribonema* sp. Bioresour Technol 172:169–173

Wang Y, Zhou P, Tong J, Gao R (2016) Advances in the genetic modification of *Rhodobacter sphaeroides* to improve hydrogen production. Renew Sustain Energy Rev 60:1312–1318

Weiland P (2010) Biogas production: current state and perspectives. Appl Microbiol Biot 85:849–860. http://www.biogas-renewable-energy.info

Author Biographies

Luisa Gouveia, Ph.D. is a Senior Research and head of Autotrophic Microalgae Unit at National Laboratory of Energy and Geology-Bioenergy Unit (LNEG–UB). She graduated in Chemical Engineering at Instituto Superior Tecnico (IST), University of Lisbon in 1986, MSc in Food Science and Technology in 2002 and PhD in Biotechnology (IST) 1996. Coordinator of UB-Research Program: Microalgae as an energy vector. Vice-chair, National Delegate of the COST Action 1408 EUALGAE-European network for algal-bioproducts (2015–2018). More than 28 years of experience in Microalgae Biotechnology. Coordinator of 4 projects related to Microalgae and Biofuel production and team member of 20 projects in the same area. Author of 1 book, 10 book's chapters, 70 international peer-reviewed papers and more than 150 papers in international conferences. ResearcherID H-4046-2011; Scopus Author ID 7004135834.

Paula Passarinho, Ph.D. is a Senior Researcher working in the Bioenergy group of the National Laboratory for Energy and Geology (LNEG) in Lisbon. With a degree in Chemistry from the Faculty of Sciences of Lisbon University and a PhD in Biotechnology from Instituto Superior Técnico (Lisbon; 2002), she has been working in photocatalytic solar technologies development and in liquid biofuels production from different raw materials, namely from microbial sources. Having participated in RD&D national and international projects she has also supervised MSc students and is author of several peer-reviewed papers and presentations in international conferences.

Chapter 5
Biomass Conversion Technologies: Catalytic Conversion Technologies

Juan Carlos Serrano-Ruiz

Abstract Diminishing fossil fuel reserves and global warming issues are driving society toward the search for new renewable sources of energy. Lignocellulosic can significantly displace petroleum in the production of fuels. Oxygenated fuels such as ethanol and biodiesel currently dominate the fuel market although they have important limitations. The production of liquid hydrocarbon fuels from biomass is a paradigmatic transformation allowing the production of fuels chemically identical to those currently used in the transportation sector. The present chapter describes some of the catalytic strategies used to transform biomass-derived molecules into liquid hydrocarbon fuels. These strategies are first focused on decreasing the oxygen content of the original molecule such that its reactivity can be more easily controlled. In a subsequent step, the less oxygenated intermediates are upgraded via C–C coupling reactions to increase the length chain to produce a final product suitable for diesel and jet fuel applications. The present chapter offers a number of examples on biomass-derived acids such as lactic and levulinic acids and biomass sugars such as glucose catalytically transformed into liquid hydrocarbon fuels.

5.1 Introduction

Fossil fuels (e.g., petroleum, coal and natural gas) accounts for roughly 90% of the global energy consumption (BP Statistical Review of World Energy 2009). The usage of fossil fuels is however associated with a number of important environmental and security issues worldwide. These important concerns have spur society to search for new renewable sources of energy and chemicals. In this sense, biomass is considered to be the only sustainable source of organic carbon for our industrial society (U.S. National Science Foundation 2008) with potential to replace petroleum in the production of fuels and chemicals. As shown in Fig. 5.1, the enormous and complex petrochemical industry is constructed over a few molecules (benzene,

J.C. Serrano-Ruiz (✉)
Universidad Loyola Andalucía, Energía Solar, 1. Edifs E, F y G, 41014 Sevilla, Spain
e-mail: jcserrano@uloyola.es

M. Rabaçal et al. (eds.), *Biorefineries*, Lecture Notes in Energy 57,
DOI 10.1007/978-3-319-48288-0_5

toluene, xylene, ethylene, propylene, 1,3-butadiene and methanol) serving as building blocks of a multitude of chemicals and specialty products. Consequently, a biomass-based industry can be envisaged through the processing of a few carefully selected biomass derivatives or platform molecules. The USDOE selected 12 compounds as key starting materials based on a number of indicators (Werpy and Petersen 2004) and more recently reviewed this list including new molecules such as ethanol and furans (Bozell and Petersen 2010).

As clearly appreciated in Fig. 5.1, renewable platform compounds are chemically opposed to petroleum feeds since biomass derivatives possess a high degree of oxygenated functionalities whereas petroleum feeds are unfunctionalized. Heterogeneous catalysis, in the same way as it did with the petrochemical industry, will play a fundamental role in the growing biomass industry. However, the different chemical composition of biomass feeds forces to design new strategies that will involve low temperature approaches (to control reactivity) and aqueous phase processing derived from the high polarity of biomass compounds (Chheda et al. 2007). Furthermore, water is the classical medium in which biomass platform molecules are obtained after pretreatment of biomass.

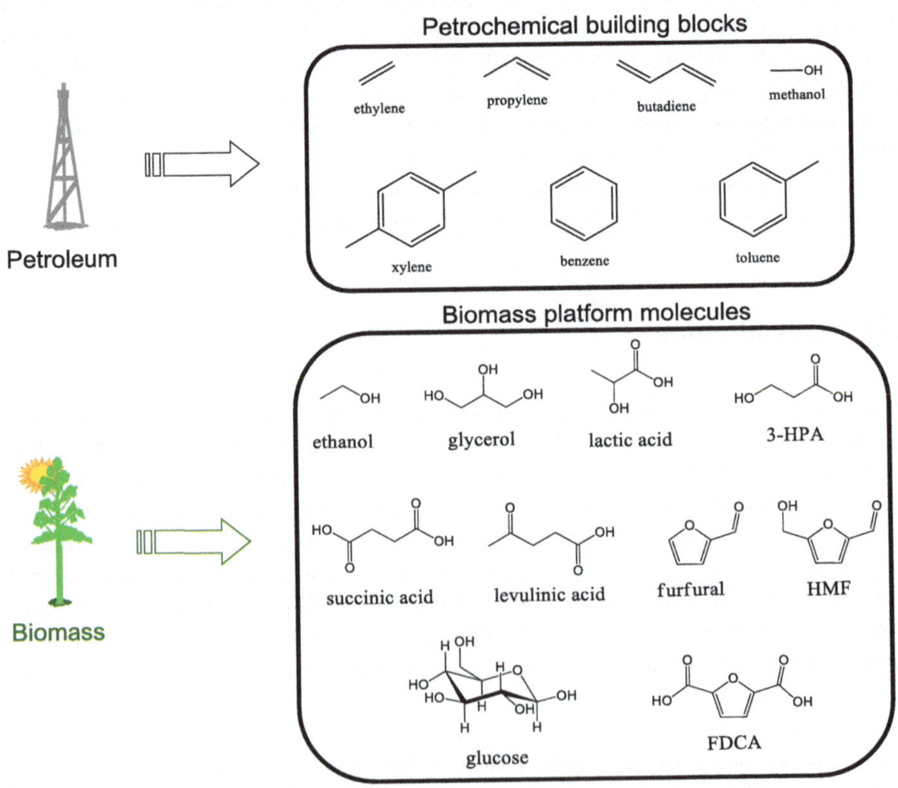

Fig. 5.1 List of petrochemical and biomass-derived building blocks

5.2 Aqueous-Phase Catalytic Processing of Biomass Platform Molecules

5.2.1 Ethanol

The fermentation of edible biomass-derived aqueous sugars to ethanol is the main technology for the production of liquid biofuels in the world, with annual productions around 90 billion L (Ethanol Producer Magazine, International Ethanol Report 2010). Ethanol has an enormous potential as a platform chemical as well, with a good number of interesting products potentially being produced via heterogeneous catalysis (Fig. 5.2). Thus, ethanol can serve as a source of renewable hydrogen by steam reforming processes (Ni et al. 2007) over metal catalyst at high temperatures, although the economics of this route are not competitive with classical methane steam reforming approaches. Ethylene, the most produced organic compound worldwide (annual production 120 million tons) and one of the seven primary building blocks of the petrochemical industry (Fig. 5.1) can be easily produced by dehydration of ethanol over an acidic catalyst at moderate temperatures (Bedia et al. 2011). The production of green polyethylene plastics is an important driver of this route, and currently companies such as Braskem are deploying this technology. Butadiene, another important building block of the petrochemical industry, can be catalytically produced from ethanol by an old and well known technology (Talalay and Magat 1945). Propylene can be also produced from biomass-derived ethanol although the process is a complex sequence of reactions involving dehydration to ethylene, partial dimerization of the latter to butene and subsequent metathesis of both C_2 and C_4 olefins (Mitsui Chemicals Inc. 2008). Aromatics such as benzene, toluene, and xylene can also be derived from aqueous ethanol in a simple process involving acidic zeolites (Inaba et al. 2006). Ethanol, consequently, can be truly considered a bridge molecule as 90% of the petrochemical building blocks can be derived from this resource. Important oxygenates such as acetaldehyde and acetic acid can be derived from ethanol by dehydrogenation and air oxidation over Cu and noble metals-based catalysts, respectively (Tu and Chen 1998; Christensen et al. 2006).

5.2.2 Hydroxymethylfurfural (HMF)

Furans such as hydroxymethylfurfural (HMF) are important intermediates in biomass conversion processes. HMF is produced by dehydration of biomass-derived hexoses under mineral acids such as H_2SO_4 and HCl. Abundantly available glucose is the desirable starting material for HMF production, although yields are typically low for this feedstock, and thus additional isomerization steps to fructose are typically required.

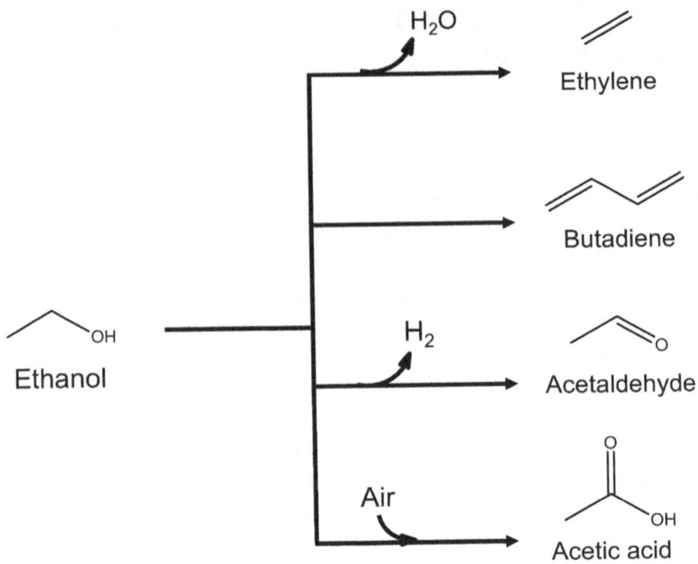

Fig. 5.2 Scheme of the different products catalytically derived from ethanol

Figure 5.3 shows the most relevant catalytic approaches to process HMF.

Dimethylfuran (DMF), an hydrophobic liquid with excellent energy-density and boiling point characteristics to be used as transportation fuel, can be produced by hydrogenation of the carbonyl group and subsequent C–O hydrogenolysis of the intermediate dihydroxymethyl furan (DHMF) (Román-Leshkov et al. 2007). Copper-based catalysts are typical materials to carry out these transformations. HMF can be upgraded using the carbonyl functionality as reactive center for C–C coupling reactions (e.g., aldol condensation). Thus, linear alkanes with molecular weights appropriate for diesel and jet fuel applications (C_9–C_{15}) can be derived from HMF by a cascade process involving dehydration, hydrogenation and aldol-condensation reactions (West et al. 2008). The catalytic transformation of HMF into furandicarboxylic acid (FDCA) represents an interesting alternative for the production of bio-PET since FDCA possesses chemical structure and properties similar to petroleum-derived terephthalic acid. The catalytic transformation of HMF into FDCA is carried out under pressurized air/oxygen at mild temperature and controlled alkaline pH with noble metal catalysts (Mäki-Arvela et al. 2007).

5.2.3 *Lactic Acid*

Lactic acid (2-hydroxypropanoic acid) is the most widely occurring carboxylic acid in nature produced by bacterial fermentation of sugars and more recently with solid

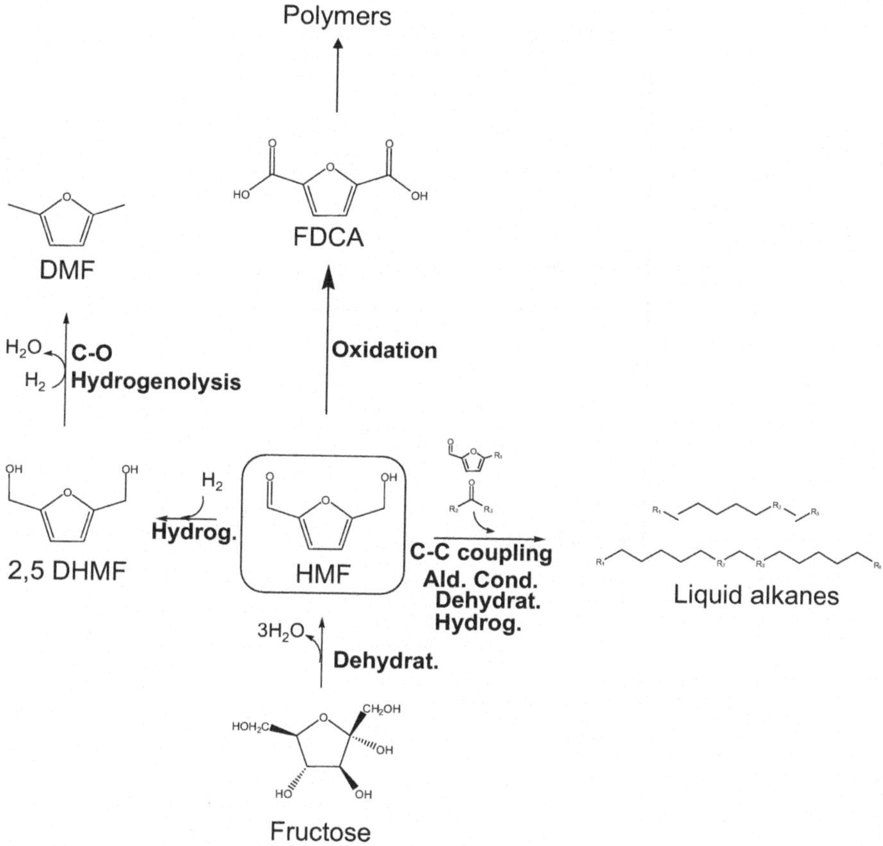

Fig. 5.3 Catalytic routes for the upgrading of HMF

zeolites (Holm et al. 2010). The bifunctional character of lactic acid (e.g., –OH and –COOH groups) allows a variety of catalytic transformations (Fig. 5.4). Thus, acrylic acid via dehydration and 2,3-pentanedione via condensation are important chemicals than can be readily derived from lactic acid using catalytic approaches (Wadley et al. 1997; Gunter et al. 1994). Polylactic acid (PLA), a new biodegradable polymer with excellent properties for the fabrication of plastics and related materials. A partial oxygen removal approach has been recently applied to catalytically process lactic acid into C_4–C_5 alcohols suitable as high energy-density liquid fuels for the transportation sector involving two key intermediates (e.g., acetaldehyde and propanoic acid) and over a single bed of Pt/Nb_2O_5 (Serrano-Ruiz and Dumesic 2009).

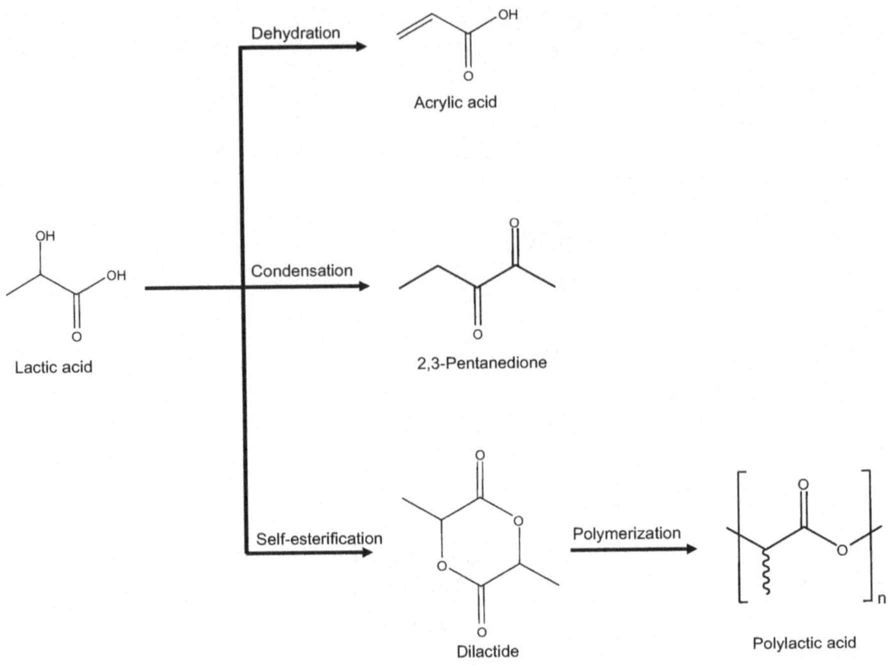

Fig. 5.4 Catalytic routes to upgrade lactic acid

5.2.4 *Levulinic Acid*

Levulinic acid (4-oxopentanoic acid, LA) is a high-boiling point (520 K), water-soluble acid (pK_a = 4.59) that crystallizes at room temperature. Levulinic acid can be produced using acid catalyzed chemical processing, e.g., dehydration of inexpensive lignocellulosic feeds (Bozell et al. 2000). Levulinic acid is typically synthetized with a maximum yield of 40% by weight, because it is coproduced along with equimolar amounts of formic acid and unwanted black insoluble materials denoted as humins. Levulinic acid has been suggested for the production of specific commodity chemicals (Fig. 5.5).

The key intermediate in these transformations is γ-valerolactone (GVL). GVL is a stable and water-soluble interesting biomass-derivative that has been proposed to be a potential gasoline additive as well as a solvent for biomass deconstruction processes (Luterbacher et al. 2014). GVL possesses a high versatility to synthesize liquid transportation fuels. It can be readily converted to MTHF via hydrogenation (Serrano-Ruiz et al. 2010). Alternatively, aqueous GVL can be transformed into 5-nonanone in high yields (90%) via ketonization reactions of hydrophobic pentanoic acid (PA) using a bifunctional Pd/Nb_2O_5 catalyst (Serrano-Ruiz et al. 2010). The C_9 ketone serves then as a platform molecule for the production of hydrocarbon fuels of different classes suitable for diesel and jet fuels applications.

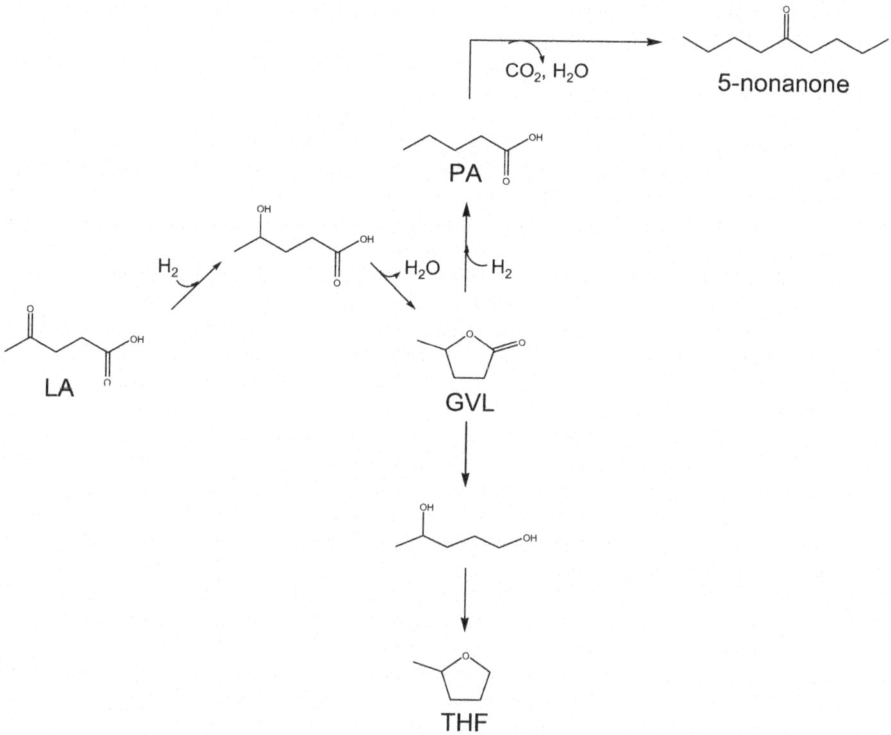

Fig. 5.5 Scheme of the different routes to catalytically upgrade levulinc acid

References

Bedia J et al (2011) Ethanol dehydration to ethylene on acid carbon catalysts. Appl Catal B 103 (3):302–310

Bozell JJ, Petersen GR (2010) Technology development for the production of biobased products from biorefinery carbohydrates—the US Department of Energy's "top 10" revisited. Green Chem 12(4):539–554

Bozell JJ et al (2000) Production of levulinic acid and use as a platform chemical for derived products. Resour Conserv Recycl 28(3):227–239

BP Statistical Review of World Energy (2009) http://www.bp.com/productlanding.do?categoryId=6929&contentId=7044622

Chheda JN, Huber GW, Dumesic JA (2007) Liquid-phase catalytic processing of biomass-derived oxygenated hydrocarbons to fuels and chemicals. Angew Chem Int Ed 46(38):7164–7183

Christensen CH et al (2006) Formation of acetic acid by aqueous-phase oxidation of ethanol with air in the presence of a heterogeneous gold catalyst. Angew Chem Int Ed 45(28):4648–4651

Ethanol Producer Magazine, International Ethanol Report (2010) http://www.ethanolproducer.com/articles/6696/international-ethanol-report-2010

Gunter GC, Miller DJ, Jackson JE (1994) Formation of 2,3-pentanedione from lactic acid over supported phosphate catalysts. J Catal 148:252–260

Holm MS, Saravanamurugan S, Taarning E (2010) Conversion of sugars to lactic acid derivatives using heterogeneous zeotype catalysts. Science 328(5978):602–605

Inaba M et al (2006) Ethanol conversion to aromatic hydrocarbons over several zeolite catalysts. React Kinet Catal Lett 88(1):135–141
Luterbacher JS et al (2014) Nonenzymatic sugar production from biomass using biomass-derived γ-valerolactone. Science 343(6168):277–280
Mäki-Arvela P et al (2007) Recent progress in synthesis of fine and specialty chemicals from wood and other biomass by heterogeneous catalytic processes. Catal Rev 49(3):197–340
Method of producing propylene containing biomass-origin carbon (2008) European Patent Application EP1953129 (A1) to Mitsui Chemicals Inc.
Ni M, Leung DYC, Leung MKH (2007) A review on reforming bio-ethanol for hydrogen production. Int J Hydrogen Energy 32(15):3238–3247
Román-Leshkov Y et al (2007) Production of dimethylfuran for liquid fuels from biomass-derived carbohydrates. Nature 447(7147):982–985
Serrano-Ruiz JC, Dumesic JA (2009) Catalytic upgrading of lactic acid to fuels and chemicals by dehydration/hydrogenation and C–C coupling reactions. Green Chem 11(8):1101–1104
Serrano-Ruiz JC, Wang D, Dumesic JA (2010) Catalytic upgrading of levulinic acid to 5-nonanone. Green Chem 12(4):574–577
Talalay A, Magat M (1945) Synthetic rubber from alcohol. Interscience, New York
Tu Y-J, Chen Y-W (1998) Effects of alkaline-earth oxide additives on silica-supported copper catalysts in ethanol dehydrogenation. Ind Eng Chem Res 37(7):2618–2622
U.S. National Science Foundation (2008) In Breaking the chemical and engineering barriers to lignocellulosic biofuels: next generation hydrocarbon biorefineries. http://www.ecs.umass.edu/biofuels/Images/Roadmap2-08.pdf
Wadley DC, Tam MS, Kokitkar PB, Jackson JE, Miller DJ (1997) Lactic acid conversion to 2,3-pentanedione and acrylic acid over silica-supported sodium nitrate: reaction optimization and identification of sodium lactate as the active catalyst. J Catal 165:162–171
Werpy T, Petersen G (2004) In Top value added chemicals from biomass. Vol. 1. Results of screening for potential candidates from sugars and synthesis gas. US Dep Energy, Off Sci Tech Inf. http://www.nrel.gov/docs/fy04osti/35523.pdf
West RM et al (2008) Liquid alkanes with targeted molecular weights from biomass-derived carbohydrates. ChemSusChem 1(5):417–424

Author Biography

Dr. Juan Carlos Serrano-Ruiz, Ph.D. graduated in Chemistry (University of Granada, Spain) in 2001 and earned his Ph.D. in Catalysis and Materials Science (University of Alicante, Spain) in 2006. In 2007, he moved to Madrid for a post-doc position at the Spanish National Council of Research (CSIC) and in 2008 he was awarded a 2 years Fulbright-MEC fellowship to conduct studies on catalytic biomass conversion to fuels and chemicals in Professor James Dumesic's research group at the University of Wisconsin-Madison (USA). Dr. Serrano-Ruiz is co-author of about 60 papers in SCI journals including one in Science (considered one of the 30 biggest chemical breakthroughs of the year according to Chemistry World), 5 book chapters, and a good number of high-impact reviews on catalytic biomass conversion (Energy and Environmental Science, Chemical Society Reviews, Annual Reviews). He has a cover paper in Green Chemistry, 3 hot papers in this prestigious journal, one of the most cited papers in Applied Catalysis B within 2010–2011, and 4 patents (one of them dealing with catalytic conversion of cellulose into diesel and gasoline which is licensed). He is only editor of three books on biomass conversion and biorefineries and regular reviewer of international scientific journals on Biofuels, Catalysis, and Green Chemistry. Dr. Juan Carlos Serrano-Ruiz has an h index of 25 and accumulates over 3000 citations (90% of them within the last 5 years). From 2012 to 2016, he worked as a Senior Researcher at Abengoa Research, the R&D branch of the multinational company Abengoa developing technologies for advanced biofuels production via catalytic conversion of biomass. In 2017, he was awarded a Ramón y Cajal fellowship to conduct his own research at the Universidad Loyola Andalucía.

Chapter 6
Biorefinery Modeling and Optimization

Abel Sanz, Ana Susmozas, Jens Peters and Javier Dufour

Abstract Most biorefinery processes are still in an early stage of development. Some pilot and demonstration plants exist, but little or no information is available from real installations at commercial scale, which is needed to determine their economic and environmental feasibility. Process simulation is a powerful tool to address this issue, since it is possible to determine mass and energy balances without the necessity of those industrial facilities. From this information, consumption of biomass and other chemicals or auxiliary services can be estimated, and plant equipment can be sized, allowing the identification of the main drawbacks and bottlenecks, the necessity of layouts modification and their optimization. This chapter reviews the different stages to carry process simulation out. As well, the main thermochemical (combustion, pyrolysis, and gasification), biochemical (fermentation) and chemical (fractionation, lignin depolymerization, and platform molecules obtaining) processes for biomass processing are discussed in terms of best approaches to simulate them. Finally, some common aspects like pinch analysis, process optimization, and upscaling are studied.

6.1 Introduction

Traditionally, fossil resources (coal, natural gas, and mainly oil) have been the most important feedstock to produce fuels, energy, and a wide variety of chemical products. Environmental concerns, such as greenhouse gases emissions due to the

A. Sanz · A. Susmozas · J. Peters · J. Dufour (✉)
Systems Analysis Unit, IMDEA Energy Institute, 28935 Móstoles, Spain
e-mail: Javier.dufour@imdea.org

J. Peters
Helmholtz Institute Ulm (HIU), Karlsruhe Institute of Technology (KIT), 76133 Karlsruhe, Germany

J. Dufour
Chemical and Environmental Engineering Group, Rey Juan Carlos University, 28935 Móstoles, Spain

M. Rabaçal et al. (eds.), *Biorefineries*, Lecture Notes in Energy 57,
DOI 10.1007/978-3-319-48288-0_6

combustion of fossil fuels and their future shortage are increasing the interest in their substitution by renewable energy sources (Sarma et al. 2015). The European Commission has been very active on this side, releasing two European Directives during the past years: the Directive on the promotion of biofuels (2003/30/EC) (2003), which specified the replacement targets to be reached, the Renewable Energy Directive (2009/28/EC) (2009), which amends the previous one and, most recently, the Directive 2015/1513 (2015). According to this one, the transport sector has to reach a 10% share of renewable energy and the greenhouse gas emissions saving must be at least 60% for biofuels and bioliquids produced in installations in which production started on or after October 5, 2015 in comparison with fossil ones. For installations already operating on this date, the reduction must be at least 35% until December 31, 2017, and 50% from January 1, 2018.

In this scenario, biomass arises as an option for the production of alternative fuels. Bioethanol, biodiesel, and biogas are the main renewable fuels obtained from biomass, but it can also be used for the production of other high value chemicals. In this way, biorefining appears as a potential alternative to petroleum-based refineries. Among the existing definitions of biorefinery, the most accepted one has been proposed by the IEA Bioenergy Task 42 (2009), Cherubini (2010): "Biorefining is the sustainable processing of biomass into a spectrum of marketable products and energy." Then, a biorefinery comprises a range of technologies to separate biomass into their building blocks (carbohydrates, proteins, triglycerides, etc.), which can be converted in other products, biofuels, and chemicals (Cherubini 2010). One of the most important challenges for biorefineries is to reach a cost-effective and efficient production of these energy and chemical products, and process integration is an interesting strategy to reduce costs and increase the overall efficiency.

6.2 Biomass Conversion Processes

A variety of biomass conversion processes may be involved in the operation of a biorefining facility. They may be divided into three main categories: thermochemical, biochemical, and chemical processes (Fig. 6.1).

Thermochemical biomass conversion processes include the direct combustion, pyrolysis, gasification and torrefaction of biomass feedstock (Kücük and Demirbas 1997; Kihedu 2015). Direct combustion provides heat, which can be used for steam production and electricity generation. Pyrolysis is the thermal decomposition of biomass in the absence of oxygen (Bridgwater 2012). This decomposition leads to the production of bio-oil, a brown liquid which has a heating value of about a half of conventional fuel oil. In this process, also char and a gaseous fraction are formed (Bridgwater and Peacocke 2000). Gasification is the conversion of biomass into a combustible gas mixture at high temperature by means of a partial combustion, with a controlled amount of oxygen and/or steam. The resulting gas is called syngas, and it can be used as a fuel or to produce other chemicals, such as methanol, hydrogen, or synthetic fuels. Torrefaction is a biomass treatment technology to improve

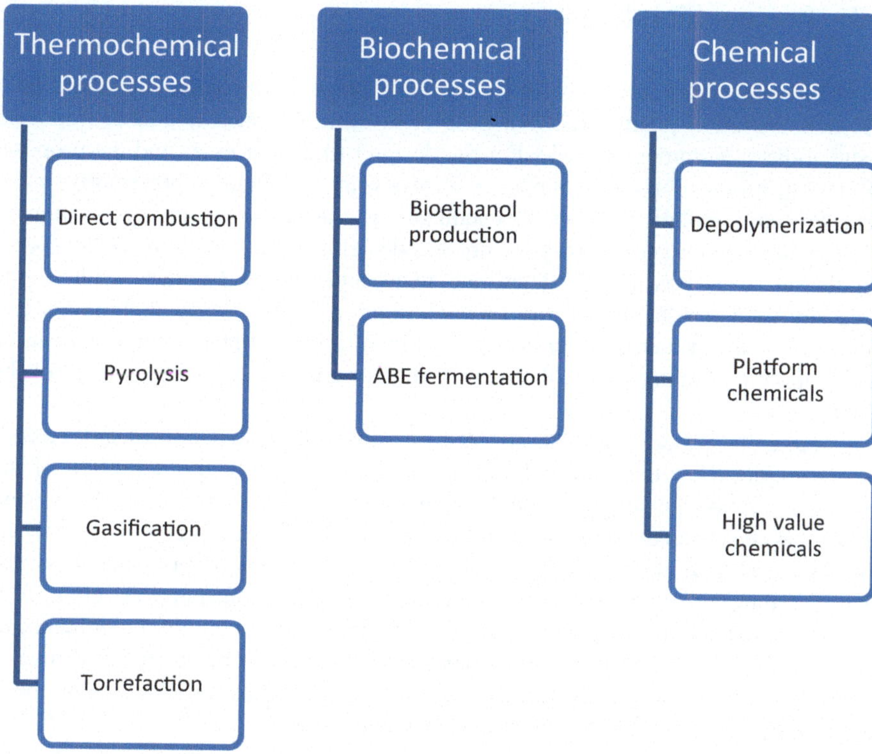

Fig. 6.1 Available biomass conversion processes involved in a biorefinery

its properties and allow easier processing. This treatment consists of a mild thermal treatment under moderate temperatures, around 200–300 °C under very low or no oxygen environment (Kihedu 2015). The moisture content is reduced, what improves the heating value and grindability of biomass and makes it easier to store, because torrefied biomass is hydrophobic. Torrefaction processes may be intended as pretreatment for other thermochemical processes.

Chemical conversion processes are commonly based on the acid degradation of lignocellulosic material, which leads to hexoses, pentoses, and lignin. The depolymerization of biomass macromolecules may lead to the production of platform molecules or other high value chemicals (Gallezot 2012). These products may be processed in conventional chemical facilities to obtain a variety of end-products, such as alcohols, ketones, acids, hydroxymethylfurfural, levulinic acid, polyols, phenol derivatives, hydrocarbons, etc. (Kücük and Demirbas 1997).

Biochemical processes convert biomass into liquid products (ethanol, butanol, acetone), gases (carbon dioxide, methane) and a solid waste which can be used as fertilizer, compost or other by using microorganisms (Kücük and Demirbas 1997). Biomass fermentation may be aerobic (in the presence of oxygen) or anaerobic (in the absence of oxygen).

6.3 Process Simulation

Most biorefinery processes are still in an early stage of development. Some pilot and demonstration plants exist, but little or no information is available from real installations at commercial scale. For the design of this processes and the proposal of modifications or improvements, it is necessary to obtain representative plant data. Process simulation allows data acquisition in these cases, and it is a common strategy in research and industry for process design and optimization. With process simulation, it is possible to determine mass and energy balances without the necessity of a real facility. From this information, consumption of biomass and other chemicals or auxiliary services can be estimated, and plant equipment can be sized, allowing the economic, thermodynamic, or environmental assessment of the plant.

Commercial process simulation software allows the calculation of any considered scenario. The first step to build a simulation model is the definition of the involved components and the process layout (process flowsheet). Once the process flow sheets has been defined and specifications of each block have been entered, the software is able to perform all the calculations and give the output data (mass and energy balances, equipment sizing, auxiliary services consumption, etc.). It is possible to run the model with variations in the input data, to study the influence of each variable on final results. Thus, sensitivity analysis can be easily performed. It is also possible to define the value of an output variable as a design requirement, and modify an input variable to meet this specification.

A variety of commercial process simulation software is available, and the final election depends on the kind of process that is being studied. For instance, Aspen Plus and Aspen HYSYS have been specifically developed for the petrochemical industry, and they are powerful tools for the simulation of chemical processes.[1] There are other simulation software suites such as ProSim,[2] CHEMCAD, [3] and TRNSYS,[4] which may also be used for the modeling and simulation of chemical processes. These commercial suites contain databases of chemical compounds and property estimation modules, to calculate stream properties and chemical reactions. Usually, nonconventional products, such as biomass or lignin, are not included in databases. However, they can be manually defined and properties can be specified to use them in a simulation model.

[1]Aspen Technology—Aspentech [Internet]. Available from: http://www.aspentech.com/products/aspen-plus.aspx.

[2]ProSim [Internet]. Available from: http://www.prosim.net/.

[3]CHEMCAD [Internet]. Available from: http://www.chemstations.com/.

[4]TRNSYS [Internet]. Available from: http://www.trnsys.com/.

6.4 Simulation in Biorefinery Processes

In this section, a general description of the necessary steps for the building of a simulation model is given, starting with the model and components selection, and ending with the custom calculations and the introduction of design specifications.

6.4.1 *Property Model Selection*

To build a simulation model of any kind of process in a simulation software suite, the first step is the selection of the thermodynamic model. This is an important decision, because this model is used to perform all calculations related to phase equilibrium and component properties. The involved components and process conditions have to be taken into account in order to choose the most adequate model. Usually, a guide for the model selection may be found in the software manual.

6.4.2 *Components Specification*

The next stage toward the simulation model is the selection of the components or chemical compounds involved in the simulation. These components are selected from the software databanks, where conventional components and its physical properties are listed. However, not all the possible components are included in these databanks. This problem is especially important when working with biomass. It is a complex material, mainly composed of three biopolymers (cellulose, hemicellulose, and lignin), whose structure and composition vary widely, depending of the origin and the pretreatment of the feedstock. Thus, these products do not exist in the databanks and, to obtain the required physical properties, are entered as user-defined components. In these cases, the component structure must be defined manually, and required data include chemical formula, boiling point, or vapor pressure. A frequent assumption is that all C6 polysaccharides have the same properties as cellulose, and C5 polysaccharides have the same properties as xylan.

It is also possible to specify components without the knowledge of its molecular structure or composition, defining them as nonconventional components. These components are heterogeneous solids that do not participate in chemical or phase equilibrium, and the only physical properties that are calculated are enthalpy and density (2010).

For instance, in Aspen Plus, properties of nonconventional components are estimated based on the component attributes, which usually are proximate, ultimate, and sulfur analysis. Proximate analysis details biomass (or any other nonconventional component) composition according to the next categories: moisture, fixed carbon, volatile matter, and ash content. For ultimate analysis, ash content, and elemental composition are specified, and sulfur analysis details the type of sulfur functional groups present in the component.

6.4.3 Flow Sheet Definition

Once the property model has been selected and the involved components have been specified and defined (if necessary), the next step is the definition of the layout of the process. The flow sheet is built using a library of unit operation blocks (mixers, separators, pumps, heat exchangers, reactors, columns, etc.), which may be combined and connected with material and energy streams in order to build the process flow sheet.

6.4.4 Streams Specification

Once the process flow sheet has been built and components have been specified, it is necessary to supply data for the inlet streams. Total flow and composition must be detailed, and pressure, temperature and vapor fraction (only two of the last three variables are independent, the third one is automatically fixed when two of them are specified).

6.4.5 Blocks Specification

Individual unit operations are represented by blocks in process simulation software. Each type of block demands a different set of specifications. For instance, in a distillation column it is possible to specify the number of equilibrium stages, position of inlet and outlet streams, reflux rate, etc.; or it is possible to define a reactor by specifying the chemical reactions involved and fractional conversions or calculate the product distribution by minimizing the free Gibbs energy.

6.4.6 Inline Calculations and Design with Specifications

At this point, it is possible to run the process model and get the simulation results. Since all inlet data has been specified, the simulation software is able to run the model and complete the calculation of each block. However, it is possible to refine the simulation. For instance, some inlet variables may be coupled, so it would be possible to determine some of them from the values of others. Although all inlet values are specified in order to run the model, some of them can be recalculated according to some conditions. Then, the recalculated value overwrites the initially entered value, fulfilling the specified conditions. For instance, if the mass flow ratio of two streams must be equal to a certain value, it is possible to calculate one of the mass flows from the other taking into account this ratio. Thus, the mass flow of the second stream will be recalculated and overwritten before running the simulation model.

It is also possible to define a value for an outlet stream as a design specification. Then, an input value (block or stream variable) can be modified to tune the model until the fulfillment of that specification.

The combination of these strategies allows the definition of highly complex models, where different variables are related according to the real process needs, making this simulation software a powerful tool to study the behavior of a plant or to design and optimize a plant meeting the product and process specifications.

6.5 Simulation of Biorefinery Processes

In this section, a variety of biorefinery processes will be described. Due to the high amount of possible biorefinery processes, a representative selection of them has been selected.

6.5.1 Direct Combustion of Biomass

Direct combustion of biomass maybe the simplest process which can be considered in a biorefinery. It consists on the oxidation of a biomass feedstock with air, and this reaction provides heat for the generation of steam and/or electricity as the final products. Figure 6.2 shows a block diagram of the direct combustion of biomass.

To build a simulation model of a direct biomass combustion process, some information about the biomass characterization is needed. Depending on this information, different approaches may be employed.

One of the scenarios is the definition of the biomass feedstock as a nonconventional component. In this case, the attributes of this component (proximate analysis, ultimate analysis , and sulfur analysis) must be manually defined. Then, the heating value of the considered biomass may be estimated from these data, or if this value is available it may be supplied as an input data instead.

If biomass composition and structure is well known, it is possible to define the chemical formula and structure of its components, such as cellulose, hemicellulose, lignin, ash, etc. Known components may be defined as solids, specifying its composition and structure, and properties may be calculated from these data. However, for some fractions of the biomass feedstock (ash, for instance), nonconventional components may be defined if detailed structure is unknown.

For this section, a biomass feedstock whose detailed composition and structure is not known will be considered, and component attributes are the only available

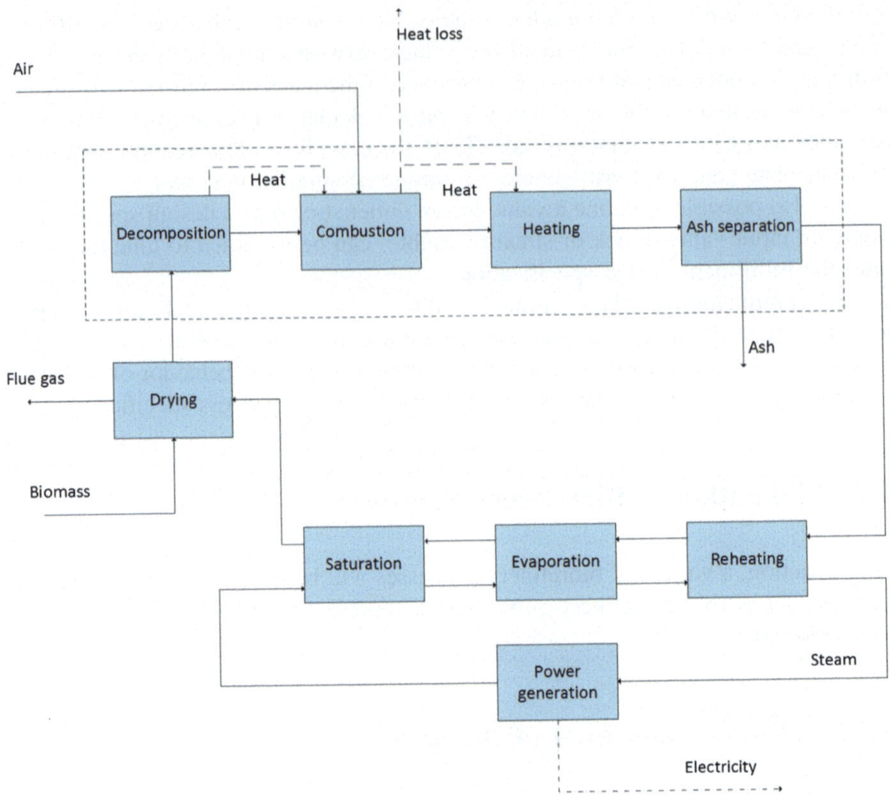

Fig. 6.2 Block diagram of the direct combustion of biomass to produce steam and/or electricity

information. Thus, biomass will be defined as a nonconventional component, and its attributes (proximate, ultimate, and sulfur analysis) will be manually defined by the user. Biomass properties will then be estimated with simulation software from this information. It is also necessary to define another nonconventional component to model the ash which is formed in the combustion process.

Before the combustion, a drying process is needed to reduce moisture content of biomass. This drying process is modeled as a chemical reaction, where raw biomass reacts to form dry biomass and water, and the moisture content of biomass is adjusted to 12%. The amount of released steam is determined by a mass balance, and heat for the drying process is provided by the exhaust gases from the combustion process.

The combustion process has been modeled using an equilibrium reactor, where the product distribution is calculated by minimizing the Gibbs free energy. This kind of reactor is not able to deal with nonconventional components. Thus, before this block has been included a yield reactor, where biomass is decomposed into its constituting elements. In this reactor, the product distribution of the outlet stream is specified by the user, but it is necessary to fulfill the atomic balance. It is possible to specify the product distribution fulfilling the atomic balance by calculating the product yields from the composition of biomass. This composition is known and is detailed in the ultimate analysis attribute. This calculation may be done with a Fortran subroutine or an Excel spreadsheet, whose input data is the atomic composition of the biomass, and the output data are the product yields for this reactor. A pseudo-code for the determination of the product yields and the description of how it works is shown below:

```
X = (100 - MOIST)/100
H2O = MOIST/100
ASH = ASH.IN/100 * F
C = C.IN/100 * F
H2 = H.IN/100 * F
N2 = N.IN/100 * F
CL2 = CL.IN/100 * F
S = S.IN/100 * F
O2 = O.IN/100 * F
```

In this code, MOIST, ASH.IN, C.IN, H.IN, N.IN, CL.IN, S.IN, and O.IN are the input variables. MOIST is the moisture content of the biomass feedstock, and ASH.IN, C.IN, H.IN, N.IN, CL.IN, S.IN, and O.IN are the values for the ash content and atomic composition (C, H, N, Cl, S, and O) of the biomass, which are obtained from ultimate analysis.

X is the conversion factor used to convert the atomic composition from a dry basis to a wet basis, calculated from the moisture content of biomass feedstock. Output variables are H_2O, ASH, C, H_2, N_2, CL_2, S, and O_2, which represent the mass yields of each product of the reactor. In this way, product yields are recalculated from biomass composition and the atomic balance is always fulfilled. Carbon dioxide is not included in this code, because it is not necessary for the decomposition of biomass. This component is later formed in the combustion reactor where the decomposed biomass is fed.

Biomass combustion has been modeled using a combination of unit operation blocks. The first one is the biomass decomposition process, and the product stream of this block is sent to the combustion reactor, where it is mixed with an air stream and the combustion reactions take place. The temperature of the gases stream has been fixed at 800 °C, but to model boiler efficiency, the reactor has been considered to work in isothermal conditions. It is important to include a heat stream from the decomposition reactor to the combustion one, because the heat of reaction involved in the decomposition has to be taken into account. Another heat stream, which represents the total heat of combustion, exits the combustion reactor and enters a splitter, which represents thermal losses. This heat of reaction is then used to raise the temperature of the products of the combustion reactor until the fixed temperature (which is controlled with the air flow fed to the combustion reactor). This step is modeled using a heat exchanger, whose inlet streams are the combustion products (material stream) and the heat of reaction (energy stream). Finally, a separator block is used to obtain a gas stream and a solid stream, which contains the ash, defined as a nonconventional component as detailed above.

Combustion gases are fed to a set of heat exchangers to generate high pressure steam. Heat exchangers are a reheater, where saturated steam is reheated, an evaporator, where saturated liquid is converted to saturated vapor, and an economizer, where the liquid water is heated until its saturation temperature. The hot combustion gases stream enters the reheater, then the evaporator and then the economizer, and finally the flue gases are released to the atmosphere.

The high pressure steam generated is then fed to a steam turbine, where electricity is generated and the pressure of the steam decreases. Then, steam is condensed, pressurized and a make-up of water is added in order to compensate the possible leaks. This liquid stream is sent again to the heat exchangers to generate steam.

6.5.2 *Biomass Pyrolysis and Bio-Oil Refining Modeling*

In this section, a description of a biomass fast pyrolysis simulation model is presented. This model has been developed in the working group of the authors, using the simulation software suite Aspen Plus. The process is divided in several sections (pyrolysis, hydrotreating, distillation and hydrocracking and steam reforming), which are described in the next paragraphs.

6.5.2.1 Pyrolysis Section

The modeling of a biofuel production process based on fast pyrolysis of hybrid poplar and hydroupgrading of the bio oil (Peters et al. 2014a) is described in this section, with the layout of current existing pyrolysis plants (Anex et al. 2010; Peters et al. 2014b; Jones et al. 2009; Laclaire and Barrett 2002; Ringer et al. 2006). Raw biomass is converted into bio-oil and char in the pyrolysis section of the plant. Figure 6.3 shows a block diagram of the pyrolysis plant, and the main unit operation blocks are specified.

Biomass is defined as a mix of cellulose, hemicellulose, lignin, ash, moisture and other components, such as extractives. Lignin is assumed to be formed by three different monomers with variations in atomic composition. The process begins with the pretreatment of the raw biomass, adjusting the moisture of the feedstock and the particle size to the pyrolyzer requirements. The first step of the pretreatment is a drying process, where moisture content of biomass is adjusted to 7%, as required by the pyrolysis reactor. This operation is modeled using a heat exchanger, where the heat is provided by the exhaust gases from the gas and char combustor. The next step in the pretreatment is the particle size reduction, which is modeled with a

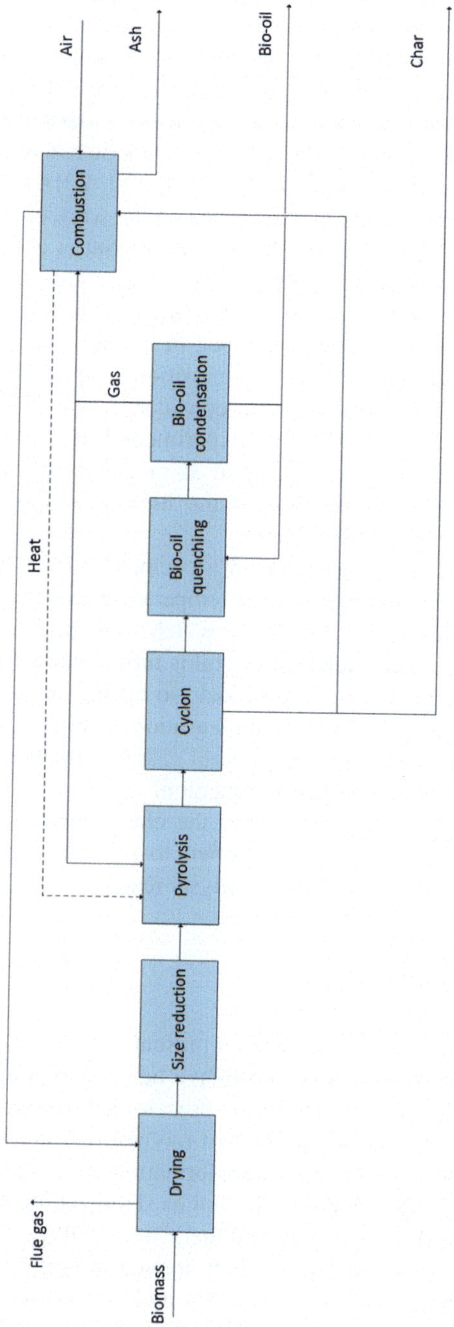

Fig. 6.3 Block diagram of the pyrolysis plant

grinder. This unit operation block needs a parameter to characterize the biomass (grindability index), which is obtained from bibliography to estimate the energy consumption of this step. Biomass particle size is reduced to 3 mm in this block.

Then, it enters the pyrolysis reactor, where it is decomposed into bio-oil, char, and gas. Typical conditions for a circulating fluidized bed reactor (520 °C, 2 s bed residence time, 0.5 s vapor residence time) are selected (Jones et al. 2009; Iribarren et al. 2012). The pyrolysis reactor is defined with a kinetic reaction model with over 150 decomposition reactions implemented (Peters et al. 2014a). The implementation of the primary pyrolysis reactions was based on the work of Ranzi et al. (2008), Faravelli et al. (2010), Calonaci et al. (2010) and Dupont et al. (2009). Secondary reactions (mainly condensation and polymerization) are modeled in another reactor before the previous one, where yields are determined and fixed according to the work of Hoekstra et al. (2012), due to the lack of kinetic data. This model allows the predictive calculation of the bio-oil composition when different kinds of biomass feedstock and reactor conditions are considered, and it has been validated against literature (Peters et al. 2013) and experimental data (Peters et al. 2014c), showing a high degree of correlation. A set of 31 model compounds are considered in order to give a detailed description of the bio-oil composition, and char is modeled as a nonconventional component.

After the pyrolyzer, char is separated in a cyclone and the gaseous fraction is quenched with a stream of cold bio-oil. The temperature reduction must be quick in order to stop the decomposition reactions, which may lead to the formation of secondary products. Later, the quenched bio-oil is further cooled with cooling water (modeled with a standard heat exchanger block) to maximize the liquid recovery in a flash separator (described as bio-oil condensation in Fig. 6.3). In this unit, the condensable fraction is separated from the light gases. A fraction of these gases are recirculated to the reactor for the bed fluidization.

A fraction of the light gases stream and the char which is obtained as a side product are burned in a combustion chamber, to provide the necessary thermal energy for the biomass drying and the pyrolysis reactor.

6.5.2.2 Hydrotreatment Section

The bio-oil produced in the pyrolysis section presents a high oxygen content, which cannot be used directly because it is a corrosive liquid with a high viscosity and it tends to polymerize. So, the next step of the process is the reduction of this oxygen content, to convert the bio-oil into an almost oxygen-free product. The layout for this section is shown in Fig. 6.4. This transformation is carried out in a two-stage catalytic hydrotreatment (described in Fig. 6.4 as stabilization and deoxygenation), using commercial Co–Mo catalyst (Choudhary and Phillips 2011; French et al. 2010; Venderbosch and Heeres 2011). Hydrotreated bio-oil has been modeled by 52 compounds with chain lengths between C4 and C18 (Peters et al. 2014a).

The first step of the hydrotreatment is the stabilization of the bio-oil, which takes place at 270 °C and 170 bar, and produces a bio-oil stream with an oxygen content

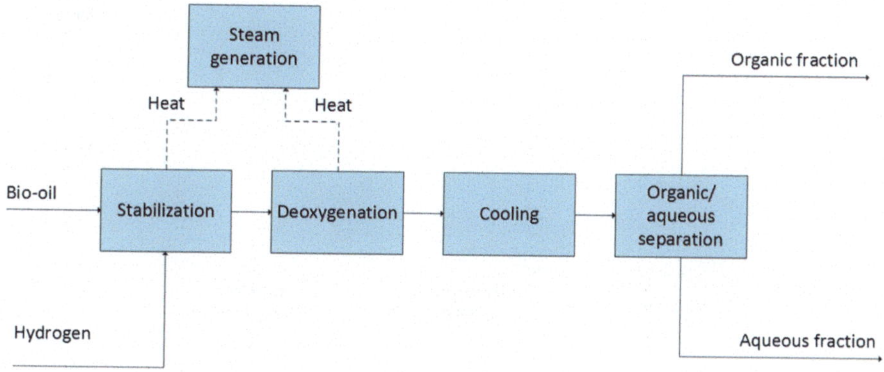

Fig. 6.4 Block diagram of the hydrotreatment section

of about 30%. The second reactor, operating at 350 °C and 150 bar, yields a deoxygenated bio-oil, with an oxygen content below 2%. Both blocks are modeled as yield reactors, and product distribution is determined by a linear regression algorithm based on literature data.

The stabilized bio-oil is cooled down in a heat exchanger after the reaction steps, and an aqueous and an organic fraction are separated in a flash unit (described in Fig. 6.4 as organic/aqueous separation).

6.5.2.3 Distillation and Hydrocracking Section

The organic fraction of the bio-oil is separated into different fuel fractions by distillation, and the heavier products are sent to a hydrocracking reactor to increase the yield in light products, as it is shown in Fig. 6.5.

Distillation products of the first column are a gas fraction, gasoline ,and a heavier product, which is sent to a vacuum distillation column, where it is separated into a diesel fraction and a heavy fraction in the bottom, which is sent to the hydrocracker reactor. Simulation software allows the detailed modeling of these distillation columns, to determine number of stages, reflux ratio and energy consumption of boilers and condensers.

The hydrocracking unit has been modeled as a stoichiometric reactor based on literature data (Peters et al. 2014a). The product is recirculated to the organic fraction and sent to the distillation columns to be separated into different fuel fractions.

6.5.2.4 Steam Reforming Section

In this process, a hydrogen supply is needed by the hydrotreaters and the hydrocracker. This hydrogen is produced by a steam reforming process of the light

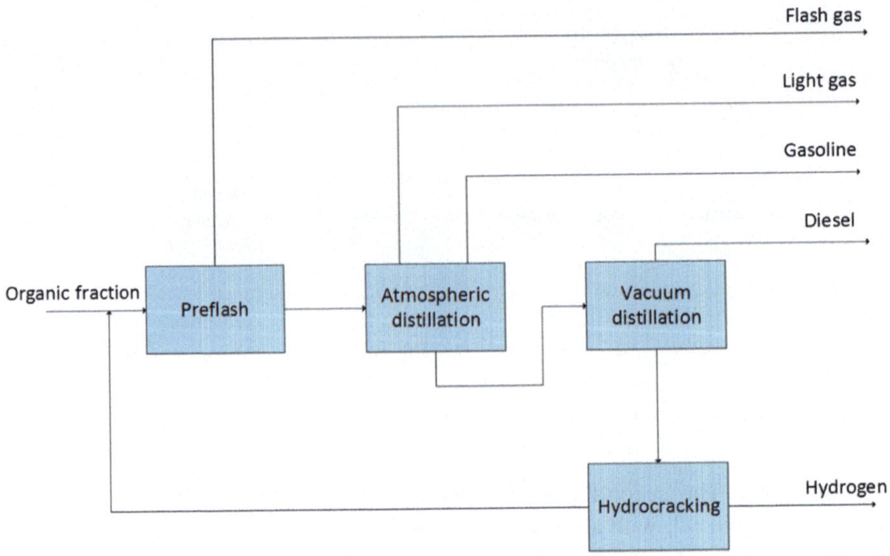

Fig. 6.5 Block diagram of the distillation and hydrocracking section

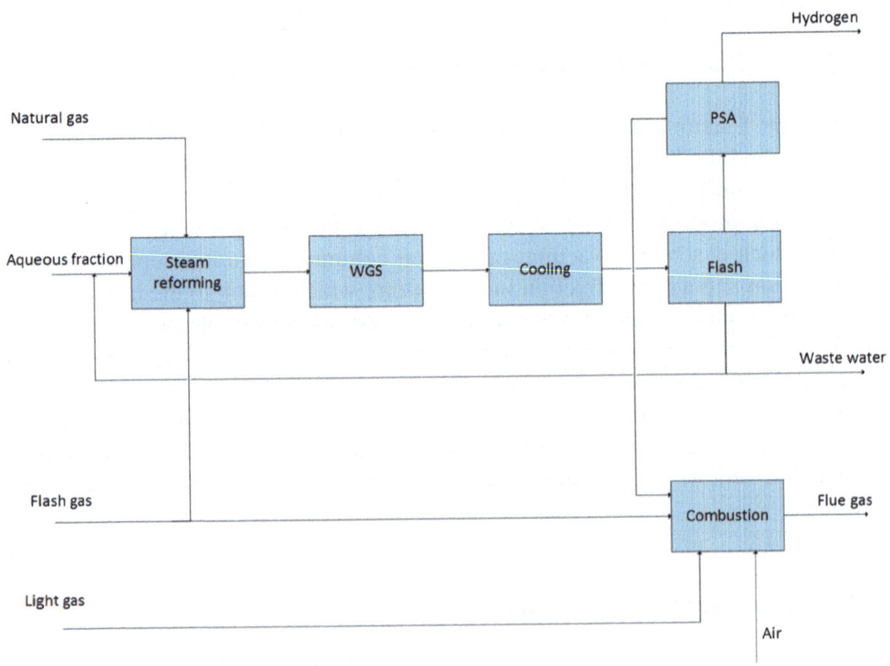

Fig. 6.6 Block diagram of the steam reforming section

hydrocarbons of the gas stream from the distillation section. A block diagram of the steam reforming section is shown in Fig. 6.6. The gas stream is mixed with water and natural gas, heated and fed to the steam reforming reactor, which has been modeled as an equilibrium reactor. It calculates the product distribution by minimizing the Gibbs free energy.

The stream that exits the steam reforming reactor is cooled down and sent to a water gas shift reactor, which has also been modeled to calculate the product distribution by minimizing the Gibbs free energy. This reactor increases the hydrogen produced by converting carbon monoxide and water to carbon dioxide and hydrogen. The products are cooled and sent to a flash, where water is separated from the gases, and a PSA, which has been modeled as a selective separator, separates hydrogen from the gases stream. The off-gas stream from the PSA is mixed with the gas stream from the distillation section and sent to a combustor, to generate the heat required by the steam reforming reactor.

6.5.3 Biomass Gasification

Gasification is a thermochemical conversion process for the generation of syngas from a solid fuel. It takes place at high temperature, and the main involved reactions include partial oxidation with a limited amount of oxidant, which can be air, oxygen or steam (Heidenreich and Foscolo 2015; Iribarren et al. 2014). The composition and properties of the syngas depend on the biomass feedstock, the gasifier type, and the gasification conditions.

Gasification is considered one of the most promising technologies for the conversion of biomass into electricity, heat, and chemical compounds (Iribarren et al. 2014). For instance, this process may be used to produce hydrogen and liquid fuels. Hydrogen is usually produced by steam methane reforming (from natural gas), but biomass gasification is attracting scientific interest due to sustainability concerns. In this section, a simulation model for a biomass gasification process using a typical fluidized bed gasifier is presented, based on the experience of the authors. For the construction of this model, Aspen Plus simulation software has been used.

6.5.3.1 Fluidized Bed Gasifier

Biomass gasification occurs at high temperature (500–1400 °C), and it is a two-stage process: pyrolysis takes place in first place, and then gasification. As shown, pyrolysis consists of the thermal decomposition of the biomass into volatile hydrocarbons and char, while gasification is the conversion of these products into syngas. Gasification pressure may vary between atmospheric pressure and 33 bar (Ciferno and Marano 2002), and depending of this pressure the size and the efficiency of process equipment may vary.

Depending of the heat demand, it is possible to distinguish between two types of gasification: direct (or autothermal) gasification, and indirect (or allothermal) gasification. In the first case, the partial oxidation of biomass is able to cover the heat demand of the process, and it is the most common scenario. In the second one, an external energy supply (e.g., natural gas or biomass) is used to cover the energy needs.

The modeling of a biomass gasification process in an indirect fluidized bed gasifier is described below. Gasification reactions take place in a circulating fluidized bed, and the used catalyst enters a combustion chamber where char is burned and the catalyst is regenerated. Then, it is sent back to the gasifier, providing energy from the combustion of the char. A block diagram of this process is shown in Fig. 6.7.

The first stage of the process is the biomass pretreatment section, where the moisture of the feedstock is decreased to meet the gasifier requirements. Biomass is defined as a nonconventional component, specifying its attributes to estimate the component properties. Biomass drying is defined as a chemical reaction, where biomass forms more biomass (with lower moisture content) and water, and the moisture content of the new biomass is adjusted to fulfill mass balances. Then, free water is separated from biomass, which is fed to the gasifier.

As it has been stated before, Aspen Plus software is not able to deal with chemical equilibrium of nonconventional components such as biomass. In order to model the gasification reactions, it is necessary to include a decomposition step in the gasifier, where biomass is decomposed into its elemental constituents. The methodology for the modeling of this decomposition is similar to the one described in the direct combustion process. The decomposition products are fed to the gasification reactor, where the product distribution is determined according to literature (Spath et al. 2005), by means of a Fortran subroutine. Products from gasifier are char and a syngas stream, which enters a tar reforming unit, where tars are reformed in order to increase hydrogen production. The tar reforming unit is modeled as an equilibrium reactor, where product distribution is determined by the minimization of the Gibbs free energy. Char is sent to a combustion chamber, which has been also modeled as an equilibrium reactor, where it is burnt with air in order to regenerate the catalyst and provide the necessary heat to the gasification reactions.

After the tar reforming, the syngas stream is cooled down and washed with water, to remove particles, chlorine compounds, ammonia, and residual tar. The clean syngas is then compressed and fed to a LO-CAT unit, where hydrogen sulfide is absorbed in an iron catalyst solution and oxidized to elemental sulfur.[5] The LO-CAT reactor is modeled as a stoichiometric reactor, where hydrogen sulfide reacts with oxygen and is completely converted to solid sulfur and water.

Once sulfur has been removed, the temperature of the syngas is increased and it is sent to a water gas shift reaction process. This stage is modeled considering two

[5]Merichem Company. LO-CAT process for cost-effective desulfurization of all types of gas streams.

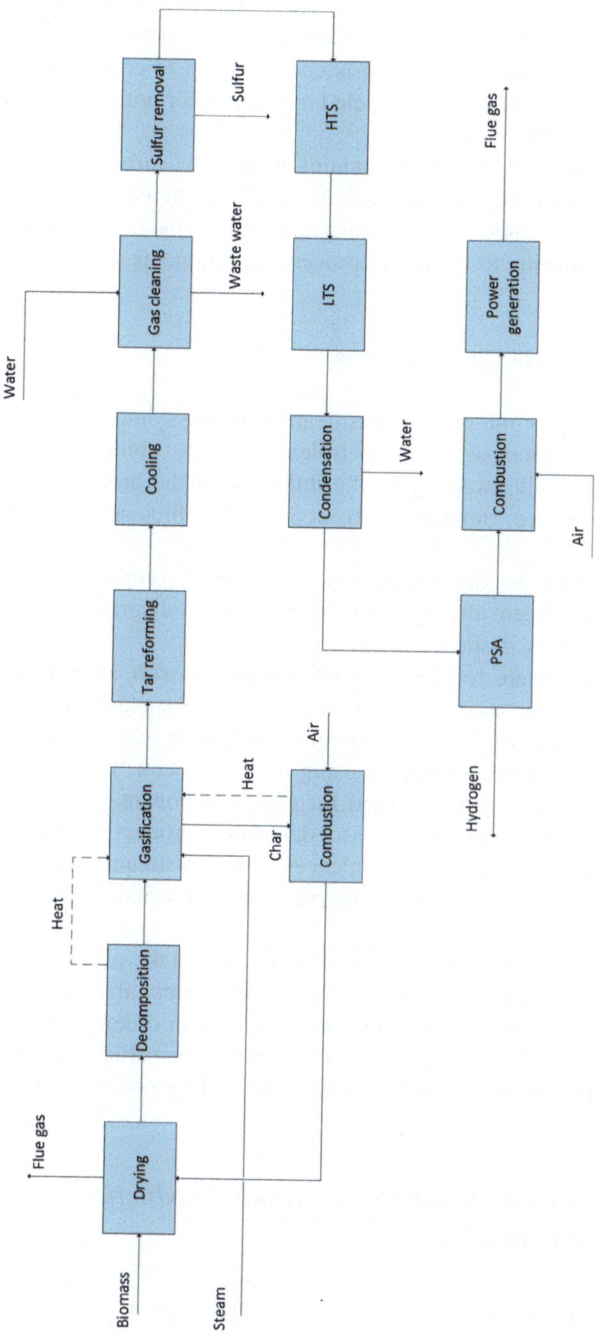

Fig. 6.7 Block diagram of a biomass gasification plant (fluidized bed gasifier)

steps at different temperatures (the former at 350 °C and the latter at 220 °C) in order to increase hydrogen production. These reactors have been modeled as equilibrium reactors, where the products distribution is determined by minimizing the Gibbs free energy. Finally, syngas is cooled down to condense water and the gaseous fraction is sent to a PSA, modeled as a separator with hydrogen purity and recovery from literature.

The off-gas from the PSA unit contains hydrogen, carbon monoxide, carbon dioxide, methane, and other combustible gases. This stream is fed to a combustor and the flue gases are used to generate steam. With this steam, electricity is produced in a steam turbine to cover the process requirements.

6.5.3.2 Biomass-Coke Co-gasification

Another possibility to integrate a gasification process into a biorefinery is the co-gasification of biomass with petroleum coke. Simulation model of a co-gasification plant allows studying the influence of the biomass to coke ratio in the plant performance, attending to economic, efficiency or environmental concerns.

Biomass and coke are defined as nonconventional components, with the same procedure that has been already specified. A block diagram of the considered co-gasification process is shown in Fig. 6.8.

Biomass and coke are fed to the gasifier together with oxygen and water and steam. In this operation unit, gasification reactions take place at a temperature of 868 °C and a pressure of 2.5 bar (Hannula and Kurkela 2012). In this case, gasification reactions have been modeled using a yield reactor. In this kind of reactor, individual product yields are specified for each component. A Fortran subroutine based on bibliographic data is used to determine product yields (Hannula and Kurkela 2012), fulfilling the atomic balances. The gas stream that exits the gasifier enters a tar reforming step, similar to the previously described one, where hydrogen content of the syngas is increased.

After the tar reforming, the syngas cleaning steps take place, in the same way described in the previous process (scrubber, sulfur removal) and a two-step water gas shift reactors are modeled (equilibrium reactors) in order to increase hydrogen content of syngas before entering the PSA unit, where high purity hydrogen is separated, and the off-gas is burnt to cover the heating needs of the process.

6.5.4 Biochemical Processes: Biofuels Production via Fermentation

The most important product from biochemical processes in a biorefinery is ethanol (bioethanol), which is usually obtained by hydrolysis and fermentation of the biomass feedstock, but also other chemicals can be synthesized via fermentation

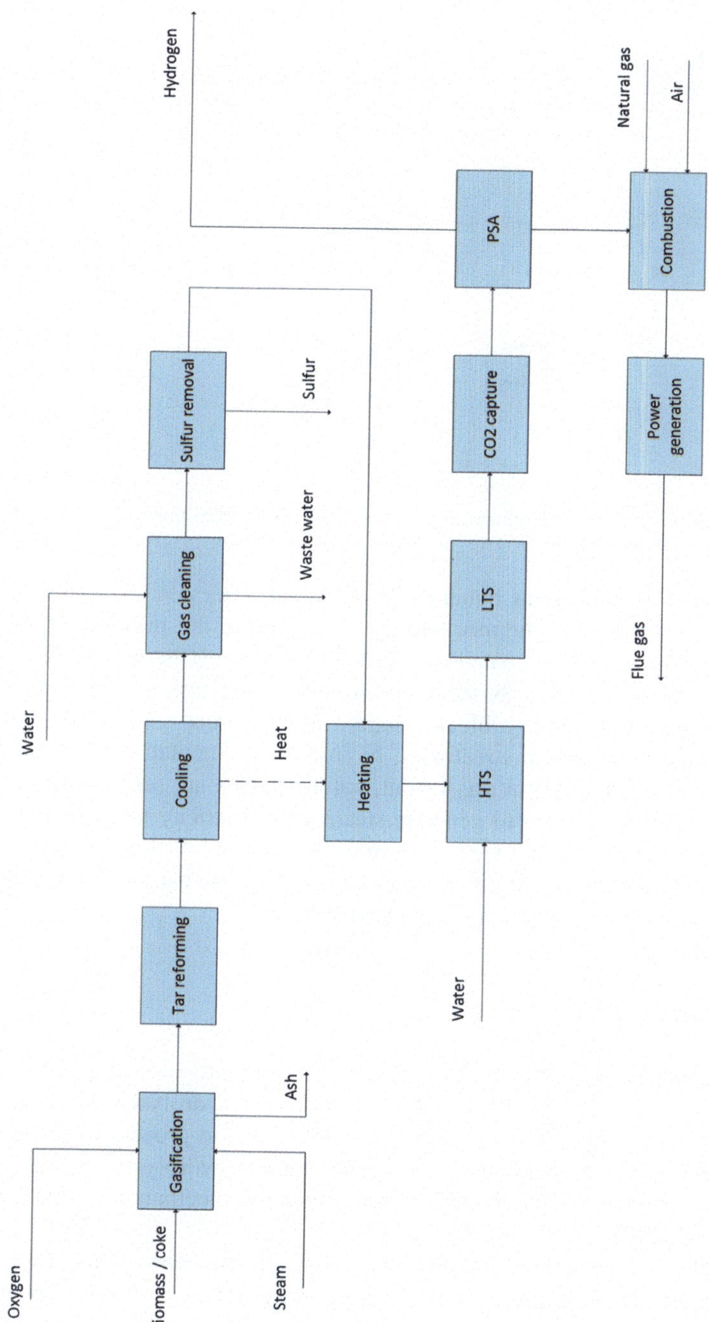

Fig. 6.8 Block diagram of a biomass and coke co-gasification process

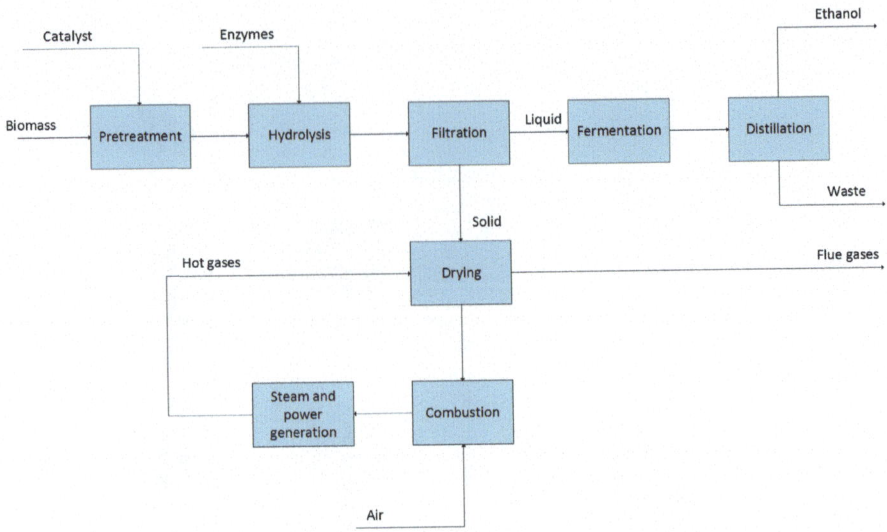

Fig. 6.9 Block diagram of a bioethanol production process via fermentation

processes, such as acetone or butanol. Biochemical processes take place at lower temperatures compared to thermochemical ones, due to the presence of enzymes and microorganisms, and so, reaction rates are also lower (Cherubini 2010). In this section, the modeling of a bioethanol production process via fermentation is described. Figure 6.9 shows a block diagram of this process, which includes the following stages: biomass pretreatment, hydrolysis, fermentation, and bioethanol purification via distillation. In this layout, solid residues are sent to a combustion unit, and hot gases are used to generate steam and electricity to cover the energy needs of the process. This hot gases stream is further employed in the drying of the solid fraction that exits the filtration stage before the combustion. A more detailed description of the involved stages is presented below.

6.5.4.1 Biomass Pretreatment

In biochemical processes, the main objective of the biomass pretreatment is to make carbohydrates more accessible to enzymes used for hydrolysis, by opening the lignocellulosic structure. The economic feasibility of the process depends on the ability to achieve a high yield in the carbohydrates hydrolysis, because this is necessary to achieve a high efficiency and reduce the bioethanol production costs (Jacquet et al. 2011).

One of the most extended pretreatments is steam explosion. This process consists of treatment of biomass with high-pressure steam, followed by a rapid decompression (Martín-Sampedro et al. 2012). These steps lead to the explosion of the lignocellulosic material into separated fibers, and to hydrolysis reactions which result in a loss of hemicellulose and, to a lower extent, cellulose. These sugars can be recovered in the liquid fraction. The steam explosion process has low environmental impact, investment costs, and energy consumption compared with other pretreatment methods (Li et al. 2007). However, the degradation of sugars leads to the formation of side products which can act as inhibitors in the subsequent steps of the process.

An alternative to steam explosion for the pretreatment of lignocellulosic biomass is liquid hot water (LHW). Biomass is treated with water at high temperature (140–180 °C) and pressure (20 bar) for a relatively short time of reaction (5–20 min) (Imman et al. 2014). Reactions taking place with this pretreatment lead to improved cellulose hydrolysis. The LHW pretreatment presents some advantages compared with other options, as the reduced production of inhibitory by-products (furfural and hydroxymethylfurfural), and the addition of extra chemical compounds is not necessary.

Most publications dealing with biomass pretreatment are based on experimental results, but the reaction mechanisms are not well established, and there is a lack of publications dealing with the modeling of these steps. These data are necessary to build a simulation model able to predict reaction yields and product distribution, but it is possible to develop a model based on experimental data (not predictive). Thus, experiments must be carried out with the selected biomass feedstock in order to determine the pretreatment conditions. From product distribution, it is possible to define the chemical reactions which are taking place and fractional conversions. Pretreatment can be modeled with a stoichiometric reactor, defining chemical reactions and conversions, or with a yield reactor, where product distribution must be specified (it is important to take care to fulfill the atomic balance of each element). When biomass is specified as a nonconventional component, an equilibrium reactor is not able to simulate the reactions that occur in this step.

6.5.4.2 Carbohydrates Fermentation

The main components of the biomass feedstock are carbohydrates and lignin. Bioethanol is obtained from carbohydrates present in biomass, which may be converted to sugar via hydrolysis, and later to bioethanol via a fermentation step.

The hydrolysis reactions may be modeled in stoichiometric reactors, where chemical reactions and conversions are manually defined. The modeling of the fermentation step is more complex, mainly due to the kinetic equations used to describe this process (Jin et al. 2012; Slininger et al. 1990, 2014).

Simulation software suites contain a variety of templates to simulate chemical reactors with kinetic data. However, due to the particularities of the kinetic expressions of biochemical processes, their use cannot be always possible.

A strategy for the simulation of processes involving microorganisms growth and fermentation reactions is the definition of a completely customized kinetic model externally, for instance using an Excel spreadsheet, and then linking this model to simulation software (Fig. 6.10).

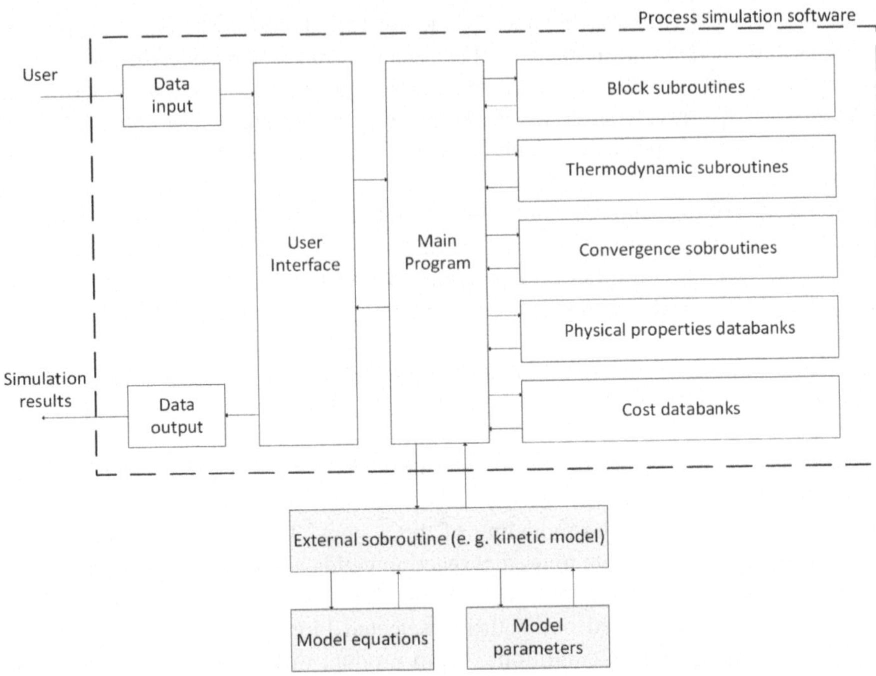

Fig. 6.10 Linking of simulation software with an externally defined model

The external model must read the input variables from the simulation software, perform the calculations, and return the output values to the simulation software, which uses them to calculate product distribution. The chemical reactor may be modeled as a stoichiometric reactor, defining the chemical reactions that take place. Reaction conditions and the composition of the feed streams are sent to the external model, and after calculating the product distribution, fractional conversion of each reaction is determined. These values are sent to simulation software, and they are used for the modeling of the fermentation reactions. Thus, product distribution obtained in simulation software fits the external kinetic model. However, some information can be obtained from simulation software: since the involved components are included in component databanks of simulation software and its properties are known, it is possible to estimate the heat duty of the reactor, and the outlet stream conditions.

Once the reactor has b een modeled, further separation and purification stages are needed in order to obtain a product which fulfills the specifications. Unit operations as distillation, filtration, absorption, etc. may be modeled conventionally, using the operation models present in the software libraries.

As it has been already mentioned, other chemical compounds can be produced via fermentation. For instance, the ABE production process is utilized to obtain acetone, butanol, and ethanol (Morales-Rodriguez et al. 2014). This process consists of various stages, and the general layout is similar to the one described for bioethanol production: initially, a pretreatment step is needed to leave the cellulose available for enzymatic hydrolysis. Then, enzymatic hydrolysis takes place to release the glucose. After the hydrolysis, glucose enters the ABE fermentation, where it is transformed into acetone, butanol, and ethanol. The last step of the process is product purification and reactant recovery. A variety of mathematical models describing the ABE fermentation step can be found in the literature (Leksawasdi et al. 2001). Pretreatment and enzymatic hydrolysis stages can be easily modeled in simulation software by specifying reaction conversions in stoichiometric reactors, while the ABE fermentation step may be modeled externally. Product distribution may be calculated from kinetic models using an Excel spreadsheet or any software able to solve numeric models, such MATLAB or Fortran. The output values of these models must be fermentation reactions conversions, which are then sent to simulation software to perform the whole process simulation. Subsequent purification steps consist of mainly distillation processes, which can be modeled in a conventional way. Process simulation software allows the rigorous design and optimization of these steps.

6.5.4.3 Lignin Valorization via Combustion in Biochemical Processes

Just sugars are used for bioethanol manufacturing. Lignin is not hydrolyzed nor fermented in this process, so it remains as a solid residue after the fermentation stage. A common pathway to integrate the lignin into the process is the combustion of this solid residue, to provide heat requirements of the biochemical process (biomass preheating, drying, reheaters of distillation columns, etc.) without an external energy source, reducing the total production cost of ethanol. This process may be modeled in simulation software in a way similar to that described for the direct biomass combustion (Fig. 6.2). Thus, lignin attributes must be specified and a decomposition reactor (yield reactor) may be used to transform lignin into its constituent elements. After this stage, a combustion reactor (equilibrium reactor) is added, and biomass is burnt with air until complete combustion. The hot gas stream which exits the reactor is used to generate steam, reducing its temperature, and with this steam it is possible to generate electricity in a steam turbine and cover power and heat needs of the process.

6.5.5 Chemical Processes

An alternative pathway to thermochemical and biochemical processes for biomass valorization in the framework of a biorefinery is chemical processing.

6.5.5.1 Biomass Fractionation

As it has been already stated, biomass is a complex material mainly formed by cellulose, hemicellulose ,and lignin. These components are polymers, constituted by the repetition of different units (monomers).

A usual pathway to increase the efficiency of the biomass feedstock processing is biomass fractionation. This process allows a more efficient use of each of the constituent fractions of the biomass, obtaining valuable products from each fraction in the framework of a biorefinery concept. Among the most usually used pre-treatments, organosolv is one of the most important ones. In this process, an aqueous organic solvent mixture is used as the pretreatment medium, and biomass is fractionated into its main components (Sannigrahi and Ragauskas 2013). Usually, ethanol is used as the solvent and sulfuric acid as the delignification catalyst, especially for the production of bioethanol. Biomass components are fractionated into different streams: hemicellulose and sugar degradation products are obtained in a liquid stream, since these components are water-soluble. A solid lignin fraction and a cellulose-rich solid are also obtained in this process (Sannigrahi and Ragauskas 2013; Johansson et al. 1987). Figure 6.11 shows a block diagram of the organosolv pretreatment using ethanol as solvent.

The organic solvent is used to solubilize both lignin and hemicellulose fractions, and they are then separated from the cellulose fraction which remains as a solid. Later, the liquid fraction is diluted with acidified water in order to precipitate and recover the lignin fraction.

The organosolv process leads to a lower average molecular weight of cellulose fragment, resulting in high yields in subsequent hydrolysis and fermentation processes for the production of bioethanol fuel.

The main chemical reactions which take place in the organosolv process are the hydrolysis of lignin-hemicellulose linkages and internal lignin bonds, which lead to lignin and hemicellulose solubilization, the degradation of monosaccharides to furfural, hydroxymethylfurfural ,and other products such as levulinic and formic acids, and lignin condensation reactions (Sannigrahi and Ragauskas 2013). Due to the high complexity level of the feedstock and the variety of operating conditions, solvents and catalysts, the modeling of this step must be supported by experimental results. Thus, the process model is not able to predict the yields and the products distribution, but they can be manually entered into the simulation software using data from experimentation, and use the model to determine mass and energy balances (Kautto et al. 2013). With these data, it is possible to perform the economic, environmental or energy analysis of this pretreatment process (García et al. 2011).

Cellulose and hemicellulose obtained via biomass fractionation are suitable for its use in bioethanol production via fermentation. Once lignin fraction has been separated from them, it can be used for the synthesis of a variety of chemical products, with applications in a wide range of processes.

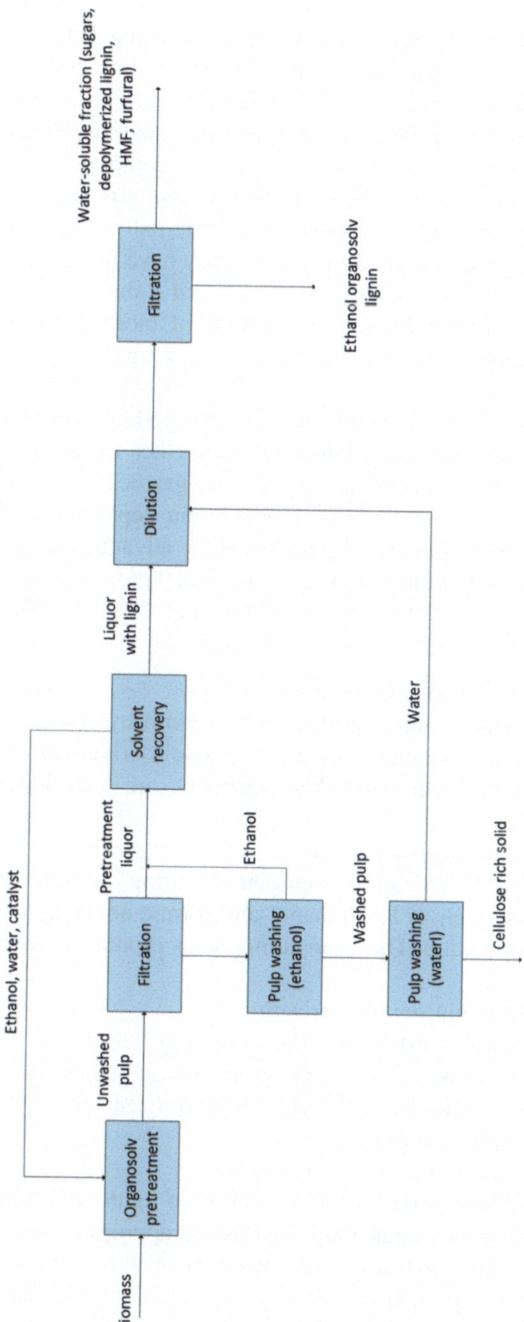

Fig. 6.11 Block diagram of the organosolv process with ethanol

6.5.5.2 Lignin Depolymerization

There has been research into lignin valorization since the 1980s, but there have been very few successful examples of its conversion into high value aromatic products (Bugg and Rahmanpour 2015). Lignin valorization processes must face a series of challenges, such as difficult hydrolysis, low solubility, heterogeneous and variable structure, repolymerization reactions or the complex composition of the depolymerization product stream. The utilization of microorganisms with capability to break down lignin has been proposed as an alternative to convert polymeric lignin into chemicals. Modified microorganisms may have the capability to degrade lignin by producing high value chemicals (Bugg and Rahmanpour 2015).

Lignin separated from the remaining fractions of biomass feedstock is still a material with a polymeric structure. Nowadays, most lignin is used as an energy source in combustion processes (Wang et al. 2013; Chávez-Sifontes and Domine 2013), but it is also a potential resource of chemical products via depolymerization. Its main objective is the conversion into small molecules for the synthesis of other chemical products. Thermochemical (pyrolysis, gasification) and biochemical (fermentation) processes can be intended as a lignin depolymerization pathways. However, the chemical treatment of lignin has some advantages related to reaction control and selectivity. A wide variety of chemical lignin depolymerization processes are available, and a brief overview (Wang et al. 2013) of them will be given in this section.

- Base catalyzed lignin depolymerization.
  This process consists on the treatment with sodium hydroxide at high temperature (270–330 °C), obtaining phenol and phenol derivatives. The most abundant components formed with this treatment are catechol, syringol and derivatives.
- Acid catalyzed lignin depolymerization.
  Acid catalyzed depolymerization is carried out using different combinations of acids and alcohols, such as HCl/ethanol and formic acid/ethylene glycol and a wide range of temperature. The main components produced are methoxyphenol, catechol and phenol.
- Metallic catalyzed lignin depolymerization.
  The objective of metallic catalysts is the increase of selectivity. The introduction of these catalysts decreases the activation energy, allowing milder reaction conditions, and the major products vary depending of the origin of the treated lignin and the selected catalyst.
- Ionic liquids assisted lignin depolymerization.
  Ionic liquids have been used for the separation of lignin and cellulose from raw lignocellulosic biomass. Some ionic liquids are appropriate solvents for lignin dissolution, and Bronsted acid associated with ionic liquid catalyzes the depolymerization reaction. However, the use of ionic liquids for these reactions is limited due to its high cost and difficulty in separation of ionic liquid and

lignin derivative molecules. Major products vary in function of the selected ionic liquid, being the most frequent guaiacol and benzoquinones.

- Supercritical fluids-assisted lignin depolymerization.
  Lignin depolymerization using a supercritical fluid as the medium for the reaction may be interesting due to its good properties as a lignin solvent. Hydrogen sources for the hydrolysis usually are acids and alcohols. Despite the good performance of these fluids, its application is limited due to the high cost. Major products obtained from lignin depolymerization with supercritical fluids are catechol, phenol, cresol, syringol or guaiacol, but product distribution depends on the specific biomass, supercritical fluid and reaction conditions used.

Due to the wide variety of depolymerization processes available, depending of the solvent and catalysts used and the reaction conditions (pressure, temperature, pH, contact time, etc.), the proposal of a general mechanism to build a simulation model able to predict product distribution is not affordable. Information about these processes can be found in literature, but it is essentially based on experimental results. However, simulation is a powerful tool to calculate mass and energy balances of a process, with reaction data gathered from experimental results, allowing the energy integration and process optimization in order to increase the overall efficiency.

6.5.5.3 Applications of Polymeric Lignin

Polymeric lignin has some industrial applications, and the development of more applications is an interesting research field, since lignin is produced at high scale not only in biorefineries, but in the paper industry. For instance, it is possible to obtain phenol-formaldehyde adhesives or phenolic resins from lignin (Ramires et al. 2010). Other applications include the fabrication of polyurethane foams (Pan and Saddler 2013), carbon fiber (Baker and Rials 2013) or porous carbon (Jeon et al. 2015). The conversion of lignin over alumina supported molybdenum catalyst has also been studied, leading to C6–C11 molecules, such as alcohols, esters, phenols, benzyl alcohols and arenes (Ma et al. 2015).

However, all this alternatives are currently in a very early stage of development, and the construction of predictive simulation models is not still viable. With the advance in the knowledge of lignin depolymerization and transformation reactions, it will be possible to propose kinetic or equilibrium based reaction schemes. These schemes would allow the construction of more powerful simulation models, with predictive capabilities.

6.5.5.4 Biomass Valorization via Chemical Processes. DIBANET Process

DIBANET process is a pathway for the production of levulinic acid from lignocellulosic biomass (Hayes 2013). Biomass is mixed with a sulphuric acid catalyst

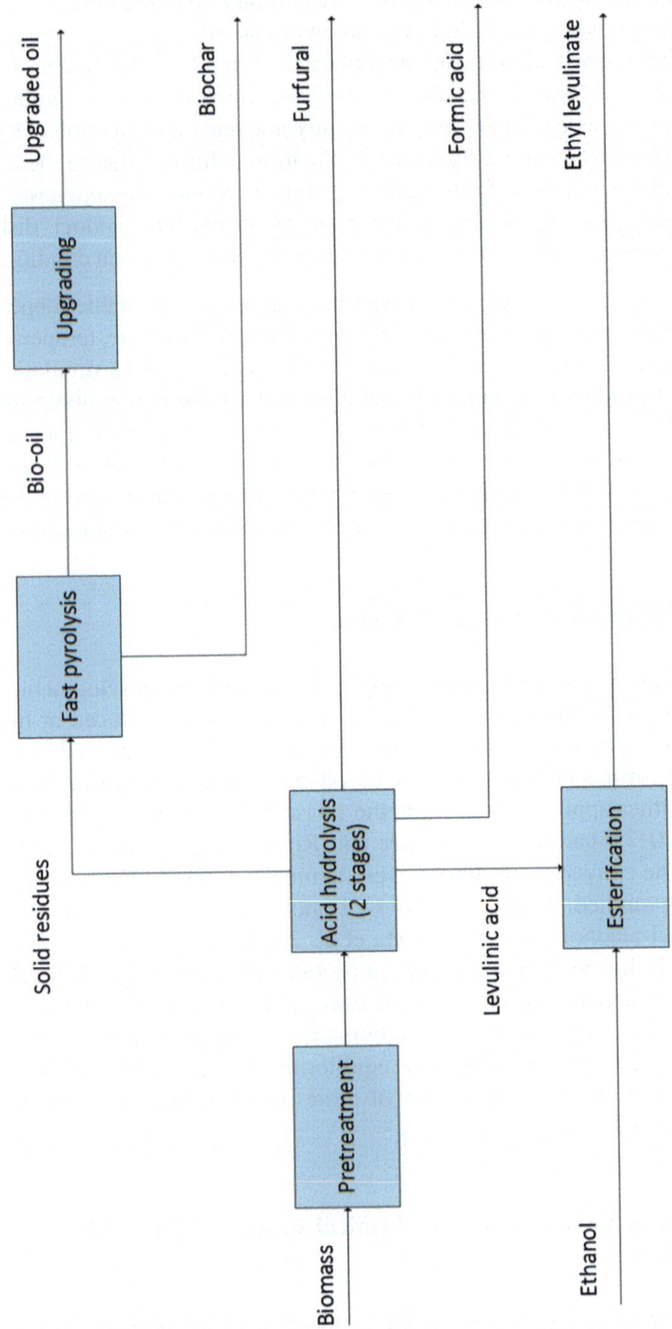

Fig. 6.12 Block diagram of the DIBANET process

solution, and this mixture is sent to a tubular reactor, which operates at a temperature of 210–220 °C and a pressure of 30 atm, with a residence time of 12 s. In this reactor, the carbohydrates are hydrolyzed into their monomers. The product of this reactor is fed to a second one, which operates at a temperature of 190–200 °C and a pressure of 12–14 atm, with a residence time of 20 min. In this reactor, formic acid an furfural are obtained in a vapor product stream, which is condensed to recuperate them, and levulinic acid is obtained in the liquid outlet stream. Finally, solid products are separated from levulinic acid by means of a filtration step. Figure 6.12 shows a block diagram of the DIBANET process.

Several improvements of the DIBANET process have been studied, as the effective and sustainable utilization of the solid residues of the hydrolysis steps, integrating it with fast pyrolysis for the production of biofuels or proposing the use of biochar as a plant growth promoter (Hayes 2013).

Levulinic acid and furfural are valuable platform chemicals, which have a wide range of applications directly or through the catalytic conversion to other chemicals (Patel 2006). For instance, levulinic acid may be used for the production of ethyl levulinate, an ester of levulinic acid and ethanol, which may be used as a diesel additive.

Process simulation software allows the determination of mass and energy balances of this process, to perform energy, environmental or economic assessments. It is possible to model the purification steps in a strict way, based on phase equilibria, but the modeling of the reactors must lean on experimental results to determine reaction conversion and product yields. In this way, the reaction steps of the DIBANET process may be modeled with stoichiometric reactors, where reaction conversions are defined by the user. These values must be calculated from experimental data, because equilibrium-based reactors are not able to deal with biomass reactions due to its complexity.

6.6 Process Optimization

Process optimization is a very important task in the design of a chemical process, and for the transition from laboratory to commercial scale. Optimization methodologies are widely employed in all areas of engineering. Efficiency is a key issue in the development of any chemical process, and this affirmation includes biorefinery processes. The goal of the optimization of any given process consists of the determination of the best possible solution, taking into account the existing constraints. Some elements are needed to perform an optimization task (Biegler 2010):

- An essential aspect in any system optimization is the selection of an objective function, whose result is the value that must be maximized or minimized. Depending on the optimization criteria, this function can be the unitary production cost, the process efficiency, product purity, emissions to the environment, etc. It is also possible to define the objective function as a combination of some indicators, giving each one a specific weight.

- A model able to describe the behavior of the studied process. This model allows the evaluation of the objective function under different conditions, to obtain characteristic curves describing the change in the objective function when the process conditions are changed. Preferably, this model must be predictive, but optimization can also been performed with experimental data based models.
- The model calculates the objective function in terms of the values of a set of variables, which affect the final result. It is necessary to identify these variables and its valid value ranges, to maximize or minimize the objective function within these ranges.

The development of a chemical process involves a sequence of steps, beginning with the laboratory scale experimentation, and leading to the construction of a commercial scale plant. In each stage of this work, some variables are studied. For instance, reaction conditions (pressure, temperature), feed composition, catalysts, solvents and other major variables are fixed at laboratory scale, and kept constant during the scale-up process.

Once these variables have been fixed, the next step of the development may be the pilot-plant scale, where a small size plant of the proposed process is constructed, allowing the study of the performance of a real plant. Other option is the direct development of a mathematical model of the process, without the construction of the pilot-plant scale facility.

Information obtained from laboratory or pilot-plant scale may be used for the development of a mathematical model of the process, to describe the observed phenomena and replicate the plant behavior before its construction. This process may be implemented using process simulation software, and simulation results may be used to carry out the optimization of the plant performance. Depending of the selected objective function, different optimization scenarios may be considered. For instance, it is possible to define as an objective the maximization of the energy or exergy efficiency of the plant, the minimization of the environmental impacts, the minimization of the unitary production cost, or a combination of them. The optimization methodology consists of the performance of a set of simulations with variations in the studied input variables, evaluating the objective function. Depending of the definition of this function, the optimum point may correspond to a maximum or to a minimum of its value.

One of the most accepted criteria for process optimization is exergy analysis. With this analysis, the objective function is the exergy efficiency of the process, which must be maximized, reducing exergy destruction and losses. The authors' group has developed an exergy analysis software, which has been named ExPE (Exergy Performance Evaluation),[6] which may be linked with Aspen Plus simulation software to analyze the exergy performance of any simulated process. This

[6]ExPE software. Exergy Performance Evaluation. Systems Analysis Unit, IMDEA Energy Institute.

program is able to read the conditions and compositions of every single stream of the process, and perform exergy balances. In exergy balances, exergy of fuel (inlet streams), exergy of product (outlet streams), and exergy destruction are determined at a unit operation block level. It is also possible to perform the global exergy balance of the process, determining the exergy of fuel, exergy of product, exergy destruction and exergy loss. Thus, it is possible to optimize the exergy performance of a single unit operation block or for the whole process, selecting the appropriate objective (which would be the exergy efficiency of the considered unit operation block or the global exergy efficiency, respectively).

6.6.1 Process Integration

Process integration is one of the key actions in order to improve the efficiency of any chemical process, in economic, environmental ,or energetic terms.

Process integration is defined as follows (Gundersen 2000): "*Systematic and General Methods for Designing Integrated Production Systems, ranging from Individual Processes to Total Sites, with special emphasis on the Efficient Use of Energy and reducing Environmental Effects.*" Thermal integration of a plant is one of the most important ways to increase its efficiency. In a biorefinery (or any other conventional plant), there is a set of material streams, transporting gases, liquids, or solids between different unit operations, and these streams have different conditions (pressure, temperature, state of aggregation). Most unit operations of the plant are carried out at fixed conditions. Therefore, heating and cooling steps are necessary to adapt the outlet stream of a unit operation to the conditions of the next one. The temperature change of the process streams is associated to an energy consumption, which may be provided by cooling water (or any other refrigeration service) in the case of temperature decrease, or by steam (or any other heating service, such as oil or hot gases) for the temperature increases.

The simplest design of the plant is the consideration of all these heating or cooling steps independently. Thus, for instance, if in the considered plant the only refrigeration service is cooling water and the only heating service is steam, all cooling steps are carried out using cooling water and all heating steps are carried out with steam. Due to its simplicity, the first version of a model is usually built in this way, which is called "linear," because there is no energy integration between the different streams of the process. The main disadvantage of this design is the high energy consumption associated with temperature changes, because all steps involving a cooling or heating process represent additional energy consumption, reducing the efficiency of the process, and increasing the unitary production cost.

Once the conditions of each unit operation block have been established, the necessary changes in the process layout are introduced to the model, to take advantage of the temperature difference between the material streams of the plant. For instance, it is possible to cool down a hot stream, and use this thermal energy for the heating of a cold stream. This modification reduces both cooling and heating needs, because energy is being transferred between streams into the plant, but it needs an additional heat exchanger to perform this operation. Thus, the global process efficiency is increased (by the reduction in its energy needs), but equipment cost is also increased (due to the additional heat exchanger).

Process integration scheme may be trivial in a simple model, with a relatively low number of unit operations and temperature change steps, but this problem increases in complexity as the number of unit operations considered increases. As the simulation model complexity increases, the amount of possible alternatives of thermal integration also increases. Thus, the simulation of all the possibilities in order to determine the optimum configuration is not an affordable problem. In this kind of processes is necessary the application of a normalized methodology, in order to propose a heat exchangers network configuration to increase the process efficiency.

Pinch analysis is a methodology for the development of process integration schemes in relatively complex plants, where the most appropriate configuration may be not obvious.

6.6.2 Pinch Analysis

Pinch analysis provides a systematic methodology for energy saving in chemical processes (March 1998), proposing a thermal integration scheme. A pinch analysis starts with the mass and energy balances of the considered process or system, and its objective is the design of a heat exchanger network in order to minimize the global energy consumption of the process (heating and cooling).

The first step of the pinch analysis is the identification of the hot streams, that need cooling, and the cold streams, that need heating. Necessary data are the initial and final temperature of each stream and its heat capacity, needed to determine the enthalpy change of each stream.

Once the streams have been identified, pinch analysis continues with the construction of the composite curves (temperature-enthalpy change), representing the enthalpy change of each stream against its initial and final temperature (Fig. 6.13).

Two composite curves must be constructed in order to perform the pinch analysis: the hot curve (with hot streams) and the cold curve (with cold streams). Both curves can be represented in the same graph, being separated by the minimum temperature difference selected (a lower value leads to a higher process efficiency and lower energy needs, but also to an increase in the total area of the needed heat exchangers). The point where the minimum temperature difference occurs is called the pinch point, and divides the process into two separate systems: above the pinch point, which requires a net heat input, and below this point, which is a net heat source and needs a cooling service (March 1998; Klemeš and Kravanja 2013). Figure 6.14 shows an example of hot and cold composite curves.

After the construction of the composite curves, the determination of the pinch point and the selection of the pinch temperature, it is possible to propose a heat

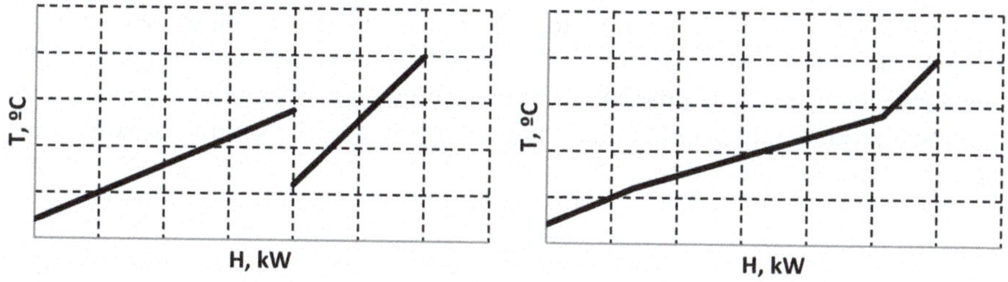

Fig. 6.13 Example of the construction of a composite curve

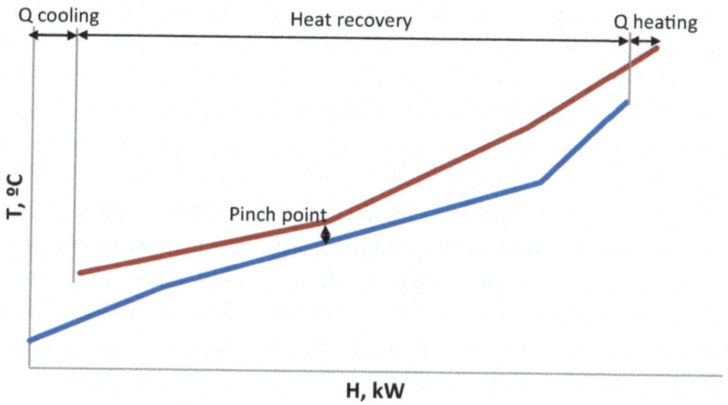

Fig. 6.14 Example of hot and cold composite curves in a Pinch analysis

exchanger network in order to reduce the global heating and cooling needs, taking into account the next considerations (March 1998):

- Heat must not be transferred across the pinch point.
- There must be no external cooling above the pinch point.
- There must be no external heating below the pinch point.

When using pinch analysis to optimize the energy consumption of any given process, it is important to notice that the technical optimum may not be the same as the economic one. The technical optimum corresponds to the point where both cooling and heating services consumption is minimal, but to reach these conditions, a high number of heat exchangers can be needed. This situation results in a decrease in the global energy consumption, which means a decrease in production cost, but capital costs are increased due to the additional necessary equipment, and the overall effect in the total production cost may be positive or negative. Thus, pinch analysis may be applied to increase the efficiency of a process, but it must not be applied in a too much strict way, because depending of the considered objective function for the optimization, the optimum point may vary.

As explained, the application of the Pinch analysis methodology results in the proposal of a heat exchangers network to reduce the energy consumption of the considered plant. So, it is possible to refine the thermal integration scheme of the simulation model, implementing the proposed changes in the simulation software to improve its overall efficiency.

6.6.3 Process Scale-up

Scale-up is defined as the process of increasing the plant capacity of a considered plant, moving to one stage of development for the next one. When moving from a research and development (laboratory scale) to a commercial scale, usually it is necessary to have an intermediate step (pilot or demonstration scale), in which the full commercial process that is proposed is replicated but with a small plant capacity (Levin 2001). This intermediate step is representative of the commercial scale process, and allows the production of the final products in order to make analysis and check if they are under specifications. It also gives information about the plant behavior, and makes possible the building of simulation models closer to the real commercial plant. These models may be implemented using process simulation software, in order to perform economic, environmental ,or efficiency assessments before the plant construction. Thus, it is possible to improve the process integration in order to increase the overall efficiency and reduce the production cost by introducing modifications in the simulation model, and take these modifications into account for the construction of the commercial scale plant. In this way, process scale-up reduces the costs of the development of a new process, allowing the implementation of a simulation model to describe the process gathering data from a plant with a small capacity.

References

Anex RP, Aden A, Kazi FK, Fortman J, Swanson RM, Wright MM et al (2010) Techno-economic comparison of biomass-to-transportation fuels via pyrolysis, gasification, and biochemical pathways. Fuel 89:S29–S35

Aspen Plus Help (2010) Getting started modeling processes with solids

Baker DA, Rials TG (2013) Recent advances in low-cost carbon fiber manufacture from lignin. J Appl Polym Sci 130(2):713–728

Biegler LT (2010) Introduction to process optimization. In: Nonlinear programming: concepts, algorithms, and applications to chemical processes. Society for Industrial and Applied Mathematics—Mathematical Optimization Society, Philadelphia, pp 1–16

Bridgwater AV (2012) Review of fast pyrolysis of biomass and product upgrading. Biomass Bioenergy 38:68–94

Bridgwater AV, Peacocke GVC (2000) Fast pyrolysis processes for biomass. Renew Sustain Energy Rev 4(1):1–73

Bugg TD, Rahmanpour R (2015) Enzymatic conversion of lignin into renewable chemicals. Curr Opin Chem Biol 29:10–17

Calonaci M, Grana R, Barker Hemings E, Bozzano G, Dente M, Ranzi E (2010) Comprehensive kinetic modeling study of bio-oil formation from fast pyrolysis of biomass. Energy Fuels 24 (10):5727–5734

Chávez-Sifontes M, Domine ME (2013) Lignina, estructura y aplicaciones: Métodos de despolimerización para la obtención de derivados aromáticos de interés industrial. Av en Ciencias e Ing 4(4):15–46

Cherubini F (2010) The biorefinery concept: using biomass instead of oil for producing energy and chemicals. Energy Convers Manag 51(7):1412–1421

Choudhary TV, Phillips CB (2011) Renewable fuels via catalytic hydrodeoxygenation. Appl Catal A Gen 397(1–2):1–12

Ciferno JP, Marano JJ (2002) Benchmarking biomass gasification technologies for fuels, chemicals and hydrogen production

Directive 2003/30/EC D. On the promotion of the use of biofuels or other renewable fuels for transport

Directive 2009/28/EC D. On the promotion of the use of energy from renewable sources and amending and subsequently repealing Directives 2001/77/EC and 2003/30/EC

Directive (EU) 2015/1513. On the promotion of the use of energy from renewable sources and amending Directive 98/70/EC relating to the quality of petrol and diesel fuels and amending Directive 2009/28/EC

Dupont C, Chen L, Cances J, Commandre J-M, Cuoci A, Pierucci S et al (2009) Biomass pyrolysis: kinetic modelling and experimental validation under high temperature and flash heating rate conditions. J Anal Appl Pyrolysis 85(1–2):260–267

Faravelli T, Frassoldati A, Migliavacca G, Ranzi E (2010) Detailed kinetic modeling of the thermal degradation of lignins. Biomass Bioenergy 34(3):290–301

French RJ, Hrdlicka J, Baldwin R (2010) Mild hydrotreating of biomass pyrolysis oils to produce a suitable refinery feedstock. Environ Prog Sustain Energy 29(2)

Gallezot P (2012) Conversion of biomass to selected chemical products. Chem Soc Rev 41 (4):1538–1558

García A, Alriols MG, Llano-Ponte R, Labidi J (2011) Energy and economic assessment of soda and organosolv biorefinery processes. Biomass Bioenergy 35(1):516–525

Gundersen T (2000) A process integration PRIMER. International energy agency report

Hannula I, Kurkela E (2012) A parametric modelling study for pressurised steam/O_2-blown fluidised-bed gasification of wood with catalytic reforming. Biomass Bioenergy 38:58–67

Hayes D (2013) DIBANET process report. Report on optimal use of DIBANET Feedstocks and Technologies

Heidenreich S, Foscolo PU (2015) New concepts in biomass gasification. Prog Energy Combust Sci 46:72–95

Hoekstra E, Westerhof RJM, Brilman W, Van Swaaij WPM, Kersten SRA, Hogendoorn KJA et al (2012) Heterogeneous and homogeneous reactions of pyrolysis vapors from pine wood. AIChE J 58(9):2830–2842

IEA Bioenergy Task 42 (2009) Biorefineries: adding value to the sustainable utilisation of biomass

Imman S, Arnthong J, Burapatana V, Champreda V, Laosiripojana N (2014) Influence of alkaline catalyst addition on compressed liquid hot water pretreatment of rice straw. Chem Eng J 278:85–91

Iribarren D, Peters JF, Petrakopoulou F, Dufour J (2012) Well-to-wheels comparison of the environmental profile of pyrolysis based biofuels. In: 20th European biomass conference and exhibition

Iribarren D, Susmozas A, Petrakopoulou F, Dufour J (2014) Environmental and exergetic evaluation of hydrogen production via lignocellulosic biomass gasification. J Clean Prod 69:165–175

Jacquet N, Quiévy N, Vanderghem C, Janas S, Blecker C, Wathelet B et al (2011) Influence of steam explosion on the thermal stability of cellulose fibres. Polym Degrad Stab 96(9):1582–1588

Jeon J-W, Zhang L, Lutkenhaus JL, Laskar DD, Lemmon JP, Choi D et al (2015) Controlling porosity in lignin-derived nanoporous carbon for supercapacitor applications. ChemSusChem 8 (3):428–432

Jin H, Liu R, He Y (2012) Kinetics of batch fermentations for ethanol production with immobilized saccharomyces cerevisiae growing on sweet sorghum stalk juice. Procedia Environ Sci 12:137–145

Johansson A, Aaltonen O, Ylinen P (1987) Organosolv pulping—methods and pulp properties. Biomass 13(1):45–65

Jones SB, Valkenburg C, Walton CW, Elliott DC, Holladay JE, Stevens DJ et al (2009) Production of gasoline and diesel from biomass via fast pyrolysis, hydrotreating and hydrocracking : a design case. Energy

Kautto J, Realff MJ, Ragauskas AJ (2013) Design and simulation of an organosolv process for bioethanol production. Biomass Convers Biorefinery 3(3):199–212

Kihedu J (2015) Torrefaction and combustion of ligno-cellulosic biomass. Energy Procedia 75:162–167

Klemeš JJ, Kravanja Z (2013) Forty years of heat integration: pinch analysis (PA) and mathematical programming (MP). Curr Opin Chem Eng 2(4):461–474

Kücük MM, Demirbas A (1997) Biomass conversion processes. Energy Convers Manag 38 (2):151–165

Laclaire CE, Barrett CJ (2002) Technical, environmental and economic feasibility of bio-oil in New Hampshire's North Country

Leksawasdi N, Joachimsthal EL, Rogers PL (2001) Mathematical modelling of ethanol production from glucose/xylose mixtures by recombinant *Zymomonas mobilis* 1087–1093

Levin M (2001) Pharmaceutical process scale-up. Marcel Dekker, Inc., New York

Li J, Henriksson G, Gellerstedt G (2007) Lignin depolymerization/repolymerization and its critical role for delignification of aspen wood by steam explosion. Bioresour Technol 98(16):3061–3068

Ma X, Cui K, Hao W, Ma R, Tian Y, Li Y (2015) Alumina supported molybdenum catalyst for lignin valorization: effect of reduction temperature. Bioresour Technol 192:17–22

March L (1998) Introduction to pinch technology. Northwich, Cheshire, England

Martín-Sampedro R, Eugenio ME, García JC, Lopez F, Villar JC, Diaz MJ (2012) Steam explosion and enzymatic pre-treatments as an approach to improve the enzymatic hydrolysis of eucalyptus globulus. Biomass Bioenergy 42:97–106

Morales-Rodriguez R, Rodriguez-Gomez D, Sales-Cruz M, de los Reyes-Heredia JA, Pérez Cisneros ES (2014) Model-based analysis for acetone-butanol-ethanol production process through a dynamic simulation. Comput Aided Chem Eng 33:133–138

Pan X, Saddler JN (2013) Effect of replacing polyol by organosolv and kraft lignin on the property and structure of rigid polyurethane foam. Biotechnol Biofuels 6(1):12

Patel M (2006) BREW project report. Medium and long-term opportunities and risks of the biotechnological production of bulk chemicals from renewable resources—The Potential of White Biotechnology

Peters JF, Iribarren D, Dufour J (2013) Predictive pyrolysis process modelling in Aspen Plus. In: 21st European biomass conference and exhibition, Copenhagen

Peters JF, Iribarren D, Dufour J (2014a) Simulation and life cycle assessment of biofuel production via fast pyrolysis and hydroupgrading. Fuel 139:441–456

Peters JF, Petrakopoulou F, Dufour J (2014b) Exergetic analysis of a fast pyrolysis process for bio-oil production. Fuel Process Technol 119:245–255

Peters JF, Banks SW, Susmozas A, Dufour J (2014c) Experimental verification of a predictive pyrolysis model in Aspen Plus. In: 22nd European biomass conference and exhibition, Hamburg, Germany

Ramires EC, Megiatto JDJ, Gardrat C, Castellan A, Frollini E (2010) Valorization of an industrial organosolv-sugarcane bagasse lignin: characterization and use as a matrix in biobased composites reinforced with sisal fibers. Biotechnol Bioeng 107(4):612–621

Ranzi E, Cuoci A, Faravelli T, Frassoldati A, Migliavacca G, Pierucci S et al (2008) Chemical kinetics of biomass pyrolysis. Energy Fuels 22(6):4292–4300

Ringer M, Putsche V, Scahill J (2006) Large-scale pyrolysis oil production: a technology assessment and economic analysis

Sannigrahi P, Ragauskas AJ (2013) Fundamentals of biomass pretreatment by fractionation. Aqueous pretreatment of plant biomass for biological and chemical conversion to fuels and chemicals, pp 201–222

Sarma SJ, Pachapur V, Brar SK, Le Bihan Y, Buelna G (2015) Hydrogen biorefinery: potential utilization of the liquid waste from fermentative hydrogen production. Renew Sustain Energy Rev 50:942–951

Slininger PJ, Branstrator JE, Lomont JM, Dien BS, Okos MR, Ladisch MR et al (1990) Stoichiometry and kinetics of xylose fermentation by Pichia Stipitis. Ann N Y Acad Sci 589 (1):25–40

Slininger PJ, Dien BS, Lomont JM, Bothast RJ, Ladisch MR, Okos MR (2014) Evaluation of a kinetic model for computer simulation of growth and fermentation by Scheffersomyces (Pichia) stipitis fed D-xylose. Biotechnol Bioeng 111(8):1532–1540

Spath P, Aden A, Eggeman T, Ringer M, Wallace B, Jechura J (2005) Biomass to hydrogen production detailed design and economics utilizing the Battelle Columbus Laboratory indirectly-heated gasifier

Venderbosch R, Heeres H (2011) Pyrolysis oil stabilisation by catalytic hydrotreatment. Biofuel's Engineering Process Technology, pp 385–410

Wang H, Tucker M, Ji Y (2013) Recent development in chemical depolymerization of lignin: a review. J Appl Chem 2013:1–9

Author Biographies

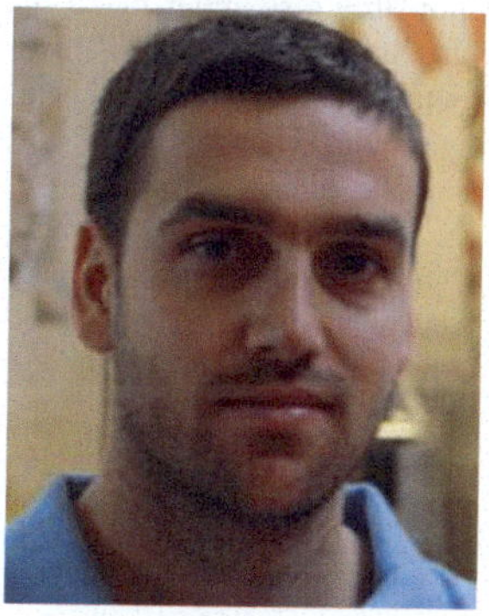

Abel Sanz de Juan, Ph.D. is Chemical Engineer (2006), Master in Industrial Process Engineering (2007), and Ph.D. in Chemical Engineering (2012) by the Complutense University of Madrid. He has been recipient of a FPU grant (2009–2012), and his doctoral research was focused in the field of gas adsorption and carbon dioxide capture by PSA. His postdoctoral investigation in Institute IMDEA Energy was focused in the simulation, design, optimization, and economic analysis of processes in the energy field. He is co-author of several research articles in peer-reviewed international journals, communications in international conferences and two book chapters. He has collaborated in 9 research projects and contracts at national and international levels.

Ana Susmozas, Ph.D. studied Chemical Engineering at Rey Juan Carlos University of Madrid (2010) and obtained her Ph.D. in Chemical and Environmental Engineering at the same university (2016). Currently, she works as Researcher in the Biofuels Unit of CIEMAT and her research focuses on the simulation and techno-economic analysis of bioenergy systems. She has participated in 5 research projects at both national and international levels, and has published 12 contributions to national and international conferences, 7 articles in peer-reviewed international journals, and 3 book chapters.

Jens Peters, Ph.D. holds a diploma degree (Dipl. Ing.) in electrical engineering (communication technologies) from the Technical University of Munich. He worked several years as R&D engineer and project leader in the automotive industry (Ingolstadt, Barcelona) in the development of electronic components. After finishing M.Sc. in renewable energies and fuel cells at UIMP/CSIC in Madrid, he started working in the field of system analysis of energy processes at Instituto IMDEA Energía, Madrid. In 2015, he finished his dissertation at Universidad Rey Juan Carlos (Madrid) about the environmental, economic, and thermodynamic assessment of pyrolysis processes for the production of biofuels and biochar. Since 2015, he is part of the research group 'Resources, Recycling, Environment & Sustainability' at HIU, where he is working on the modeling and assessment of novel electrochemical energy storage technologies, with a special focus on material issues and sustainability of new battery systems (eco-design).

Javier Dufour, Ph.D. is Senior Researcher and Head of the Systems Analysis Unit at IMDEA Energy Institute, and Associate Professor at Rey Juan Carlos University. His research lines are focused in the analysis of energy systems, design, simulation and optimization of processes, and life cycle sustainability assessment He is the author of more than 75 papers published in international journals, 140 contributions to conferences, and 3 patents. He has collaborated in 55 research projects (8 international ones), being the responsible researcher in 28 of them. Currently, he is Chairman of the Spanish Network of Life Cycle Assessment and Operating Agent of Task 36 "Life Cycle Sustainability Assessment of Hydrogen Energy Systems" of the Hydrogen Implementing Agreement of the International Energy Agency.

Chapter 7
Biorefinery Sustainability Analysis

Carla A.M. Silva, Remus M. Prunescu, Krist V. Gernaey, Gürkan Sin and Rocio A. Diaz-Chavez

Abstract This chapter deals with sustainability analysis of biorefinery systems in terms of environmental and socio-economic indicators. Life cycle analysis has methodological issues related to the functional unit (FU), allocation, land use and biogenic carbon neutrality of the reference system and of the biorefinery-based system. Socio-economic criteria and indicators used in sustainability frameworks assessment are presented and discussed. There is not one single methodology that can aptly cover the synergies of environmental, economic, social and governance issues required to assess the sustainable production and use of bioenergy systems. The perfect metric for environmental issues is not yet established and some researchers prefer to avoid high levels of uncertainty in life cycle assessment (LCA) methodology and adopt more physically quantifying methods like the annual basis carbon (ABC) method presented here. In addition to establishing the perfect metric, there are three types of uncertainty when building scenarios with biorefinery-based systems that must be regarded to have a more holistic point of view. This uncertainty is at the level of the concept, of the configuration and of the operation.

C.A.M. Silva (✉)
Instituto Dom Luiz, Faculdade de Ciências, Universidade de Lisboa, 1749-016 Lisboa, Portugal
e-mail: carla.silva@fc.ul.pt

R.M. Prunescu · K.V. Gernaey · G. Sin
CAPEC-PROCESS Research Center, DTU Chemical Engineering, Department of Chemical and Biochemical Engineering, Søltofts Plads, Building 227 (Postal Address: Building 229), 2800 kgs Lyngby, Denmark
e-mail: kvg@kt.dtu.dk

G. Sin
e-mail: gsi@kt.dtu.dk

R.A. Diaz-Chavez
Instituto Centre for Environmental Policy, ICEPT, Imperial College London, 14 Princess Gardens, South Kensington, London SW7 1NA, UK
e-mail: r.diaz-chavez@imperial.ac.uk

M. Rabaçal et al. (eds.), *Biorefineries*, Lecture Notes in Energy 57,
DOI 10.1007/978-3-319-48288-0_7

7.1 Life Cycle Assessment

As part of a sustainability assessment, life cycle assessment (LCA) is commonly used to compare two or more pre-defined systems or to identify environmental burdens and identify bottlenecks and possible improvements of a pre-defined system. The system must be clearly defined in terms of boundaries, mass and energy inputs and outputs and its function clearly acknowledged.

ISO 14040:2006 (ISO 2006a) describes the principles and framework for LCA including: definition of the goal, scope, boundary and functional unit (FU); the life cycle inventory analysis (LCI), which accounts the mass and energy flows within the defined boundary; the life cycle impact assessment (LCIA), which aggregates inventory data in impact categories; the interpretation, reporting; and, finally, a critical review of the LCA bearing in mind its limitations.

The boundary is essential to identify the mass flows and energy flows in and out of the system. For example a biorefinery for sugar and ethanol production is shown, and the function of the biorefinery is to produce ethanol and sugar, so the FU could be 1 L of ethanol produced, 1 kg of sugar produced, or both. Other option would be to consider that the biorefinery function is to transform 1 tonne of biomass and so the FU would then be 1 tonne biomass transformed. All the mass and energy flows would be related to this FU (Figure 7.1).

From the impact categories that can be analyzed, the carbon footprint is the most common, usually called global warming potential.

For example, policies for supporting biofuels, such as the Renewable Energy Directive, RED (for Europe), the Renewable Fuel Standard (for US), and the

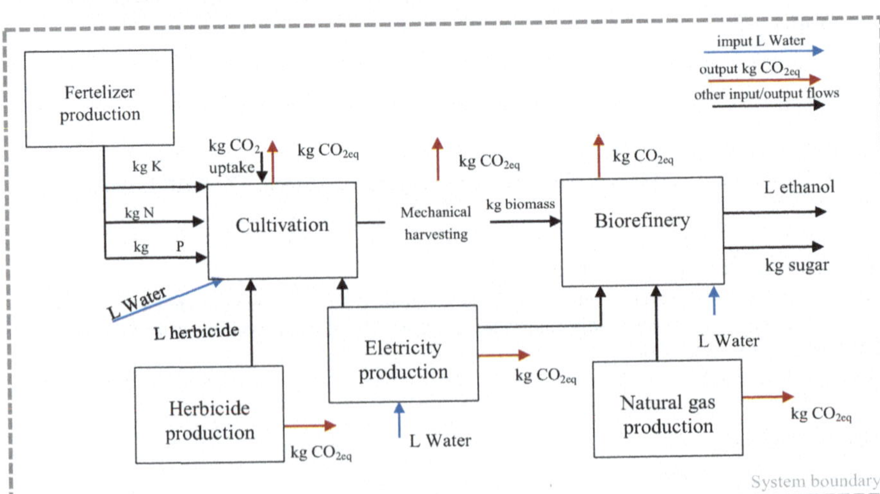

Fig. 7.1 Example of a system boundary for a biomass conversion to sugar and ethanol. If the feedstock is sugar cane usually the electricity and natural gas are avoided and produced within the biorefinery from by-products

Renewable Transport Fuel Obligation, RTFO (for UK), require life cycle carbon reporting to ensure that biofuels achieve greenhouse gas (GHG) reductions relative to fossil fuels.

Most recently, the water footprint (Hoekstra et al. 2009) is frameworked and could be part of the LCI, throughout the supply chain of a product, where green, blue and grey water quantities are accounted for. The blue water footprint refers to consumption of surface and groundwater water resource. The green water footprint refers to consumption of for example rainwater. The grey water refers to water resources appropriation through pollution. Water vapour emissions could potentially increase global warming but anthropogenic contribution is usually disregarded (Silva 2011). In a global context, the water footprint is a relevant indicator of how much of the globe's scarce freshwater resources are used for a certain product.

As part of the carbon footprint inventory the GHG protocol corporate standard (Schmitz et al. 2000) was launched in 1998 and covers the following GHG emissions: carbon dioxide (CO_2), methane (CH_4), nitrous oxide (N_2O), hydrofluorocarbons (HFCs), perfluorocarbons (PFCs), and sulphur hexafluoride (SF_6) and as of May 2013—nitrogen trifluoride (NF_3). The inventory for GHG emissions has three distinct scopes:

- Scope 1: Direct GHG emissions. Direct CO_2 emissions from the combustion of biomass or fossil fuel;
- Scope 2: Electricity indirect GHG emissions. Scope 2 emissions physically occur at the facility where electricity is generated;
- Scope 3: Other indirect GHG emissions, is an optional reporting category that allows for the treatment of all other indirect emissions. Some examples of scope 3 are extraction and production of purchased materials; transportation of purchased fuels; and use of sold products and services.

Other authors divide energy and mass flows balances in three levels, e.g. (Macedo et al. 2008). The Level 1 direct mass and energy flows are considered. Level 2 (indirect energy, water and emissions due to energy and chemicals mass consumption) and Level 3 (indirect energy, water and emission due to manufacture, construction, maintenance and end of life of equipment and buildings) are strongly dependent of the electricity mix of the country where the plant is placed, the origin of the consumables and materials for the equipment and the plant building itself.

Biogenic CO_2 emissions are related to the natural carbon cycle, as well as those resulting from the combustion, digestion, fermentation, decomposition or processing of biologically based materials. Fossil CO_2 is derived from fossil fuel combustion. For the majority of studies biogenic CO_2 is disregarded and only CO_2 from fossil hydrocarbon burning is accounted. For example the RED (for Europe), do not consider biogenic emissions when comparing the GHG of petrol fuels (83.4 g CO_2eq/MJ). Figure 7.2 shows CO_2eq for each life cycle stage. The FU is 1 MJ of ethanol produced. Other issue that may arise in terms of methodological issues is the allocation used for the fossil system (e.g. mass based) being different

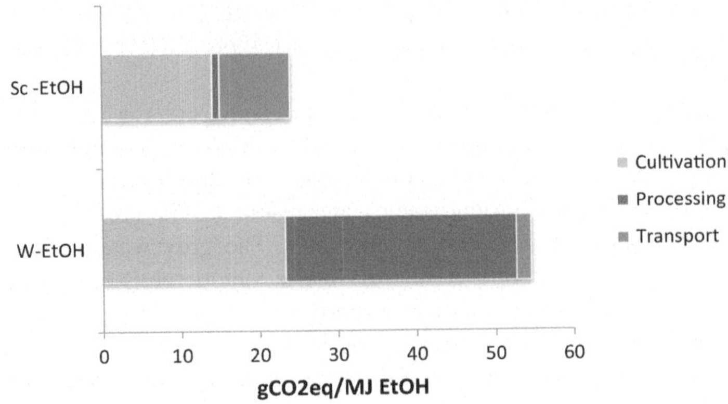

Fig. 7.2 Renewable energy directive for Europe applied to Ethanol from sugarcane (*Sc-EtOH*) and Ethanol from wheat (*W-EtOH*) with Natural Gas boiler in the processing step. Biogenic CO_2 not accounted, land use change not accounted nor materials of the factory (Hennecke et al. 2013)

than the allocation choose for the ethanol system (e.g. energy based). Having the same allocation method for the reference fossil and biorefinery-based system would be preferable.

7.1.1 Methodological Issues

Results of LCA are highly variable due to different assumptions regarding, indirect land use change (iLUC), allocation method, boundaries assumed, geographical differences and disregarded items in the feedstock processing chain due to limited information. Also a distinction can be made in two types of approaches: consequential LCA (CLCA) and attributional LCA (ALCA) (Brander et al. 2008). ALCA use processes and material flows directly used in the production, consumption and disposal of the product, to clearly identify the total emissions. On the other side, CLCA uses processes and material flows which are directly or indirectly affected by a marginal change in the output of a product (e.g. through market effects, substitution, use of constrained resources etc.), being therefore more affected by presumed economic scenarios.

Land use change (LUC): Soil organic carbon includes plant, animal and microbial residues in all stages of decomposition. The loss of soil organic carbon by conversion of natural vegetation to cultivated use is discussed in (Post and Kwon 2000). LUC result from giving a different use to a portion of land previously with a certain carbon storage potential, what will cause a carbon debt. Direct LUC can be determined from a comparison of the carbon balances of the previous land use with those after the land has been used to produce biomass crops. iLUC adds more uncertainty by assuming for example that the used land will cause that forestland or wetlands somewhere else in the world will be converted to cropland. According to (Searchinger et al. 2008) these adds carbon emissions that occur as farmers worldwide respond to higher prices and convert forest and grassland to new cropland to replace the grain (or cropland) diverted to biofuels. The iLUC use is part of a CLCA analysis.

Allocation is inherent to a process with several product, by-product and co-product outcome. The choice of allocation method (mass, energy content, economic, system expansion, preferences of the LCA practitioner or requests from the study's commissioner) can introduce differences in results by several orders of magnitude and therefore affecting decision making (Sandin et al. 2015). ALCA allocates emissions to co-products based on economic value, energy content or mass. CLCA uses system expansion to quantify the effect of co-products on emissions. In a biorefinery, the environmental impact of the selected main product can be calculated as the emissions from the main production system minus the avoided emissions from the use of the by-products. This is called system expansion or substitution.

The FU is the function of the system under study and serves as a basis for the calculations. It can be difficult to identity in a biorefinery system that produces multiple outputs with different functions (Ahlgren et al. 2013), and therefore being hard to choose one main product or function. A FU that avoids allocation between coproducts could be, e.g. 1 tonne of biomass input, 1 biorefinery or a combination of all outputs (i.e. production of 1 kg product of A, 2 MJ of product B, 100 kg of product C). For example (Cherubini 2010), uses the amount of biomass treated per year to compare a biorefinery concept which produces bioethanol, electricity, heat, and phenols from switchgrass, to a fossil reference system that delivers the same amount of products.

Direct carbon mass flows are usually not fully accounted. It is usually assumed that biofuel burning emits carbon that was uptake during cultivation of the feedstock. Carbon gains in soil due to harvesting leftovers, direct carbon emissions during fermentation or within leftovers are usually disregarded. This carbon flows accounting are defended in the annual basis carbon (ABC) analysis has a less uncertain way of comparing two systems with the same function (DeCicco and Krishnan 2015). In the report it is claimed that: the biofuel carbon neutrality assumption built into LCA models does not hold up for real-world biofuel production. How to treat the timing of sequestration and emission of biogenic carbon may be an issue as reported in (Ahlgren et al. 2013).

Carbon in soil may be measured (Slepetiene et al. 2008; Degerickx et al. 2015), carbon in cultivated biomass and processing leftovers may be known using elemental analysis, e.g. for *Jatropha Curcas*, soil organic carbon (Degerickx et al. 2015), husk, seed shell, tree blanch (Murata et al. 2012), leaves (Méndez et al. 2014), whole seed and seed pressed cake (Jourabchi et al. 2014). GHG emissions due to N_2O and CH_4 may occur: 0.042 g N_2O per g N fertilizer applied (Cherubini 2010); the reduction in methane uptake is equivalent to an emission of methane from cultivated soils of 10 g CH_4/kgN (Cherubini 2010).

7.1.2 *Biorefinery Case Studies*

In this section examples of biorefinery analysis are presented to illustrate the several possible approaches. It is not meant to be an exhaustive list of case studies.

7.1.2.1 Comparing Multiple Biorefinery Configurations

A problem often arising is whether to convert a certain biomass feedstock or feedstocks in a biorefinery concept, maximizing its profitability (Sukumara et al. 2014). Other objectives that can be added are minimizing the carbon footprint and the water footprint (Giarola et al. 2011; Bernardi et al. 2013). The former proposes a multiobjective Mixed Integer Linear Programming modelling framework to optimize the environmental (i.e. the carbon and water footprints) and economic performances of bioethanol supply chains in Italian context. The Pareto curve of optimal solutions reveal a conflict between environmental and economic performance in dealing with biofuels production: minimum carbon <5 kg CO_2eq/GJ) and water footprint (−0.06 m^3/GJ) but negative net present value (−7.5 €/GJ); supply chain based on the standard dry–grind process with the distiller's dried Grains with solubles sold as animal fodder has the best net present value: 1.17 €/GJ of ethanol produced. The environmental impact of this configuration is high on water footprint (9.98 m^3/GJ, corresponding to 212 L of water/L of fuel) as well as on carbon footprint (77.2 kg CO_2/GJ).

Bradley et al. (2015) presents a common LCA methodology for the comparison of three different algae biofuel demonstration facilities created in the Algae Cluster, a group of three European Commission funded projects, each building a different demonstration algae biofuel facility up to 10 ha in size. The idea is to produce biodiesel from the feedstocks, and assure that the LCA of the biorefinery within each project is comparable. Otherwise a wide range of 0.75 kg CO_2eq/MJ–5.34 kg CO_2eq/MJ, depending on LCA assumptions, can be found for the same biorefinery. The FU chosen was "combustion of 1 MJ (Lower Heating Value) of algal biofuel in a car engine". Twenty one impact categories were selected:

- Ozone depletion (kg CFC-11 equivalent);
- Terrestrial acidification (kg SO_2 equivalent to air);
- Freshwater eutrophication (kg P equivalent to freshwater);
- Marine eutrophication (kg N equivalent to freshwater);
- Human toxicity (kg 1,4 dichlorobenzene to urban air) and (DALY/PDF);
- Photochemical oxidant formation (kg NMVOC compound equivalent to air);
- Particulate matter formation (kg PM10 to air);
- Terrestrial ecotoxicity (kg 1,4 dichlorobenzene to industrial soil) and (DALY/PDF);
- Freshwater ecotoxicity (kg 1,4 dichlorobenzene to freshwater) and (DALY/PDF);
- Marine ecotoxicity (kg 1,4 dichlorobenzene to marine water) and (DALY/PDF);

- Agricultural land occupation (m^2 × year of agricultural land);
- Urban land occupation (m^2 × year of urban land);
- Natural land transformation (m^2 × year of natural land);
- Mineral resource depletion (kg Fe equivalent);
- Fossil resource depletion (kg oil equivalent);
- Primary energy consumption [MJ];
- Land occupation [m^2];
- Blue water consumption [m^3];
- Climate Change over a 100-year period (kgCO_2eq);
- Climate Change over a 20-year period (kgCO_2eq);
- Land use change (LUC) (100-year and 20-year based kgCO_2eq).

Allocation is replaced by system expansion by using credits or avoided burdens. The average energy mix to be included is the EU28 average for 2013. The geographical differences between sites affecting affect the growth of the algae were accounted by using the Photovoltaic Geographical Information System, as produced by the Joint Research Centre (JRC), the total yearly irradiance, average daytime temperature and average daily temperature were compared for each site.

7.1.2.2 Comparing a Biorefinery with Fossil Based Counterpart

Other problem that may arise is how to choose the reference fossil reference system that delivers the same amount of products.

Cherubini and Jungmeier (2010) focuses on a biorefinery concept which produces bioethanol, bioenergy and biochemical from switchgrass, a lignocellulosic crop. Results are compared with a fossil reference system producing the same products/services from fossil sources trough an LCA aproach. The main impact categories are greenhouse gas (GHG) emissions and cumulative primary energy demand (distinguished into fossil and renewable). The FU of the assessment is the amount of biomass treated per year. The methodology is to build a fossil reference system that ends up with the same amount of products than the biorefinery system. It is concluded that the use of switchgrass in a biorefinery offsets GHG emissions and reduces fossil energy demand: GHG emissions are decreased by 79% and about 80% of non-renewable energy is saved. The energy return on investment is also calculated and equals 3.6: this means that the energy output (i.e. the energy content of the products) of this system contains more than three times the non-renewable energy invested (i.e. all the non-renewable energy inputs, direct and indirect, required along the full life cycle).

A scheme of this "basket" of products approach is given below in a comparison between the integrated algal biorefinery and the soybean based linear economy where the same (amounts of) functionalities were produced. System expansion was used to avoid allocation. And, a comparison between the production chain of the switchgrass biorefinery and the respective fossil reference systems (Fig. 7.3).

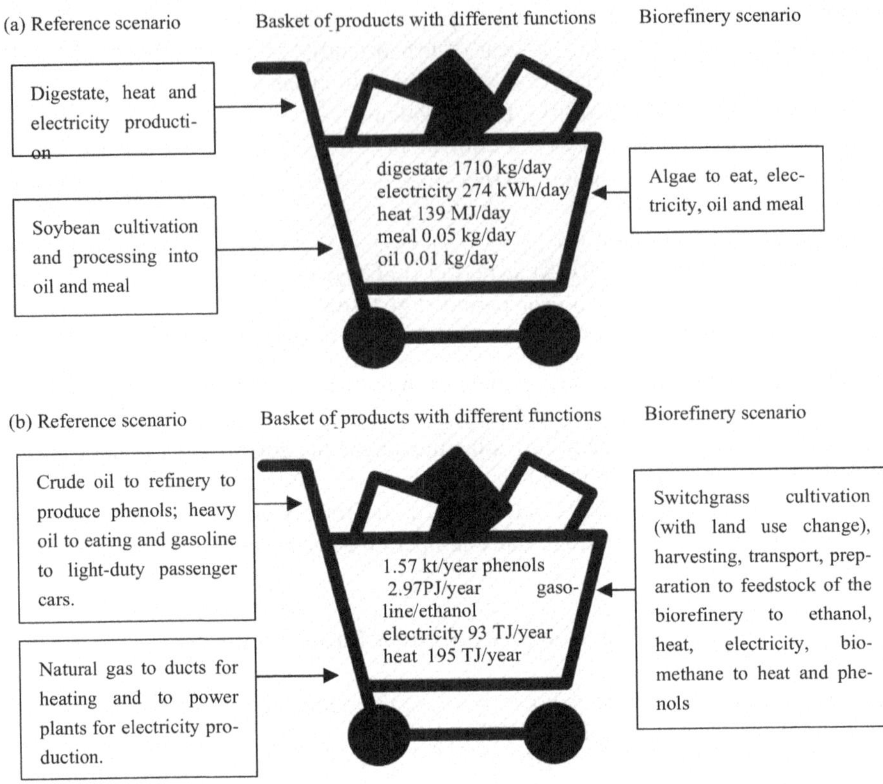

Fig. 7.3 Example of two "basket" of products approach to compare a biorefinery system with a linear economy equivalent system. **a** Comparison between an algal biorefinery with the soybean based linear economy system (Taelman and Sfez 2015). **b** Swichgrass biorefinery comparison with the equivalent products and services from fossil system (Cherubini and Jungmeier 2010)

The FU of the assessment is the amount of biomass treated per year, i.e. 477 ktdry/ha of switchgrass (corresponding to 1.45 ktdry biomass per day).

7.1.2.3 Comparing a Product from a Biorefinery with the Fossil Counterpart

DeCicco and Krishnan (2015) uses the ABC analysis instead of LCA claiming it is a less uncertain way of comparing two systems. In year zero a system with a farm to produce food and a refinery producing gasoline for vehicles, and in year one a biorefinery of corn to ethanol. The carbon-equivalent mass flows are computed and a net increase of 4 thousand metric tonnes per year (ktcarbon/yr) is found for year one. Includes biogenic process emissions, notably the CO_2 released during fermentation of corn for ethanol production. The carbon exported from the system is

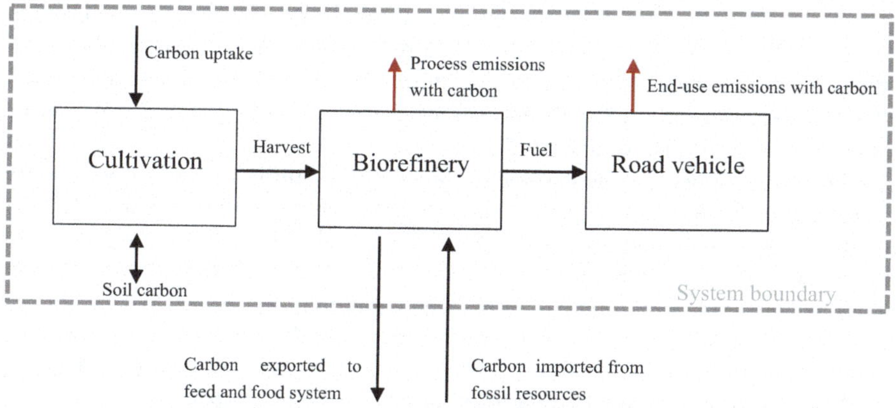

Fig. 7.4 Year 1 biomass pathway resulting in 13,524 tonnes of carbon per year, meaning a 4000 tonne increase as opposed to year 0 biomass processing for food and use of fossil fuel in the vehicles

higher in the case of year zero due to carbon embodied in food products. However, these are set aside of the final balance (see Fig. 7.4).

7.2 Socio-Economic Sustainability Assessment

The concepts of sustainable development and sustainability assessment have changed since the original ideas related to it were considered in the 1970s as part of the concerns on depletion of resources and growth limits. It was not until 1987 with the Brundtland Report that it acquired global recognition (Brundtland 1987). Nevertheless, some basic principles sill stand, such as those that relate to environmental and social impacts, along with economic issues. Many methodologies now exist that assess sustainability, ranging from traditional environmental management tools (e.g. EIA, SEA, SIA), to more complex uses of models, multi-criteria assessment and spatial referenced tools.

In the last few years, sustainability assessment, particularly for bioenergy, biofuels and, more recently, biorefineries, has driven the analysis of the viability of the supply chains, including the role of greenhouse gases (GHG), land use and biodiversity conservation. Social aspects as well as policy and governance have been brought into attention mainly by the work of NGOs and sustainability standards more focused on the impacts produced, particularly in developing countries. This section reviews extant examples of sustainability assessment of biorefineries. The review focuses mostly on indicators and methodologies that can be used to assess sustainability in these supply chains and processes, with special attention paid to socio-economic issues.

Since sustainable development became firmly established as a mainstream concept, different methodologies have been developed to assess sustainability relating to the three key pillars of the concept (environmental, social and economic), although more recently a fourth one has emerged too, policy and/or governance (Gibson et al. 2005; Diaz-Chavez 2011).

Although sustainability assessment of bioenergy is still in the early stages, there are several methodologies in place that are context contingent (Dale et al. 2013). As the overview of methodologies by Diaz-Chavez et al. (2015) has highlighted, the importance of socio-economic assessment through Social Impact Assessment (SIA), social LCA, input–output models and multi-criteria has increased. In the economic area, more specific methods include cost benefit analysis, economic modelling, whilst in the business arenas Environmental Management Systems (EMS) and Corporate Social Responsibility have been developed.

The use of lignocellulosic material, including agricultural and forestry residues, woody material from forests, and dedicated plantations provide the main feedstocks for advanced biorefineries. Therefore, the supply chains assessment for feedstock production is virtually the same as that for bioenergy supply chains. For the production and final use of products, the supply chains are also similar to those of any other chemical process. Nevertheless, the application of methods and mainly indicators and criteria may allow the sustainability assessment to be more focused not just on the feedstock but also on the process and the final products.

Some of the available methods are based on either the spatial characteristics (land use analysis), the technoeconomical characteristics of the process or the integrated assessment of the bioenergy supply chain (GBEP 2012). Other methods consider mostly mitigation measures and monitoring options that use criteria and indicators. Traditional environmental management tools used prior to the implementation of the project, or policy or plans, such as environmental and social impact assessment (ESIA), strategic environmental assessment (SEA) have been used to assess bioenergy proposals (OECD 2011).

Methodologies such as LCA, mentioned in the begining of this chapter, have also contributed to the sustainability assessment of the bioenergy sector and provide a methodology that is standardized and considers the whole supply chain, rather than just the production of the feedstock. The ISO standards 14040 (2006a, b) provide the framework and guidelines for its application (e.g. definition of systems boundaries, allocation of impacts, and choice of data sources, among others). But additional issues need to be considered, such as, for example the availability of good quality data; and a consideration of the local, national and global scales when assessing supply chains based on biomass feedstocks for bioenergy provision (Black et al. 2011). Methodologies to address LCA and sustainability have made important strides and become more sophisticated, but data availability may constitute a hindrance for certain feedstocks and impacts. LCA has traditionally been used for environmental impacts (Environmental LCA) but it has been extended to consider a full sustainability assessment (Sustainable LCA), including Life Cycle Costing (LCC) and social LCA (Wolf et al. 2012; Valdivia et al. 2013).

The use of social LCA (S-LCA) can provide key information to impact analysis (UNEP Setac Life Cycle Initiative 2009). This also applies not only to the final product, but also for green procurement. The S-LCA uses additional information on organization related aspects along the chain (Diaz-Chavez 2014a). S-LCA is still a novel methodology and few assessments exist that are based on it. Although the main benefit is to include life cycle thinking on social issues in the supply chain, its scope is still rather limited. It is recommended that a combination of methods is employed to address social and economic impacts, such as social impact assessment (SIA), social life cycle assessment (sLCA) and sustainability assessment to link it with the environmental assessment (Diaz-Chavez 2014a). An innovative aspect of the S-LCA is that it enables the use of hotspots database where integrated data on social issues can be accessed (Social hotspot organisation 2014).

Therefore, extant and emerging methodologies need to take account of issues contingent on the geographical area and level, as well as on the project. What majority of these methodologies have in common is the use of indicators.

7.2.1 Criteria and Indicators

Criteria and indicators are widely used in sustainability frameworks. Indicators can be used to organize, monitor and assess information in different contexts, but they are only useful for describing or helping to describe a given situation, rather than explaining it. International and national institutions have been using indicators to assess performance and change on a number of dimensions, such as income, education, health and welfare, both at the regional and national levels (Diaz-Chavez 2014b). Sustainability indicators can be useful in showing the ways changes in the economy, the environment and society interrelate or setting up a goal or thresholds to achieve and to aid the decision-making process. Overall frameworks of indicators are useful but in the end they need to be related to the local or regional context, and the purpose of the sustainability assessment will determine the indicators needed or selected (Efroymson et al. 2013).

In the context of bioenergy, more attention has been given to environmental indicators which are tied to the assessment of economic indicators. Social indicators had been more difficult to monitor and quantify as most of the time they require longitudinal assessment, which can be time consuming and expensive to conduct. Traditional social indicators (and indices) have been used in connection with aspects such as demography (population size) and well-being (e.g. sustainable development index, Gini). There is also the need to consider the interactions between environmental and socio-economic indicators, which have been advanced through the ecosystem services approach. Socio-economic indicators are used to analyse a particular social phenomenon or society as whole and may be derived from qualitative and quantitative data, and might be applicable to the assessment of supply chains (e.g. feedstock production and conversion).

Some of the trade-offs that need to be considered in the social sustainability assessment are shown in Fig. 7.5. These were explored in Diaz-Chavez et al. (2015), focusing mainly for bioenergy.

Each one of the topics shown in Fig. 7.5 may be a criteria or an indicator depending on the framework and goal which is to be included in the assessment. Some indicators are applicable to multiple spatial scales, as shown in Fig. 7.5 but others may only be applicable to specific context and to a particular scale (Ness et al. 2007). Indicators can be included in a framework depending on the aims. In the case of bioenergy, sustainability frameworks have focused on indicators that consider issues of social well-being and justice regarding the producers. Diaz-Chavez (2014b) put forward a set of indicators to monitor socio-economic impacts of biofuels production. These included a wide range, showing the trade-offs between environment and social attributes and a link to ecosystem services. Table 7.1 illustrates these criteria.

Dale et al. (2013) also proposed a framework for socio-economic indicators that focus on socio-economic impacts on bioenergy systems and that are feasible to measure. They proposed six categories: social well-being, energy security, external trade, profitability, resource conservation, and social acceptability. Their framework also supports links with environmental indicators, and contains a minimum of ten indicators.

Indicators can be used for monitoring activities through the voluntary application of standards. Nevertheless, standards need to be audited and an audit is a "snapshot" of the activities performed by a company, and its production system in a

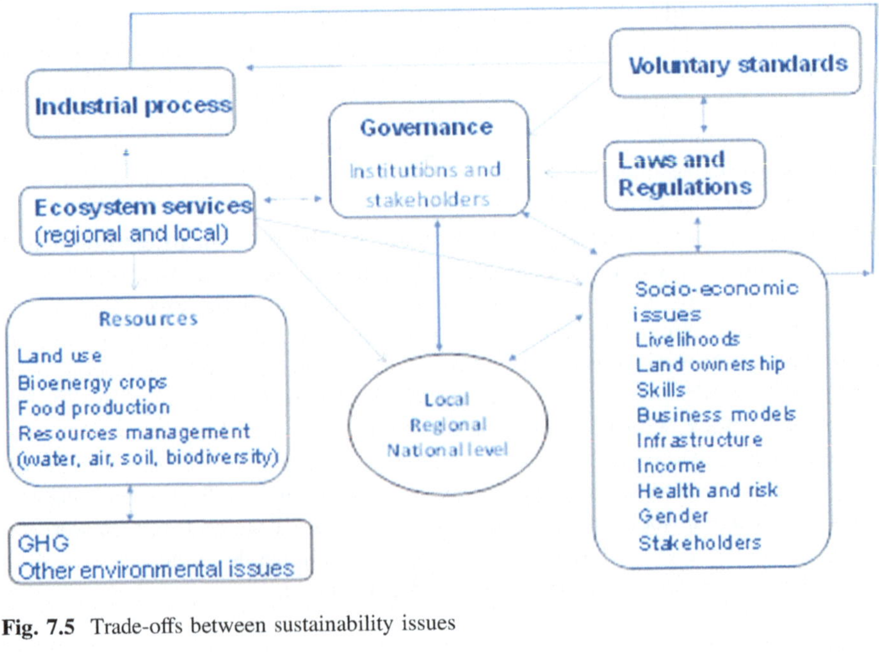

Fig. 7.5 Trade-offs between sustainability issues

Table 7.1 Indicators for monitoring socio-economic impacts of biofuels production (Diaz-Chavez 2014b)

Socio-economic	Social well-being/justice	Social–environmental
Contribution to local economy	Working conditions and rights	Air
	Health and safety	Soil
	Gender;	Water
	Land rights and conflicts	Biodiversity
	Food security	Ecosystem services

particular location at a specific time. It is an ex-post environmental management tool. However, few of these indicators can be monitored over time and within clear quantitative or qualitative parameters. As in the voluntary standards and sustainability schemes, indicators can h help monitor impacts (both positive and negative) of certain activities or over a period of time at the national level (OECD 2011; GBEP 2012). Voluntary reporting is another form of compliance and monitoring, and Corporate Social Responsibility's goals for sustainability reporting, comprises the most recent method for companies. The Global Reporting Initiative (GRI) promotes the use of sustainability reporting as a way for organizations to become more sustainable and contribute to sustainable development (GRI 2014). The process also involves auditors and experts in various fields; and regulators and governmental agencies in several countries (GRI 2014). It is also based on a set of indicators but there has been criticism on the overall number of indicators and their lack of contingency regarding specific industrial sectors or regions.

Some examples exist on the application of socio-economic indicators independent of the available voluntary standards for bioenergy. Diaz-Chavez and Vuohelainen (2014) applied a set of socio-economic indicators to a case study on sugar cane production in Brazil and one based on soy production in Argentina. The focus was on production and the final products were biofuels, but the overall review of indicators and the process of the selection was one of the main outcomes. The application of the indicators also entailed a stakeholders mapping and analysis. An important issue reported on the case studies was the challenge of obtaining relevant data, along with a lack of monitoring for some issues not regulated and a lack of public and stakeholders participation. Environmental aspects s such as soil analysis, water and air emissions had clear indicators that companies were monitoring. Other social indicators, such as number of jobs, number of women working in the plantations or the mills, the number of accidents at the work place, and average wages were all reported. However, a number of socio-economic indicators, such as income spent on basic needs, and access to ecosystem services, were never reviewed by other key social actors, such as companies, local authorities, or NGOs. Another important issue revealed by the study was the need for investment in resources (economic and human) to set up regional or local databases through private public partnerships (Diaz-Chavez and Vuohelainen 2014).

Other studies have also reviewed the socio-economic impacts of biofuels production in Latin America (e.g. Solomon and Bailis 2014) or in Africa (Johnson and

Seebaluck 2012; Rutz and Janssen 2014). Such reviews have shown that the most the most controversial social issues are land rights and food production in the context of bioenergy production. These topics have been extensively analysed (e.g. Lorenzo Cotula and Nat Dyer 2008; Rosillo-Calle and Johnson 2010; FAO 2012; Hamelinck 2013; Souza et al. 2015). Nevertheless, there very few studies have focused specifically on the use of indicators in the case of biorefineries. The reason for this is because the overall supply chain remains similar to the bioenergy ones (see Fig. 7.6). The feedstock supply remains one of the main challenges to sustainability. As seen in Fig. 7.6 land use, availability of resources (residues or dedicated crops), along with the GHG calculation are some of the key issues, but these are all environmental issues.

Few methodologies are available that look specifically at biorefineries, especially high technology biorefineries that use lignocellulosic feedstocks and focus on high value products. For example, the proposed methodology for sustainability assessment of bioerefineries by Jungmeier (2014) of the International Energy Agency Task 42 on biorefineries is based on fact sheets that explain the process of feedstock use to produce a product through different pre-treatments and chemical pathways. This methodology does focus on environmental and some economic indicators, but not on social indicators. The methodology provided by Sacramento (Sacramento-Rivero 2012) uses a sustainability scale normalized for sustainability indicators applicable to biorefineries. The framework has a set of 5 indicator categories and 14 metrics. It is suggested as an ex-post tool. This methodology could be useful for established biorefineries but the number of high technology lignocellulosic biorefineries currently working places limits on it. Dalgaard (2012) proposed a multi-criteria sustainability assessment of the socio-economic and environmental effects of different scenarios to increase agricultural production and a new biobased economy in rural landscapes. The assessment also included socio-economic indicators where competition with other activities and policy issues were assessed.

The EC funded BIOCORE project (BIOCORE 2014) conducted a sustainability assessment of process for biorefineries. The methodology is based on Keller et al. (2015), and is called Integrated Life Cycle sustainability assessment (ILCA). This considered the ex-ante assessment based on other existing frameworks and other ex-ante assessments to provide a better decision-making assessment. The methodology helps to identify barriers that decision-makers need to consider. The

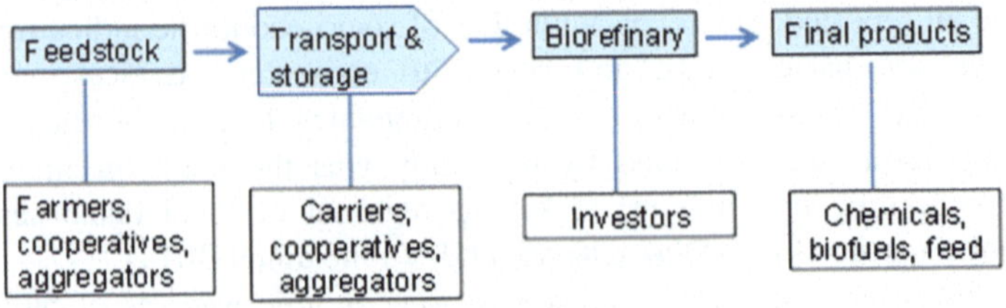

Fig. 7.6 Simplified supply chain showing some stakeholders and relevant issues in each stage

socio-economic assessment of the project focused on a methodology that combined the social life cycle methodology (UNEP Setac Life Cycle Initiative 2009) with a social impact assessment methodology (Diaz-Chavez 2014a). This is also an ex-ante methodology which includes the use of 13 indicators, the mapping of the stakeholders, a policy assessment and hotspots assessment (Benoit-Norris et al. 2012) to determine the main impacts and barriers in the selected case studies of the project. The selected cases were regions in France, Hungary, Germany and India (see Table 7.2).

The list of socio-economic indicators included:

1. Trade of feedstock
2. Identification of stakeholders along the supply chain
3. Policies and regulations
4. Potential biorefinery location Logistic
5. Land use tenure
6. Community participation
8. Rural development and Infrastructure
9. Jobs creation and wages
10. Gender equity
11. Labour conditions
12. Health and safety
13. Competition with other sectors

The stakeholder's assessment provided valuable information on some of the challenges that the bioeconomy faces regarding socio-economic issues. In depth interviews with stakeholders, mainly farmers and producers of feedstock, revealed for instance how strong is the competition of the residues or the feedstock with other uses, in the selected regions. The combined methodology is robust in that it demonstrated possible barriers and negative impacts which can be overcome through mitigation measures (Diaz-Chavez 2014a).

One area that is not well researched is health and safety in the place of work of biorefineries. This is a subject that is in general regulated at national level and the International Labour Organisation (ILO) has over 40 standards and agreements (e.g. ILO C184—Safety and Health in Agriculture Convention, 2001 (No. 184) and

Table 7.2 Case studies of the BIOCORE project (BIOCORE 2014)

Characteristics	France	Germany	Hungary	India
Location	Beauce	Midwest	South-West	Punjab
Main feedstock	Wheat straw	Hardwood	Straw	Rice/wheat straw
Other feedstock	Niche crops (*Miscanthus*)	Softwood	SRC poplar hardwood	Rice/wheat straw
Capacity (1000 t of dry feedstock)	150	150	150	150 and 500*

*Two sites in India Sangrur (150 and 500) and Faridkot (150)

C170—Chemicals Convention, 1990 (No. 170), which can be used as a proxy. Accardi et al. (2013) reviewed some safety concerns in integrated biorefineries in Italy. Their study reviewed environmental and health risks caused by integrated biorefineries processes. Nevertheless, they concluded that these are expected to be lower than with traditional chemical and petrochemical plants, but there is still need of research on safety.

Therefore, as discussed above, indicators on their own can be helpful for reviewing targets, setting up objectives and monitor performance, but they should be considered within a framework. Additionally, the variety of methodologies allows to reduce subjectivity in the assessments and to consider the context of the case study or project.

However, when assessing sustainability, stakeholder participation and good governance are a requirement that a social assessment needs to incorporate. The engagement and analysis of stakeholders has a broad range of methods, but according to Reed et al. (2009), there is little information regarding how, when and why they are effective. The following section examines what is sometimes called fourth pillar of sustainability, that is, governance.

7.2.2 Governance

In addition to the three pillars of sustainability, several other frameworks have introduced a fourth pillar on policy or good governance (Diaz-Chavez 2011). This has been incorporated in sustainability assessments, although not directly related to biorefineries. Still, stakeholders' analysis is a key consideration in sustainability principles and used in the S-LCA and other methodologies as explained above.

The discussion on green economy and good governance has raged on for quite a number of years, but it has yet to be fully acknowledged. According to Mathai and Parayil (2013), the green economy is understood as an economic arrangement that improves growth, social and environmental issues. Furthermore, these authors argue that economic growth is expected to alleviate poverty, promote equality and achieve sustainability. Nevertheless, the definitions of governance also vary according to institutions, political and decision-making processes, and management institutions. There are as well many interpretations of governance, for instance social, political, corporate or environmental (Afful-Koomson 2012).

While the green economy is broader in context, the bioeconomy is considered an economy where the basic building blocks for materials, chemicals and energy are derived from renewable biological resources, such as plant and animal sources (McCormick and Kautto 2013). According to McCormick and Kautto (2013), the bioeconomy can meet the requirements for sustainability from environmental, social and economic perspectives if designed and implemented logically. The advantage of the bioeconomy is that multi-purposes of the use of biomass can be reached including food production, fibres, material and energy. Within this framework and in the context of the EU, the bioeconomy faces a long-term goal, which is to develop a competitive, resource efficient and low carbon economy by 2050. The bioeconomy includes different sectors such as agriculture, energy and climate, transport, industry. Nevertheless, the bioeconomy's focus is on new growth opportunities in traditional and emerging bio-based sectors, while considering global challenges and resource/environmental limitations (Scarlat et al. 2015). This biobased economy may provide a way to address global problems such as climate change and food production, but will have to consider a wider form of understanding of the links between good governance and the use of natural resources (Ostrom 2000). This includes the information on benefits and risks of the bioeconomy that need to be informed to the general public and which currently is not fully understood and communicated (McCormick and Kautto 2013; Diaz-Chavez 2014a). The experience with first generation biofuels and some cases in bioenergy demonstrates that stakeholder participation, public perception and the willingness to pay or accept a technology need to be considered within the bioeconomy (Diaz-Chavez et al. 2015).

As McCormick and Kautto stated (2013), a more coherent, integrated and strategic policy approach combined with a strong emphasis on collaboration is necessary to stimulate the bioeconomy.

7.3 Uncertainty Issues in Biorefinery Sustainability Assessment

Uncertainty affects various aspects of biorefinery research including the synthesis, configuration and operation of a biorefinery. In particular, uncertainty significantly affects the choice of the right feedstock as well as the conversion path to chemicals with high added value. Uncertainty also determines the accuracy of mathematical models of the process (e.g. yield predictions), and has an impact when finding the optimal operational point for maximizing economical profit. This section presents a generic methodology for assessing the impacts of uncertainties on a biorefinery, with applications in different stages of the biorefinery design. The method quantifies uncertainty and propagates it through a mathematical model following a Monte Carlo technique, with an additional step where the results are analysed critically using sensitivity analysis. The methodology is suitable to support robust and reliable decision making in biorefinery design and operation, with the objective to contribute to realizing future profitability of the biorefinery. This part of the chapter presents a generic uncertainty analysis framework consisting of four steps: (1) identify sources of uncertainty and assign a proper distribution function; (2) Latin Hypercube Sampling with correlation control of all sources of uncertainty; (3) model evaluations for all samples; and (4) analysis of results through global sensitivity analysis with the standardized regression coefficients (SRC) method. Afterwards the methodology was applied to the three different stages of biorefinery design: (A) Biorefinery synthesis for finding the best processing pathway from raw material to product; (B) The biorefinery configuration for determining the optimal plant setup given a product pathway; and (C) Operation optimization with two examples, i.e. a dynamic pretreatment (PT) model and a supervisory optimization layer that ensures maximum profitability in production.

Nowadays society depends on an unsustainable source of energy, i.e. fossil fuels. In addition, these conventional fuels emit gases with greenhouse effects, which are responsible for global warming and other negative climate changes. Extensive research and development efforts are continuously made to seek alternative and sustainable sources of energy, such as biofuels. The second-generation bioethanol production process reached commercial reality in 2012 (Larsen et al. 2012), and the technology is exploited nowadays at large-scale facilities around the world. The first commercial scale second-generation bioethanol plant was opened by Beta Renewables in Italy in early 2014. USA followed with three other plants, i.e. POET-DSM, Abengoa, and DuPont, which use corn stover as feedstock to produce bioethanol (Tristan et al. 2015). Brazil also operates two commercial plants based on sugar cane, i.e. GranBio and Iogen. Another important player in the biofuel industry is Inbicon, a Danish company which developed the technology at lower scales, and awaits political support to build a large-scale facility.

Biorefineries transform agricultural wastes into chemicals with higher added value such as biofuels after a chain of conversion steps. There are many technological alternatives for producing biofuels, and appropriate tools need to be developed to support taking the right decision. When designing and operating a biorefinery, three main phases can be identified:

A. Conceptual design:

The starting point is a complex series of conversion pathways from different types of feedstock to possible products.

The objective of this step is to select the path that offers the best economic return considering the impact of an economic risk assessment.

B. Biorefinery configuration:

Given a selected pathway from feedstock to products from the previous step, the objective is to find the best biorefinery configuration that minimizes product costs. The term "configuration" refers to defining recycle streams, reactor hold-ups, retention times, and overall refinery throughput.

C. Operation optimization:

A large-scale biorefinery is constructed following the optimal configuration determined in the previous step. The objective now is to design a control and optimization layer for ensuring a process operation that maximizes refinery profitability considering variations in feedstocks and product prices.

Uncertainty affects all three steps. Market conditions, feedstock variability and model mismatches are all sources of uncertainty that influence the biorefinery profitability in early concept design as well as in real operation after commissioning the plant. Assessing and analysing uncertainty in a quantitative manner is an important step in making the right decision in all stages of a biorefinery design.

This section of the chapter presents a generic methodology for analysing uncertainties with applications to the three cases: evaluation of two different technological platforms for a lignocellulosic biofuel plant (A); configuration alternatives for a second-generation bioethanol plant (B); and dynamic modelling and optimization for the same type of plant (C).

7.3.1 Method

A mathematical model describes the process with different prediction objectives and levels of detail for each of the design phases A, B, and C. A generic input–output nonlinear process model consists of a series of relations $f(u, \theta)$ between the input vector u, model parameters θ and outputs y:

$$y = f(u, \theta) \tag{7.1}$$

When process dynamics—the change of one or more variables as a function of time—are of interest, such as in batch processes, control design or dynamic optimization problems, the mathematical model contains time dynamics and consists of a series of ordinary differential equations (ODE):

$$\begin{aligned} \dot{x} &= g(x, u, \theta) \\ y &= h(x, u, \theta) \end{aligned} \tag{7.2}$$

Variable x is the state vector, and u and θ have the same meaning as in Eq. (7.1); $g(x, u, \theta)$ describes the nonlinear time dynamics, and $h(x, u, \theta)$ calculates the outputs of interest.

An optimization problem then exploits the mathematical model, either in formulation (1) or (2), to reach the objectives of each design step. This will be highlighted in more detail for each case study:

A. The conceptual biorefinery analysis follows a superstructure optimization approach that incorporates uncertainty. The decision on the biorefinery synthesis comes from the result of a stochastic mixed integer linear or non-linear programming problem (MILP or MINLP) that evaluates possible pathways from different feedstocks to a wide range of products following a number of alternative conversion steps (Quaglia et al. 2013). The objective function of the optimization problem is to maximize earnings before interest, taxes, depreciation, and amortization, i.e. EBITDA (Cheali et al. 2014). The constraints of the optimization problem include a database with steady state mathematical models with a low level of detail, describing simplified mass and energy balances of the possible conversion pathways with focus on economics, sustainability, available technologies and global market conditions.

B. Biorefinery configuration studies rely on formulating a non-linear optimization problem (NLP) that evaluates different setups in order to minimize production costs for an already established conversion pathway from feedstock to product (Morales-Rodriguez et al. 2012). The mathematical models focus in greater detail on the conversion kinetics than the models from case A, but they describe the process itself at a low level of detail, assuming perfect control of key parameters, and do not include transport phenomena or the effect of delays between units. For example, such a study can then be used to determine if the enzymatic hydrolysis and fermentation should be performed separately or simultaneously, or if the plant should recycle from the fermentation reactors to liquefaction, or construct separate recycles for hydrolysis and fermentation.
C. Operation optimization deals with building an appropriate control and optimization layer for an existing plant design. The automatic control layer tracks the nominal operational point rejecting process disturbances. The optimization layer finds the optimal operational point given a certain economic objective function, and provides set points to the control layer. The mathematical models have the greatest level of detail, and include time dynamics, transport phenomena and delays between units reflecting the configuration of a large-scale plant. One can observe the plant dynamic behaviour in closed loop including overshoots, settling times, oscillations, and system responses to disturbances. For example, liquefaction pH and PT temperature are key parameters in second-generation bioethanol production, and feedback controllers ensure adequate tracking of their set points (Prunescu et al. 2013a, b). For the same type of plant, an optimization layer finds the best trade-off between the conversion steps of PT, enzymatic hydrolysis and fermentation providing optimal set points for enzyme dosage, yeast seed and PT temperature such that profitability is maximized.

A classical Monte Carlo simulation technique is used, where the uncertainty propagates through the mathematical model of the process. The algorithm for uncertainty analysis consists of four steps:

1. Identify all variables affected by uncertainty and assign a proper distribution function:

$$v \leftarrow [v_1 v_2 \ldots v_{n_v}]$$

The sources of uncertainties can propagate from either inputs u or parameters θ. Therefore, the vector v consists of a combination of elements from u and θ:

$$v \leftarrow [u_1 u_2 \ldots u_{n_u} \theta_1 \theta_2 \ldots \theta_{n_\theta}]$$

where n_u is the number of uncertain inputs and n_θ is the number of uncertain model parameters.

Statistical characterization for each source of uncertainty then follows:

$$v_i \sim D(\theta_d),$$

where v_i is the ith component in vector v and $D(\theta_d)$ is a generic distribution function for parameter vector θ_d. For example:

$$v_i \sim N\left(\mu_i, \sigma_i^2\right) \text{or } v_j \sim U(a_j, b_j)$$

says that v_i follows a normal distribution with mean μ_i and variance σ_i^2, while v_j is uniformly distributed in between lower and upper bounds a_j and b_j. A model correlation matrix **C** describes the correlations between model parameters and completes the statistical characterization of vector v.

A parameter estimation exercise requires real measurements, and provides the statistical characterization of model parameters and inputs, i.e. parameter estimates μ_i, standard deviations σ_i and correlation matrix **C**.

2. Latin hypercube sampling (LHS) with correlation control (Iman and Conover 1982) is performed using the implementation from (Sin et al. 2011):
   Vector v might contain a large number of components, in which case generating all combinations of values is impractical due to the large number of possibilities. LHS is a method that extracts a reduced but representative set of samples suitable for complex numerical analysis (Helton and Davis 2003). This procedure generates a number N of samples for each source of uncertainty, which also considers potential correlation between the sources of uncertainty.
3. Model evaluation in simulations:
   The goal is to propagate uncertainty from sources to model outputs. For each set of samples v, a simulation is performed and the predicted outputs are collected.
4. Analysis of the results:
   In this step, statistical analysis of the collected outputs from the previous simulations is performed. A common technique is the global sensitivity analysis method with Standardized Regression Coefficients (SRC). The aim is to quantify how much each source of uncertainty contributes to model output inaccuracy. The first step is to fit a linear model from the sample vector v, which represents the sources of uncertainty, to each model output of interest from the Monte Carlo simulations (Sin et al. 2011). The fitted output is denoted as y_{reg}:

$$y_{reg} = a + \sum_i b_i v_i \tag{7.3}$$

Finding the coefficients a and b_i is the result of solving a linear least squares problem. A fitting coefficient $R^2 > 0.7$ indicates a sufficient linear model, which validates the results, i.e. it indicates that the linear model explains at least 70% of the observed variance in an output. The next step is to calculate the standardized regression coefficients by scaling the fitting parameters b_i with the help of the standard deviations of the uncertainty sources v_i and the predicted model output y:

$$\beta_i = \frac{\sigma_{v_i}}{\sigma_y} b_i \tag{7.4}$$

The β_i coefficients take values between −1 and +1, demonstrating the relative contribution of each source of uncertainty to the output variance.

7.3.2 Case Study A—Conceptual Design

This case study evaluates design alternatives of a lignocellulosic biofuel plant at an early development stage in order to decide on the platform to implement, what feedstock to use and which products to make (Cheali et al. 2014). Figure 7.7 illustrates the superstructure concept that incorporates all possible pathways from

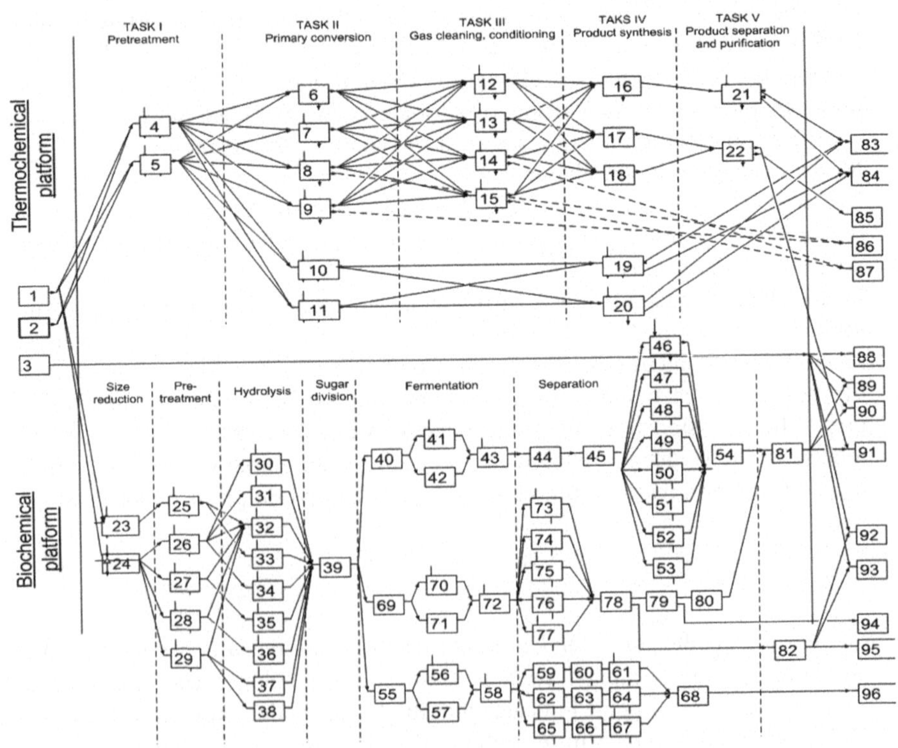

Fig. 7.7 Combined superstructure of two biorefinery conversion platforms: thermochemical (*top*) and biochemical platform (*bottom*) (Cheali et al. 2014)

feedstock to products through a combination of available technologies. This specific study compares the thermochemical against the biochemical platform. The evaluated feedstock includes corn stover and wood, while the possible products refer to gasoline, diesel and ethanol.

The methodology is applied systematically:

1. Identify sources of uncertainty and assign probability distribution functions: Feedstock and product market prices are the most important sources of uncertainty to be considered. Feedstock costs are assumed to follow a uniform distribution within a lower and upper bound, while market prices are assumed to follow a normal distribution with mean and standard deviation shown in Table 7.3. The correlation matrix between the parameters is an important element in addition to the normally distributed parameters.
2. LHS: The uncertain inputs defined in step 1 are now sampled. Figure 7.8 illustrates the 200 samples generated through LHS with correlation control. The diagonal plots show the histogram for each uncertain input, i.e. uniform distribution (rectangular shape) for corn stover and wood cost $P1_1$ and $P1_2$, and normal distribution (Gaussian bell curve) for gasoline, diesel and ethanol price $P3_{122}$, $P3_{123}$ and $P3_{131}$. The off-diagonal plots show the generated samples from the distribution of each uncertain input. Uncorrelated uniformly distributed samples give a rectangular shape while correlated and normally distributed samples appear as a skewed elliptic shape.
3. Model evaluation through simulations:
   The objective function of the MINLP problem is to maximize EBITDA. The solution of the optimization problem yields one solution for each of the 200 samples, i.e. the optimal processing path derived for one combination of uncertain inputs sampled from the distributions.

Table 7.3 Input uncertainty and correlation matrix criteria (Cheali et al. 2014)

Input uncertainty		Min.	Max.		
Corn stover cost ($/dry tonne)		60	100		
Wood cost ($/dry tonne)		60	100		
		Mean	Std.		
Gasoline price ($/gal)		3.53	0.21		
Diesel price ($/gal)		3.97	0.14		
Ethanol price ($/gal)		2.24	0.18		
Correlation matrix	Stover cost	Wood cost	Gasoline price	Diesel price	Ethanol price
Stover cost	1	0	0	0	0
Wood cost	0	1	0	0	0
Gasoline price	0	0	1	0.71	0.12
Diesel price	0	0	0.71	1	0.36
Ethanol price	0	0	0.12	0.36	1

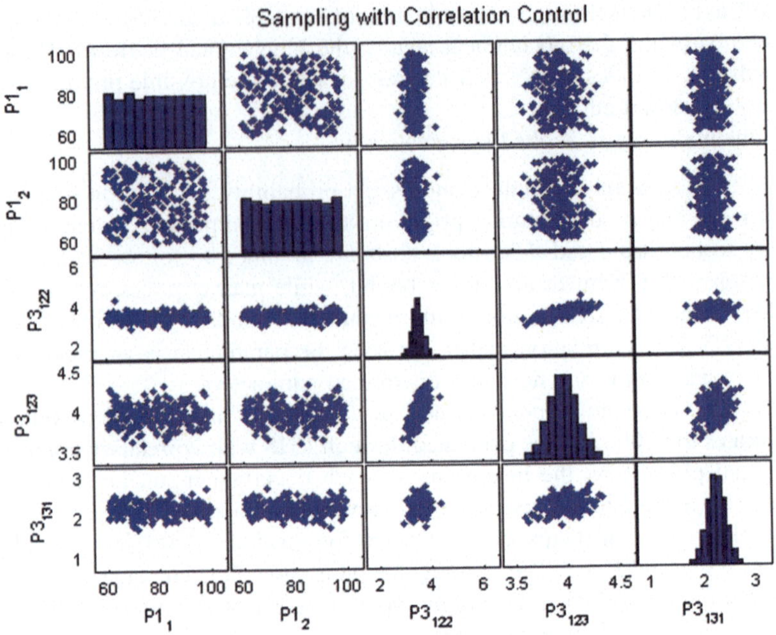

Fig. 7.8 Latin hypercube sampling results with correlation control (Cheali et al. 2014)

4. Analysis of results:
   Table 7.4 presents the selection frequency of the processing paths, and the optimized value of EBITDA. Feedstock (2) is woody biomass and appears to be the best choice due to its high carbon content. Products (83) and (84) are transportation fuels, i.e. gasoline and diesel. The intermediate production steps refer to the thermochemical platform technology.
   There are two pathways from woody biomass to gasoline and diesel. Table 7.5 compares the two networks with respect to the internal rate of return (IRR). The higher a project's IRR is, the more desirable it is to establish that design under the evaluated set of market conditions (Cheali et al. 2014). The results show that network 1 is a safer investment compared to network 2.

Table 7.4 The frequency of selection of the optimal processing paths for 200 scenarios (Cheali et al. 2014)

Network No.	Processing path	Frequency of selection	EBITDA (MM$/a)
1	2 4 6 15 16 21 83 84	83/200	138–230
2	2 5 6 14 16 21 83 84	74/200	140–197
3	1 4 11 20 83 84	18/200	133–195
4	1 5 11 20 83 84	16/200	146–177
5	1 5 10 20 83 84	7/200	154–175
6	1 5 6 14 16 22 83 84	2/200	138–173

Table 7.5 The comparison of risk occurring under uncertainties and the distribution characterization of %IRR when comparing network 1 and network 2 (Cheali et al. 2014)

	Units	Network 1	Network 2
A quantified economic risk			
Total investment	(MM$)	575	600
Expected return	(MM$/a)	210.25	205
Risk	(MM$/a)	0.84	1.35
Data characterization			
Frequency of selection	–	83/200	74/200
Average IRR (%)	(%)	11.06	10.99
Standard deviation of IRR	–	0.94	1.19

7.3.3 Case Study B—Biorefinery Configuration

The objective of the (Morales-Rodriguez et al. 2012) case study is to minimize the cost of bioethanol production given a specific technological pathway from feedstock to biofuel. Figure 7.9 shows the evaluated configurations: (a) simultaneous saccharification and co-fermentation with recycle (SSCF-C RECY); (b) separate hydrolysis and co-fermentation (SHCF) with double recycles; and, (c) separate hydrolysis and co-fermentation with single recycle for enzymes.
The different steps in the method are again followed:

1. Sources of uncertainty:
   The results originate from a nonlinear programming problem that uses a mathematical model for describing the biomass conversion steps, i.e. PT (Lavarack et al. 2002), enzymatic hydrolysis (Kadam et al. 2004), C5 and C6 co-fermentation (Krishnan et al. 1999) and distillation and purification. Model inputs u refer to the feedstock composition, and the parameter vector θ contains over 70 reaction kinetic constants.
   The study assumes a uniform distribution for all components in θ with uncertainty bounds of 5% for PT and co-fermentation parameters, and 25% for enzymatic hydrolysis and feedstock composition. Previous publications contain real experimental data, which can be used for parameter estimation. The parameter estimates and correlation matrix for the enzymatic hydrolysis model were obtained from (Sin et al. 2010), while the statistical characterization of PT and co-fermentation parameters was recalculated based on the experimental data published in (Lavarack et al. 2002) and (Krishnan et al. 1999).
2. LHS with correlation control:
   In this case, the technique generates 150 samples for both feedstock composition and model parameters. Each sample constitutes a simulation scenario.
3. Model evaluations:
   The bioethanol production costs are calculated for all scenarios, i.e. all three configurations are evaluated for all 150 scenarios.

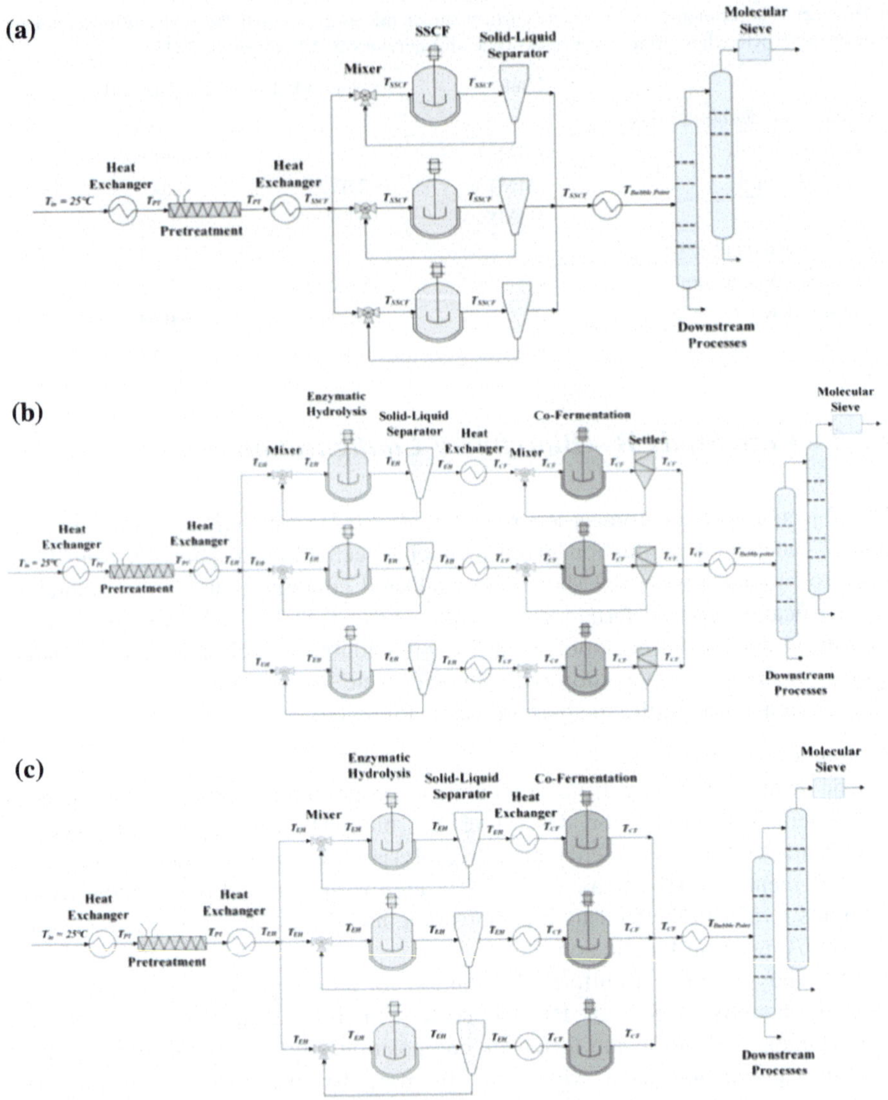

Fig. 7.9 Extended process flowsheet configurations for bioethanol production: **a** SSCF-C RECY, **b** SHCF with double recycle and **c** SHCF with single recycle (Morales-Rodriguez et al. 2012)

4. Analysis of results:
   Table 7.6 displays the results of the uncertainty and global sensitivity analysis of the bioethanol costs in all three configurations. The top table ranks all configurations with respect to bioethanol costs uncertainty, i.e. σ^2. Simultaneous saccharification and co-fermentation with recycle (SSCF-C RECY) has the

Table 7.6 Uncertainty and sensitivity analysis results for the ethanol manufacturing cost for SSCF-C RECY, SHCF with double recycle and SHCF with single recycle configurations (Morales-Rodriguez et al. 2012). Only β coefficients with an absolute value above 0.1 are shown

			SSCF-C RECY		SHCF double recycle		SHCF single recycle		
Uncertainty analysis σ^2			0.03		0.05		0.07		
Sensitivity analysis R^2			0.95		0.89		0.93		
Rank	θ_i	SRC	Section	θ_i	SRC	Section	θ_i	SRC	Section
	SSCF-C RECY			SHCF double recycle			SHCF single recycle		
1	$K^{CF}_{1X_GIG}$	−0.99	SSCF	$Y_{Et_{Xy}\mid Xy}$	−0.99	CF	$Y_{Et_G\mid G}$	0.99	CF
2	$Y_{Et_{Xy}\mid Xy}$	−0.99	SSCF	$Y_{Et_G\mid G}$	0.99	CF	$K^{CF}_{1X_GIG}$	−0.99	CF
3	$Y_{Et_G\mid G}$	0.99	SSCF	n^{PT}_{Xy}	−0.94	PT	$Ea^{PT}_{1,G}$	0.83	PT
4	$Ea^{PT}_{2,Xy}$	0.99	PT	$Ea^{PT}_{1,G}$	0.85	PT	$Y_{Et_{Xy}\mid Xy}$	−0.79	CF
5	n^{PT}_{Xy}	−0.96	PT	$K^{CF}_{2X_{Xy}\mid Xy}$	−0.83	CF	$K^{CF}_{2X_{Xy}\mid Xy}$	−0.72	CF
6	$K^{CF}_{2X_{Xy}\mid Xy}$	−0.93	SSCF	$K^{CF}_{5\mid G}$	0.78	CF	$Ea^{PT}_{2,Xy}$	0.69	PT
7	$Ea^{PT}_{1,G}$	0.87	PT	$Ea^{PT}_{2,Xy}$	0.78	PT	n^{PT}_{Xy}	−0.43	PT
8	$Ea^{PT}_{1,Xy}$	0.31	PT	$K^{CF}_{1X_GIG}$	−0.28	CF	$K^{EH}_{3,Xy}$	0.20	EH
9	C_{Gn}	0.20	FS	C_{Gn}	0.18	FS	$Ea^{PT}_{1,Xy}$	−0.17	PT
10	C_{Xn}	−0.16	FS	n^{PT}_{A}	0.18	PT	k^{EH}_{G2}	−0.11	EH
11	$K^{EH}_{3,Xy}$	0.10	SSCF	$Ea^{PT}_{1,Xy}$	0.17	PT	n^{PT}_{A}	0.11	PT
12				$K^{EH}_{3,Xy}$	0.16	EH			

lowest uncertainty, i.e. a variance of 0.03, which makes this configuration the preferred design for a large-scale plant.

The second table ranks model kinetic parameters and feedstock composition with respect to the β coefficients for all configurations. The results show that SSCF kinetic parameters are the more important ones, followed by PT and feedstock composition. Inhibition of cell biomass growth on glucose $K^{CF}_{1X_GIG}$, and ethanol yield on glucose and xylose, i.e. $Y_{Et_G|G}$ and $Y_{Et_{Xy}|Xy}$, which are all yeast dependent, demonstrate the highest sensitivity in the SSCF-C_RECY case. PT parameters are also ranked high, i.e. the activation energy for xylan hydrolysis $Ea^{PT}_{2,Xy}$ and the order of the reaction to produce xylose n^{PT}_{Xy}.

7.3.4 Case Study C—Operation Optimization

Pretreatment Model Analysis

Raw biomass consists of cellulose, hemicellulose (xylan, arabinan and acetyl groups), lignin, ash and finally other residues in negligible amounts. The purpose of

the PT process is to break the biomatrix such that cellulose becomes more accessible for the enzymatic hydrolysis process that takes place downstream from the PT. During PT, lignin relocates, and hemicellulose partially hydrolyzes forming xylooligomers, xylose and arabinose, along with sugar degradation products such as furfural, 5-hydroxymethylfurfural (5-HMF) and pseudo-lignin, which inhibit downstream processes.
The biomass PT mathematical model has the generic formulation from Eq. 7.5. The next steps illustrate the application of the methodology for assessing the model uncertainty on this case study:

1. Sources of uncertainty:

- Feedstock composition:

$$u = [C_{C_S} C_{X_S} C_{A_S} C_{Ac_S} C_{L_S}] \tag{7.5}$$

The raw biomass composition shows considerable variations due to harvest location and seasonality. Most components are considered in this step, i.e. cellulose C_{C_S}, xylan C_{X_S}, arabinan C_{A_S}, acetyl groups C_{Ac_S} and lignin C_{L_S}.

- Model parameters for reaction kinetics:

$$\theta = [E_{Xo} E_F E_{Ac} E_G E_{PL} E_H]$$

A preliminary sensitivity analysis showed that the PT process is sensitive to the activation energy of all reactions, i.e. production of xylooligomers E_{Xo}, furfural E_F, organic acids E_{Ac}, glucose E_G, pseudo-lignin E_{PL} and 5-HMF E_H.
Vectors u and θ are appended to create the array v:

$$v = [u\theta]$$

Statistical characterization of uncertainty:

- Feedstock composition:

$$u \sim U(\mathrm{a}, \mathrm{b})$$

We assume a uniform distribution of the composition of the feedstock within an interval that is $\pm 7\%$ of the nominal composition.

- Model parameters:

$$\theta \sim N(\mu_\theta, \sigma_\theta^2)$$

Parameter estimation provides the statistical information for vector θ. The kinetic parameters are assumed to follow a normal distribution with estimates μ_θ, standard deviations σ_θ and correlation matrix $\mathbf{R}_\theta$.

2. LHS: we generate 200 samples for v.

3. Monte Carlo simulations: for each sample of v, a simulation is performed and the outputs are collected. On these outputs, the 5th (lower bound), 50th (median) and 95th (upper bound) percentiles are found. Figure 7.10 illustrates model predictions with uncertainty bounds on the one hand, and real measurements taken from a demonstration scale plant on the other hand. The top plot displays the solid composition of pretreated biomass, i.e. the fraction of cellulose, xylan and lignin in the pretreated biomass, while the bottom plot refers to inhibitor formation contained in the liquid phase such as furfural and acetic acid. The plant uses a NIR instrument to determine these concentrations. The results show that most of the NIR readings are well within the uncertainty bounds of the model outputs.
4. Global sensitivity analysis with SRC:

Table 7.7 displays the β coefficients as calculated by using Eq. (7.4). The table ranks all sources of uncertainty contained in vector v with respect to the β coefficients for each model output, i.e. cellulose, xylan, lignin, organic acids and furfural in pretreated fibres.

As expected, the concentrations of cellulose, xylan, lignin and organic acids are sensitive primarily to the feedstock composition of the same components, i.e. C_{C_S}, C_{X_S}, C_{L_S} and C_{Ac_S}, which account for 82, 70, 77 and 72% of the variations in the outputs. In addition, the activation energy for furfural formation is the most sensitive parameter for the furfural content in pretreated biomass. The cellulose content in the fibers in the outlet stream is the second most sensitive input to C_{L_S}, which illustrates the protective role which the lignin has on the cellulosic fibers. More importantly, this specific β coefficient has a negative value showing that if lignin content increases in raw biomass then the cellulose concentration decreases in pretreated fibers.

Xylan hydrolysis is sensitive to the activation energy for xylooligomers production E_{Xo}. The lignin concentration changes as pseudo-lignin is produced, and the

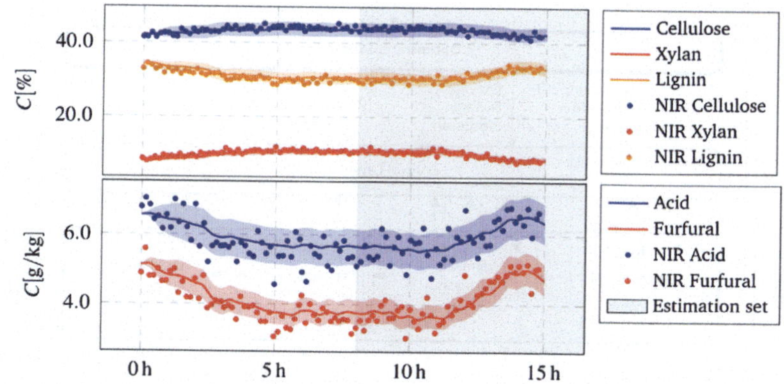

Fig. 7.10 Predictions with uncertainty interval from a biomass *pretreatment* model (Prunescu et al. 2015). A detailed explanation of the results in this figure is provided in the text

Table 7.7 The β coefficients for a biomass *PT* model (Prunescu et al. 2015)

v	Cellulose	v	Xylan	v	Lignin	v	Acid	v	Furfural
C_{C_S}	0.82	C_{X_S}	0.70	C_{L_S}	0.77	C_{Ac_S}	0.72	E_F	−1.00
C_{L_S}	−0.43	E_{Xo}	0.54	E_{PL}	−0.72	E_{Ac}	−0.60	E_{PL}	0.75
E_G	0.38	C_{C_S}	−0.38	C_{C_S}	−0.47	C_{C_S}	0.06	C_{X_S}	0.56
E_{PL}	0.36	C_{L_S}	−0.23	E_H	−0.47	C_{L_S}	0.04	E_{Xo}	−0.16
E_H	0.23	E_{PL}	0.20	E_G	−0.24	E_{PL}	−0.04	C_{A_S}	0.08
C_{X_S}	−0.23	E_H	0.14	E_{Xo}	−0.11	E_H	−0.02	E_H	0.07
E_{Xo}	−0.08	E_G	−0.12	E_F	−0.06	E_G	0.02	C_{L_S}	0.04
C_{Ac_S}	−0.05	C_{Ac_S}	−0.03	C_{X_S}	−0.04	C_{X_S}	0.01	C_{C_S}	0.03
E_F	0.02	E_F	0.02	C_{Ac_S}	−0.04	E_{Xo}	0.01	E_G	0.02
E_{Ac}	−0.02	E_{Ac}	−0.02	E_{Ac}	−0.02	E_F	0.00	C_{Ac_S}	0.01
C_{A_S}	−0.01	C_{A_S}	−0.01	C_{A_S}	0.02	C_{A_S}	0.00	E_{Ac}	0.00
R^2	1.00	R^2	1.00	R^2	0.99	R^2	1.00	R^2	0.93

activation energy for acid formation E_{Ac} significantly affects the concentration of organic acids. Furfural participates in pseudo-lignin creation, and E_{PL} influences its concentration.

The high linearity coefficient (R^2) values increase the confidence in the above results, which is also in agreement with expert knowledge in the domain.

Optimization layer

This study aims at developing an optimization layer for a second-generation biorefinery. Figure 7.11 shows the interaction between a real plant with automation and supervisory layers for optimization and system identification. The stages in

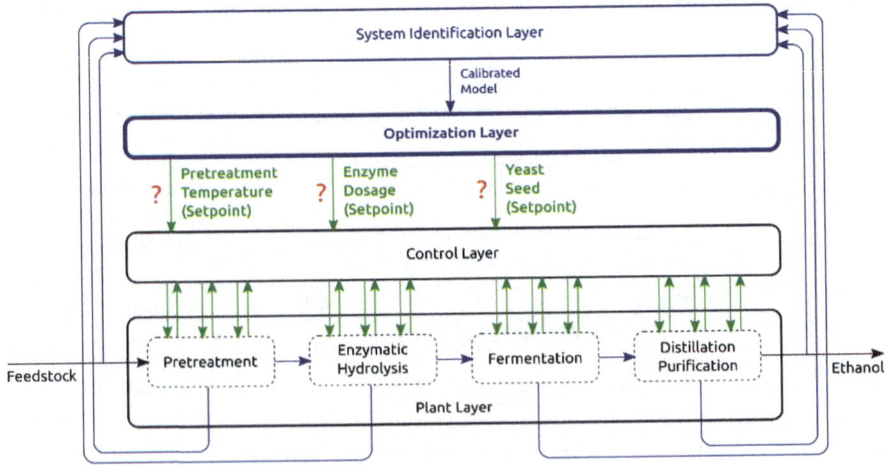

Fig. 7.11 The interaction between the physical plant and the layers for control, optimization and system identification

biomass conversion influence each other, and it is important to treat the refinery in an integrated manner in order to capture all possible trade-offs between PT, enzymatic hydrolysis and fermentation.

The optimization layer searches for the nominal operational point that maximizes the economic profit, and provides set points for PT temperature, enzyme dosage in liquefaction, and the required yeast seed in fermentation. The control layer supports the optimal operation ensuring reference tracking of the optimal set points and disturbance rejection. The system identification layer calibrates the mathematical models for achieving more accurate predictions.

The optimization layer solves the following optimization problem:

$$\begin{gathered}\max_{T_{tr},F_e,M_y} M_{Eth}(t_f)P_{Eth} - (F_S P_S + F_e P_e + M_y P_y)\\ 0 = f(x, u, \theta)\\ \dot{x} = h(x_f, u_f, \theta)\\ 150 \leq T_{tr} \leq 210\,\mathrm{C}\\ 10 \leq F_e \leq 1000\,\mathrm{kg/h}\\ 10 \leq M_y \leq 1000\,\mathrm{kg}\end{gathered} \tag{7.6}$$

We apply the same methodology for assessing the uncertainty of the optimal solution found by the optimization layer:

1. Sources of uncertainty:

We treat the second-generation biorefinery in an integrated manner connecting PT, enzymatic hydrolysis and fermentation together. The total number of model parameters is 96 but a preliminary sensitivity analysis has shown that the profit curve is sensitive to a subset of 22 parameters. Vector v includes both information on the feedstock composition with the same components as in (5), and a vector of the sensitive model parameters θ of length 22 referring to reaction kinetics.

Statistical characterization of uncertainty:

This step is identical to the one from the previous case study, i.e. the feedstock composition is assumed to have a uniform distribution while model parameters are assumed to follow a normal distribution.

2. LHS with correlation control is applied, generating 100 samples.
3. Monte Carlo simulations. Figure 7.12 is a 3 × 3 matrix of graphs containing the refinery profit (the optimized solution compared to a traditional operation without optimization), the refinery costs (split into PT, liquefaction and fermentation), and the optimal solution (for enzyme dosage and yeast seed). The first 2 columns with graphs show the propagation of separated feedstock and kinetic parameter uncertainty while the last column combines all sources of uncertainty.

   The optimization layer diminishes the uncertainty bounds around the profit curve when comparing to a traditional operation. The first two columns from Fig. 7.12 demonstrate that kinetic parameters have a lower uncertainty impact than feedstock composition, suggesting that measuring the raw biomass

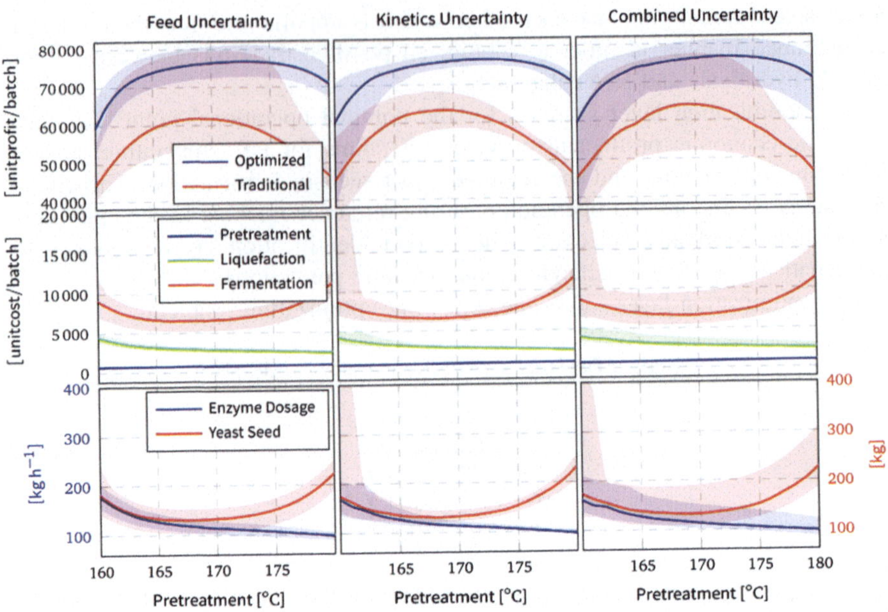

Fig. 7.12 Uncertainty analysis of a biorefinery optimal operation: profit, costs and optimal solution

composition is more important than estimating model parameters for improving the accuracy of the model predictions. Another aspect is that increasing the PT temperature is beneficial for the enzymatic hydrolysis due to a proper biomatrix opening, which reflects in a decreasing enzyme dosage curve. However, a high PT temperature degrades sugars leading to inhibitor formation, which has the disadvantage of increasing the yeast seed and fermentation costs. The optimization layer flattens the profit curve ensuring high profitability for a wide range of operation conditions with a maximum median profit peak at the following point:

$$z_0 = \begin{bmatrix} T_{tr} \\ F_e \\ M_y \end{bmatrix} = \begin{bmatrix} 172\,\mathrm{C} \\ 110\,\mathrm{kg/h} \\ 142\,\mathrm{kg} \end{bmatrix} \tag{7.7}$$

The control layer takes the optimal set points from z_0 and adjusts the biorefinery operation.

4. Global sensitivity analysis (SRC):
The aim of the global sensitivity analysis is to identify which feedstock components and model kinetic parameters influence most the optimized profit curve. Table 7.8 ranks all parameters with respect to the β coefficients. The cellulose content in dry biomass C_{C_S} appears first because it significantly affects the biorefinery profit, which is expected since cellulose is the substrate for glucose

Table 7.8 The β coefficients for the optimization layer (Prunescu, Blanke, Jakobsen, & Sin, Model-Based Plantwide Optimization of a Large Scale Lignocellulosic Bioethanol Plant, 2016)

v	c
C_{C_S}	0.91
C_{Ac_S}	−0.31
Y_{PS_G}	0.28
$q_{Ac_{Max}}$	0.14
Y_{XS_G}	0.12
Y_{PS_X}	0.11
K_2	0.10
I_{G_2}	0.07
…	…
R^2	0.99

production. The acetyl content C_{Ac_S} appears as second most influential parameter because it constitutes the substrate for formation of organic acids and acetate, which disrupt and inhibit both enzymatic hydrolysis and fermentation leading to higher costs. This β coefficient has a negative value indicating that a higher content of acetyl groups will decrease profitability. Other sensitive parameters refer to ethanol yield from glucose Y_{PS_G}, maximum acetate uptake $q_{Ac_{Max}}$, cell biomass growth on glucose Y_{XS_G}, and ethanol yield on xylose Y_{PS_X}. The results from the global sensitivity analysis are in accordance with the uncertainty bounds shown in Fig. 7.12, i.e. profitability is more sensitive to feedstock composition (cellulose and acetyls content) than to model kinetic parameters. This suggests that placing a sensor to measure accurately the feedstock composition has the potential to reduce the uncertainty on the profit curve.

7.4 Conclusions

The need is recognized to have metrics to compare systems in terms of carbon footprint, water footprint, net present value, annual carbon flows, social indicators or even using the energy return on investment. There is not one single methodology that can aptly cover the synergies of environmental, economic, social and governance issues required to assess the sustainable production and use of bioenergy systems. Much of the focus of sustainability assessment in this sector has been given to the feedstock production, particularly for biofuels production.

The perfect environmental metric is not yet established and some researchers prefer to avoid high levels of uncertainty in life cycle analysis methodology and adopt more physically quantifying methods like the ABC method presented here.

An effort in making systems/products comparable in a clear unbiased way is still a huge problem to tackle in bioenergy sustainability analysis.

Life cycle analysis and environmental management tools are a form to combine data for a sustainability assessment. This continues to be a more common form to assess projects that cover the whole supply chain, and this is much more so when the production system is involved. Both methodologies use indicators although the application and type of data may be a challenge for a combined methodology. This is particularly the case for those indicators that are based on qualitative data. As demonstrated with practical projects, some qualitative indicators could be further standardized in terms of the information requested, making them easier to measure and compare across time scales and even across regions or, still, feedstocks. More detailed research needs to be carried out on socio-economic assessment of biorefineries and as many pathways are still being developed, there are not yet concrete cases to assess finalized bioproducts. Participatory approaches may benefit more the support to bioeconomy if the benefits and also risks are better communicated to the general public and involved stakeholders.

In addition to establishing the perfect metric, there are three types of uncertainty when building scenarios with biorefinery based systems that must be considered to have a broader sustainability analysis.

The method presented for assessing uncertainty in biorefinery concept (case study A), plant configuration (case study B) and plant operation analysis (case study C) proved to be a valuable tool to support decision making both at the early design stage of a biorefinery, but also during operation after the biorefinery has been commissioned. In addition, the sensitivity analysis step helped quantify and identify the uncertainty bottlenecks by ranking all sources of uncertainty with respect to their impact on model outputs.

References

Accardi DS, Bubbico R, Di Palma L, Pietrangeli B (2013) Environmental and safety aspects of integrated biorefineries (IBR) in Italy. In: Chemical engineering transactions, vol 32. pp 169–174

Afful-Koomson T (2012) Governance challenges for promoting the green economy in Africa. In: Puppim de Oliveira JA (ed) Green economy and good governance for sustainable development: opportunities, promises and concerns. United Nations University Press, Tokyo, pp 136–160

Ahlgren S, Björklund A, Ekman A, Karlsson H, Berlin J, Börjesson P, Ekvall T, Finnveden G, Janssen M, Strid I (2013) LCA of biorefineries identification of key issues and methodological recommendations. Sweden

Benoit-Norris C, Cavan DA, Norris G (2012) Identifying social impacts in product supply chains: overview and application of the social hotspot database. Sustainability 4:1946–1965. doi:10.3390/su4091946

Bernardi A, Giarola S, Bezzo F (2013) Spatially explicit multiobjective optimization for the strategic design of first and second generation biorefineries including carbon and water footprints. Ind Eng Chem Res 52:7170–7180. doi:10.1021/ie302442j

BIOCORE (2014) Integrated assessment of overall sustainability
Black MJ, Whittaker C, Hosseini SA, Diaz-Chavez R, Woods J, Murphy RJ (2011) Life cycle assessment and sustainability methodologies for assessing industrial crops, processes and end products. Ind Crops Prod 34:1332–1339. doi:10.1016/j.indcrop.2010.12.002
Bradley T, Maga D, Antón S (2015) Unified approach to life cycle assessment between three unique algae biofuel facilities. Appl Energy 154:1052–1061. doi:10.1016/j.apenergy.2014.12.087
Brander M, Tipper R, Hutchison C, Davis G (2008) Consequential and attributional approaches to LCA: a Guide to policy makers with specific reference to greenhouse gas LCA of biofuels. Econom Press, pp 1–14
Brundtland GH (1987) Our common future: report of the world commission on environment and development. Med Confl Surviv 4:300. doi:10.1080/07488008808408783
Cheali P, Quaglia A, Gernaey KV, Sin G (2014) Effect of market price uncertainties on the design of optimal biorefinery systems-a systematic approach. Ind Eng Chem Res 53:6021–6032. doi:10.1021/ie4042164
Cherubini F (2010) The biorefinery concept: using biomass instead of oil for producing energy and chemicals. Energy Convers Manag 51:1412–1421. doi:10.1016/j.enconman.2010.01.015
Cherubini F, Jungmeier G (2010) LCA of a biorefinery concept producing bioethanol, bioenergy, and chemicals from switchgrass. Int J Life Cycle Assess 15:53–66. doi:10.1007/s11367-009-0124-2
Dale VH, Efroymson RA, Kline KL, Langholtz MH, Leiby PN, Oladosu GA, Davis MR, Downing ME, Hilliard MR (2013) Indicators for assessing socioeconomic sustainability of bioenergy systems: a short list of practical measures. Ecol Indic 26:87–102. doi:10.1016/j.ecolind.2012.10.014
Dalgaard T (2012) Concepts for a multi-criteria sustainability assessment of a new more biobased economy in rural production landscapes
DeCicco J, Krishnan R (2015) Annual Basis Carbon (ABC) analysis of biofuel production at the facility level. Ann arbor
Degerickx J, Almeida J, Moonen PCJ, Vervoort L, Muys B, Achten WMJ (2015) Impact of land-use change to Jatropha bioenergy plantations on biomass and soil carbon stocks: a field study in Mali. GCB Bioenergy. doi:10.1111/gcbb.12288
Diaz-Chavez R (2014a) WP7 Sustainability assessment and socio-economic methodology. Biocore project
Diaz-Chavez R (2014b) Indicators for socio-economic sustainability assessment. In: Rutz D, Janssen R (eds) Socio-economic impacts of bioenergy production. Springer International Publishing, Cham, pp 17–37
Diaz-Chavez RA (2011) Assessing biofuels: aiming for sustainable development or complying with the market? Energy Policy 39:5763–5769. doi:10.1016/j.enpol.2011.03.054
Diaz-Chavez R, Colangeli M, Morese M, Fallot A, Azanha M, Sibanda L, Mapako M (2015) Social considerations. In: Souza GM, Victoria R, Joly C, Verdade L (eds) Bioenergy & sustainability: bridging the gaps. SCOPE, Paris, pp 514–539
Diaz-Chavez R, Vuohelainen A (2014) Test auditing of socio-economic indicators for biofuel production. Socio-economic impacts of bioenergy production. Springer International Publishing, Cham, pp 39–58
Efroymson RA, Dale VH, Kline KL, McBride AC, Bielicki JM, Smith RL, Parish ES, Schweizer PE, Shaw DM (2013) Environmental indicators of biofuel sustainability: what about context? Environ Manage 51:291–306. doi:10.1007/s00267-012-9907-5
FAO (2012) Voluntary guidelines on the responsible governance of tenure of land, fisheries and forests in the context of national food security
GBEP (2012) The global bioenergy partnership sustainability indicators for bioenergy. Rome
Giarola S, Zamboni A, Bezzo F (2011) Spatially explicit multi-objective optimisation for design and planning of hybrid first and second generation biorefineries. Comput Chem Eng 35:1782–1797. doi:10.1016/j.compchemeng.2011.01.020

Gibson RB, Hassan S, Holtz S, Tansey J, Whitelaw G (2005) Sustainability assessment: criteria and processes
GRI (2014) Sustainability reporting database
Hamelinck C (2013) Land grabs for biofuels driven by EU biofuels policies
Helton JC, Davis FJ (2003) Latin hypercube sampling and the propagation of uncertainty in analyses of complex systems. Reliab Eng Syst Saf 81:23–69
Hennecke AM, Faist M, Reinhardt J, Junquera V, Neeft J, Fehrenbach H (2013) Biofuel greenhouse gas calculations under the European renewable energy directive—a comparison of the BioGrace tool versus the tool of the roundtable on sustainable biofuels. Appl Energy 102:55–62. doi:10.1016/j.apenergy.2012.04.020
Hoekstra AY, Chapagain AK, Aldaya MM, Mekonnen MM (2009) Water footprint manual state of the art 2009. Water Footpr Netw 131
Iman RL, Conover WJ (1982) A distribution-free approach to inducing rank correlation among input variables. Commun Stat-Simul Comput 11:311–334. doi:10.1080/03610918208812265
ISO (2006a) ISO 14040: Environmental management—life cycle assessment—principles and framework. Environ Manage 3:28. doi:10.1002/jtr
ISO (2006b) ISO 14044:2006. Environmental management – life cycle assessment. Requirements and Guidelines. International Organization for Standardization
Johnson F, Seebaluck V (2012) Bioenergy for sustainable development and international competitiveness. The role of sugar cane in Africa. Earthsan, London
Jourabchi SA, Gan S, Ng HK (2014) Pyrolysis of Jatropha curcas pressed cake for bio-oil production in a fixed-bed system. Energy Convers Manag 78:518–526. doi:10.1016/j.enconman.2013.11.005
Jungmeier G (2014) The biorefinery fact sheet. Version 1.0, 2014-09-19
Kadam KL, Rydholm EC, McMillan JD (2004) Development and validation of a kinetic model for enzymatic saccharification of lignocellulosic biomass. Biotechnol Prog 20:698–705. doi:10.1021/bp034316x
Keller H, Rettenmaier N, Reinhardt GA (2015) Integrated life cycle sustainability assessment - A practical approach applied to biorefineries. Appl Energy 154:1072–1081. doi:10.1016/j.apenergy.2015.01.095
Krishnan MS, Ho NW, Tsao GT (1999) Fermentation kinetics of ethanol production from glucose and xylose by recombinant Saccharomyces 1400 (pLNH33). Appl Biochem Biotechnol 77–79:373–388. doi:10.1385/ABAB:78:1-3:373
Larsen J, Haven MØ, Thirup L (2012) Inbicon makes lignocellulosic ethanol a commercial reality. Biomass Bioenergy 46:36–45. doi:10.1016/j.biombioe.2012.03.033
Lavarack BP, Griffin GJ, Rodman D (2002) The acid hydrolysis of sugarcane bagasse hemicellulose to produce xylose, arabinose, glucose and other products. Biomass Bioenergy 23:367–380. doi:10.1016/S0961-9534(02)00066-1
Cotula Lorenzo, Nat Dyer SV (2008) Fuelling exclusion? The biofuels boom and poor people's access to land. FAO and IIED, London
Macedo IC, Seabra JEA, Silva JEAR (2008) Green house gases emissions in the production and use of ethanol from sugarcane in Brazil: the 2005/2006 averages and a prediction for 2020. Biomass Bioenergy 32:582–595. doi:10.1016/j.biombioe.2007.12.006
Mathai M, Parayil G (2013) Towards equity and sustainability in the "green economy." In: Puppim de Oliveira JA (ed) Green economy and good governance for sustainable development: opportunities, promises and concerns. United Nations University Press, Tokyo
McCormick K, Kautto N (2013) The Bioeconomy in Europe: an overview. Sustain 5:2589–2608. doi:10.3390/su5062589
Méndez L, Rojas J, Izaguirre C, Contreras B, Gómez R (2014) Jatropha curcas leaves analysis, reveals it as mineral source for low sodium diets. Food Chem 165:575–577. doi:10.1016/j.foodchem.2014.05.124

Morales-Rodriguez R, Meyer AS, Gernaey KV, Sin G (2012) A framework for model-based optimization of bioprocesses under uncertainty: lignocellulosic ethanol production case. Comput Chem Eng 42:115–129. doi:10.1016/j.compchemeng.2011.12.004

Murata K, Liu Y, Inaba M, Takahara I (2012) Catalytic fast pyrolysis of jatropha wastes. J Anal Appl Pyrolysis 94:75–82. doi:10.1016/j.jaap.2011.11.008

Ness B, Urbel-Piirsalu E, Anderberg S, Olsson L (2007) Categorising tools for sustainability assessment. Ecol Econ 60:498–508

OECD (2011) Strategic environmental assessment and biofuel development. Paris

Ostrom E (2000) Collective action and the evolution of social norms. J Econ Perspect 14:137–158

Post WM, Kwon KC (2000) Soil carbon sequestration and land-use change: processes and potential. Glob Chang Biol 6:317–327. doi:10.1046/j.1365-2486.2000.00308.x

Prunescu RM, Blanke M, Jakobsen JG, Sin G (2015) Dynamic modeling and validation of a biomass hydrothermal pretreatment process-a demonstration scale study. AIChE J 61:4235–4250. doi:10.1002/aic.14954

Prunescu RM, Blanke M, Sin G (2013a) Modelling and L < inf > 1</inf > adaptive control of pH in bioethanol enzymatic process. In: 2013 American control conference. IEEE, pp 1888–1895

Prunescu RM, Blanke M, Sin G (2013b) Modelling and L1 adaptive control of temperature in biomass pretreatment. In: 52nd IEEE conference on decision and control. IEEE, pp 3152–3159

Quaglia A, Sarup B, Sin G, Gani R (2013) A systematic framework for enterprise-wide optimization: synthesis and design of processing networks under uncertainty. Comput Chem Eng 59:47–62. doi:10.1016/j.compchemeng.2013.03.018

Reed MS, Graves A, Dandy N, Posthumus H, Hubacek K, Morris J, Prell C, Quinn CH, Stringer LC (2009) Who's in and why? a typology of stakeholder analysis methods for natural resource management. J Environ Manage 90:1933–1949. doi:10.1016/j.jenvman.2009.01.001

Rosillo-Calle F, Johnson F (2010) Food versus fuel. An informed introduction to biofuels. ZED Books, London

Rutz D, Janssen R (eds) (2014) Socio-economic impacts of bioenergy production. Springer International Publishing, Cham

Sacramento-Rivero JC (2012) A methodology for evaluating the sustainability of biorefineries: framework and indicators. Biofuels, Bioprod Biorefining 6:32–44. doi:10.1002/bbb.335

Sandin G, Røyne F, Berlin J, Peters GM, Svanström M (2015) Allocation in LCAs of biorefinery products: implications for results and decision-making. J Clean Prod 93:213–221. doi:10.1016/j.jclepro.2015.01.013

Scarlat N, Dallemand J-F, Monforti-Ferrario F, Banja M, Motola V (2015) Renewable energy policy framework and bioenergy contribution in the European Union—an overview from national renewable energy action plans and progress reports. Renew Sustain Energy Rev 51:969–985. doi:10.1016/j.rser.2015.06.062

Schmitz S, Dawson B, Spannagle M, Thomson F, Koch J, Eaton R (2000) The Greenhouse gas protocol—a corporate accounting and reporting standard, revised edition. GHG Protoc Corp Account Report Stand 9:116

Searchinger T, Heimlich R, Houghton RA, Dong F, Elobeid A, Fabiosa J, Tokgoz S, Hayes D, Yu T-H (2008) Use of U.S. Croplands for biofuels increases greenhouse gases through emissions from land-use change. Science (80-) 319:1238–1240

Silva C (2011) Electric and plug-in hybrid vehicles influence on CO_2 and water vapour emissions. Int J Hydrogen Energy 36:13225–13232. doi:10.1016/j.ijhydene.2011.07.023

Sin G, Gernaey KV, Neumann MB, van Loosdrecht MCM, Gujer W (2011) Global sensitivity analysis in wastewater treatment plant model applications: prioritizing sources of uncertainty. Water Res 45:639–651. doi:10.1016/j.watres.2010.08.025

Sin G, Meyer AS, Gernaey KV (2010) Assessing reliability of cellulose hydrolysis models to support biofuel process design-Identifiability and uncertainty analysis. Comput Chem Eng 34:1385–1392. doi:10.1016/j.compchemeng.2010.02.012

Slepetiene a, Slepetys J, Liaudanskiene I (2008) Standard and modified methods for soil organic carbon determination in agricultural soils. Agron Res 6:543–554

Social hotspot organisation (2014). Social hotspot database. http://socialhotspot.org/. Accessed May 2014

Solomon BD, Bailis R (eds) (2014) Sustainable development of biofuels in Latin America and the caribbean. Springer New York, New York, NY

Souza SP, Gopal AR, Seabra JEA (2015) Life cycle assessment of biofuels from an integrated Brazilian algae-sugarcane biorefinery. Energy 81:373–381. doi:10.1016/j.energy.2014.12.050

Sukumara S, Faulkner W, Amundson J, Badurdeen F, Seay J (2014) A multidisciplinary decision support tool for evaluating multiple biorefinery conversion technologies and supply chain performance. Clean Technol Environ Policy 16:1027–1044. doi:10.1007/s10098-013-0703-6

Taelman SE, Sfez S (2015) Environmental Life Cycle Assessment (LCA) of algae production in NorthWest Europe (NWE)-EnAlgae project

Tristan RB, Robert B, Vonnie E (2015) SBE special section: lignocellulosic Biofuels—commercial-scale production of lignocellulosic biofuels. CEP Mag, AIChE

UNEP Setac Life Cycle Initiative (2009) Guidelines for Social Life Cycle Assessment of Products

Valdivia S, Ugaya CML, Hildenbrand J, Traverso M, Mazijn B, Sonnemann G (2013) A UNEP/SETAC approach towards a life cycle sustainability assessment—Our contribution to Rio + 20. Int J Life Cycle Assess 18:1673–1685. doi:10.1007/s11367-012-0529-1

Wolf M-A, Pant R, Chomkhamsri K, Sala S, Pennington D (2012) The International Reference Life Cycle Data System (ILCD) Handbook

Author Biographies

Dr. Carla A.M. Silva, Ph.D. received the Mechanical Engineering degree (five-year course) in 2000 and the Ph.D. degree in Mechanical Engineering in 2005, at Instituto Superior Técnico (IST), University of Lisbon, Portugal. She went for a postdoc at both IST and University of Michigan working on CO2 mitigation in road vehicles. She was a senior researcher at the Institute of Mechanical Engineering IDMEC since 2008 to 2015. She is now assistant professor at the Department of Geographic, Geophysics and Energy Engineering. She teaches, in the integrated master of Energy and Environment Engineering, the courses of Sustainable Mobility, Bioenergy, Combustion Technologies, Life cycle assessment and Electrical circuits. She has supervised more than 50 M.Sc. and Ph.D. students. She has 1 book, 2 book chapters, more than 50 papers in international peer-reviewed journals and more than 60 papers in international conferences. She has been awarded with seven prizes, including the best mechanical engineering internship and the 3M prize for innovation. Areas of interest are new fuels, biomass, alternative road vehicle simulation, energy and environment impacts, life cycle assessment, indicators for sustainable mobility.

Remus Prunescu, Ph.D. is a researcher in the area of mathematical modelling, advanced process control and optimization. He graduated in Electrical Engineering (Automation and Control) in 2012, and obtained his Ph.D. in 2016 at the Technical University of Denmark for his research on "Dynamic Modeling, Optimization and Advanced Control for Large Scale Biorefineries". His contributions refer to creating a dynamic simulator for a large-scale second-generation bioethanol plant together with an automation and optimization layer. All his work includes sensitivity and uncertainty analysis. He has a vast experience in multi-disciplinary areas including chemical, biochemical, robotics and maritime technologies, focusing on design and implementation of advanced model-based control algorithms for large-scale facilities.

Krist V. Gernaey, Ph.D. has a Master's degree (1993) and Ph. D. degree (1997) from Ghent University (Belgium). He is Professor in Industrial Fermentation Technology and Head of the CAPEC-PROCESS Research Center, Department of Chemical and Biochemical Engineering, DTU. His main research focus is on the development and optimization of fermentation, biocatalysis, pharmaceutical and biological wastewater treatment processes. His research work also includes mathematical modelling, investigation of mass transfer related issues in bioreactors, scale-up and scale-down of bioreactors, PAT and development of continuous production processes. Krist V. Gernaey has until now co-authored 195 publications in peer-reviewed journals (H-index = 34, ISI Web of Science).

Gürkan Sin, Ph.D. is Associate Professor within the Process Systems Engineering (PSE) area at the CAPEC-PROCESS Research Center at the Department of Chemical and Biochemical Engineering at Technical University of Denmark (DTU). Dr. Sin has a broad expertise in development and application of systems engineering for process design and innovation. His research aims to develop enabling technology in the form of tools, methods and software to support cost-competitive process–product development for the life sciences (food, pharma), biotechnology, biorefinery and water industries. Dr. Sin has pioneered uncertainty and sensitivity analysis for engineering use of models for process design, product design and process scale-up and optimization. Dr. Sin has supervised 15 completed Ph.D. projects and is currently supervising 11 Ph.D. students. Dr. Sin has 145+ peer-reviewed publications (Web of Science), 1700 citations and an H-index of 26.

Rocio A. Diaz-Chavez, Ph.D. is Research Fellow at the Centre for Environmental Policy of Imperial College London and tutor with the Distance learning M.Sc. of the Centre for Development, Environment and Policy at SOAS (School of Oriental and African Studies). She has extensive work and academic experience in sustainability assessment and environmental management tools. She has participated in and coordinated different EU funded projects on bioenergy, climate change and bioeconomy (CARENSA, BEST, COMPETE, Global-Bio-Pact, AfriCAN Climate, BIOCORE, COSMOS, Biotrade 2020). She has collaborated with international organizations including FAO and UNIDO. Her regional expertise extends to the EU and other regions in Africa, Asia and Latin America. Dr. Diaz-Chavez acted as Co-Chair of the Social Group developing the Bioenergy Standard for the International Organisation for Standardisation (ISO) and is vice-chair of the Bioenergy Committee at the British Standards Institute. She is the co-chair of the International Energy Agency for the UK Task 40 (Biomass trading) and chair of the Oil, Gas and Renewable Fuels at the International Association of Impact Assessment. Dr Diaz-Chavez has authored a number of scientific publications and international reports. She received the SCOPE 2010 Young Scientist Award in Environmental Management for her work on indicators and standards.

Chapter 8
Designing Integrated Biorefineries Using Process Systems Engineering Tools

Behrang Mansoornejad, Shabnam Sanaei, Banafsheh Gilani, Dieudonné R. Batsy, Marzouk Benali and Paul R. Stuart

Abstract Biorefinery concept is now well established as an attractive production system for effectively converting lignocellulosic biomass. Sustainable design of biorefinery processes requires considering multiple criteria to analyze the performance of biorefinery from different perspectives. The focus of this chapter is on introducing a systematic methodology for designing integrated biorefineries using process systems engineering tools, which include market analysis, techno-economic assessment, cost accounting, energy integration analysis, life-cycle assessment, supply chain analysis, as well as a multi-criteria decision-making framework to put forward the most effective biorefinery strategies that fulfill the needs of the forest industry. The proposed methodology, aggregating the impacts into sustainability scores, is illustrated through a case study consisting of evaluating the potential of implementing black liquor lignin recovery process or fast pyrolysis within a Canadian softwood Kraft pulp mill with an annual pulp production capacity of approximately 330,000 air-dry-tons. These options are focusing on producing one to two types of Phenol Formaldehyde resins based on functional group and/or molecular structure modifications of recovered lignin.

Highlights

- Multi-criteria methodology conceived for prefeasibility of integrated biorefineries
- Set of sustainability metrics developed for designing integrated biorefineries
- Bringing supply chain operational considerations to the strategic decision-making

B. Mansoornejad · B. Gilani · M. Benali (✉)
Natural Resources Canada, CanmetENERGY, Industrial Systems Optimization, 1615 Lionel-Boulet Blvd., P.O. Box 4800, Varennes, QC J3X 1S6, Canada
e-mail: marzouk.benali@canada.ca

S. Sanaei · D.R. Batsy · P.R. Stuart
Department of Chemical Engineering, Polytechnique Montreal, Succ. Centre-ville, P.O. Box 6079, Montreal, QC H3C 3A7, Canada

M. Rabaçal et al. (eds.), *Biorefineries*, Lecture Notes in Energy 57,
DOI 10.1007/978-3-319-48288-0_8

- Stepwise methodology developed for supply chain strategic design
- Assessment methodology was demonstrated for lignin recovery and postprocessing

8.1 Introduction

Forest products industry transformation is close to a reality. The emerging bioeconomy provides forest products industry with promising opportunities to generate new revenue streams by taking advantage of abundant, high-quality and renewable forest resources. Moving forward and establishing biorefinery strategies imply that companies have to identify which bio-based products are commercially attractive and how to produce them cost-efficiently with minimum technological, market, financial, and environmental risks. As a result, a number of key questions must be answered to support the decision-making:

- Which products are the most interesting given the biomass availability, existing infrastructure, and access to energy and water resources?
- What are the available processes to produce the candidate products, and which coproducts could also be considered?
- How should one deal with complex and interconnected range of technological, environmental and economic issues, as well as operational constraints that compete, making trade-offs unavoidable?

Implementing Integrated Forest Biorefinery (IFBR) introduces conflicting challenges that must be systematically addressed. The design of an IFBR is therefore a highly complex problem (Andiappan et al. 2015). One of the approaches that can be considered to identify the promising biorefinery strategies and their long-term implications is to evaluate decision criteria and to assess the implementation strategies using a Multi-Criteria Decision-Making (MCDM) framework, which is a panel-based decision-making technique. Process Systems Engineering (PSE) tools are applied to, first, analyze each biorefinery strategy from a specific perspective, e.g., economic, environmental, market, and then, provide relevant criteria for decision-making.

The systematic use of PSE tools for biorefinery design has gained an increased interest, as evidenced by the papers published in topical journals. Sammons et al. (2008) proposed a systematic framework for optimizing product portfolio and process configuration in integrated biorefineries. It determines the variable and fixed costs using yield, conversion and energy usage for each process model. Process integration tools optimize process performance in terms of energy usage and material consumption. An optimization model enables the framework to verify whether a certain product should be sold or processed further, or which processing route must be pursued if multiple production pathways exist for a special product. Kokossis and Yang (2010) presented a hierarchical cascade which explains the roles of systems tools in the design of biorefineries. In this cascade, systems approach supports the evaluation of techno-economic trade-offs and determines robust product portfolios to match market, process , and feedstock uncertainties. It also systematically assesses retrofit initiatives. Sharma et al. (2011) developed a decision-analysis framework for technology and product portfolio design for a biorefining enterprise. It considers the operational, economic, environmental , and social aspects of a project by utilizing flexibility, structural evaluation, environmental and social life-cycle assessment modules. A financial planning model based on an MILP is used to maximize the value of the business model and stakeholder interest by selecting appropriate feedstocks, technologies, and products. Sukumara et al. (2012) proposed a multidisciplinary methodology for the preliminary analysis of region-specific biorefinery models. It assesses the amount and type of feedstock available in the region. Process simulation estimates the maximum amount of biofuels that can be produced by the available feedstock. Given the capacity of biorefinery, potential locations, and feedstock options, the optimal transportation network is determined using a supply chain (SC) model. Capital and operating costs are estimated. The results can be used to assess environmental and social impacts. Finally, the results are analyzed based on profitability measures. Sharma et al. (2013) presented a strategic optimization model that optimizes the value of a multiproduct biorefinery to its stakeholders by making optimal decisions with respect to feedstock mix, product portfolio, technologies and process configuration, and capital structure for financing. Santibanez et al. (2014) presented a model to design and plan biorefinery SC. They incorporated economic, environmental and social performance data in the evaluation of SC design results in order to evaluate them from a sustainability perspective. Kokossis (2014) proposed a systems methodology which assists in developing integrated biorefineries. Decisions are made at three stages; chemistry and product portfolio selection, setting efficiency targets for individual processes, and developing flowsheets for selected processes. Considering the effect of design changes in retrofit and the result of LCA studies, the processing paths are screened from a background of chemistries that link core process chemicals to intermediates and end products. Selected paths are integrated to the core process. Mass and energy balances of the new configuration are developed for further LCA studies. Finally, new LCA and sustainability analyses are performed. Kasivisvanathan et al. (2014) have presented a robust MILP with input–output model to support decision-makers in addressing process synthesis problems due to uncertainties that arise from variation in feedstock supply and product demand. The goal is to maximize economic performance and to minimize environmental impacts for a polygeneration and palm oil plant, based on the concept of "multifunctional energy system". The MILP takes into account technical performance (i.e., energy efficiency, yield, etc.), sustainability and economic metrics. Zhang et al. (2014) proposed a superstructure that employs a bi-criteria MINLP model to maximize the economic performance (net present value) and minimize the global warming potential (based on LCA procedures) of a hydrocarbon biorefinery that involves fast pyrolysis and hydro-processing. Sandin et al. (2015) explored how choosing different LCA methods for allocating environmental impacts of biorefinery processes would affect the decision-making with regards to biorefinery projects. They tested six methods and investigated the decision contexts in which the choice of method is important.

The decision as to what biorefinery strategy to pursue depends on several factors, many of which cannot be well reflected in an optimization problem, e.g., understanding market strategies, emerging products and technologies, capabilities of existing supply chain assets, and potential partners. The practical way is to employ a hierarchical methodology that addresses these factors in a stepwise manner. The methodology presented in this chapter, which is the extension of a previous version introduced by Mansoornejad et al. (2010), presents a systematic approach for assessing the biorefinery alternatives using PSE tools. It includes market analysis, techno-economic study, LCA, SC analysis, and MCDM. There are similar works in the literature that have employed different PSE tools to assess biorefinery alternatives from different sustainability perspectives and then have employed MCDM to aggregate a set of conflicting decision criteria in order to identify promising biorefinery strategies (Sugiyama et al. 2008; Othman et al. 2010; Schaidle et al. 2010). However, in those studies among the employed PSE tools, SC analysis is missing and consequently the level of flexibility associated with biorefinery alternatives have not been addressed in their decision-making process. The proposed hierarchical methodology incorporates process flexibility analysis as a result of SC analysis into the decision-making process.

In this context, forest products companies envisage generating additional revenues in economic activity through producing new products and entering new markets. To achieve these goals, they plan to implement biorefinery and want to identify profitable product/process portfolios, robust SC strategies, and promising economic and environmental implementation strategies, given its competitive position. PSE tools are employed in a stepwise manner to evaluate each strategy through providing criteria and assigning score to each strategy from a specific perspective. The unpromising strategies are screened out in each step based on the scores. Finally, an MCDM approach is used to identify the most promising strategy among the remaining strategies taking into account the provided conflicting criteria.

8.2 Methodology

In biorefinery strategic design, long-term decisions should be made. Such decisions include the type of products that should be produced, the technologies that can be used, the number, location, and capacity of each type of facility, and the targeted markets. Multi-objective optimization is a tool that can be used to solve such a multifaceted problem. However, in a practical problem, it is sometimes difficult to address market strategies, emerging products and technologies, capabilities of existing SC assets, and potential partners within a single optimization framework. Involving these factors in an optimization problem complicates the solution process and increases the computational intensity of the problem. Instead, a hierarchical methodology that can address all these factors in a stepwise manner is more applicable. Because of the combinatorial aspect of such design problems, the hierarchical methodology might miss the global optimum. However, this methodology does not seek to identify a global optimum, but a set of feasible and practical biorefinery options (near-optimal solutions) that a company can strategically select. Many of these aspects can be addressed in different scenarios instead of being modeled into an optimization formulation. In this way, a simpler model with practical considerations will be solved.

The integration of sustainability indicators at early stage into biorefinery process design contributes in minimizing the possible negative impacts, rather than performing later corrective and costly modifications (i.e., retrofitting). Hence, it is important to apply a systematic and practical methodology for evaluating the various integrated biorefinery strategies from sustainability, technology, and market risk mitigation perspectives. Companies seek a set of biorefinery options that would improve their profitability. This should include the optimum and/or near-optimum solutions that can be pursued by a company in parallel with potential partners to establish mutual interests and to address most effectively the competitive disadvantages of forestry companies, such as lack of capital. This requires applying a methodology that identifies the most promising biorefinery option from a specific company's point of view, considering all the complexities involved in the industry sector.

To achieve a hierarchical methodology, the strategic decisions must be made through the integration of different methodologies. The potential set of products is determined by a product portfolio definition methodology (Chambost and Stuart 2007). Then, technologies that can be employed to produce the potential products are chosen through a techno-economic study (Hytönen and Stuart 2011). The result will be a set of product/process portfolios screened out from the non-profitable ones based on a preliminary market analysis and techno-economic assessment. From this point, two separate analysis tools, one related to SC and one related to LCA, are utilized to analyze the remaining portfolios. Each of these analyses, along with product portfolio definition and techno-economic study, provide several metrics that can ultimately be used in an MCDM. MCDM considers several criteria provided from different analysis tools and helps in making decision by analyzing biorefinery options from different perspectives. The hierarchical methodology is presented schematically in Fig. 8.1.

8.2.1 Product Portfolio Definition

There are numerous product options offered by biorefinery which will add value to the existing product portfolio of a forestry company. However, the question is which set of products is the most promising for the implementation of biorefinery in the long term. A key criterion in targeting the product portfolio of a future biorefinery is the assessment of current market needs, though considering the available technologies. Another crucial issue that must be taken into consideration in defining

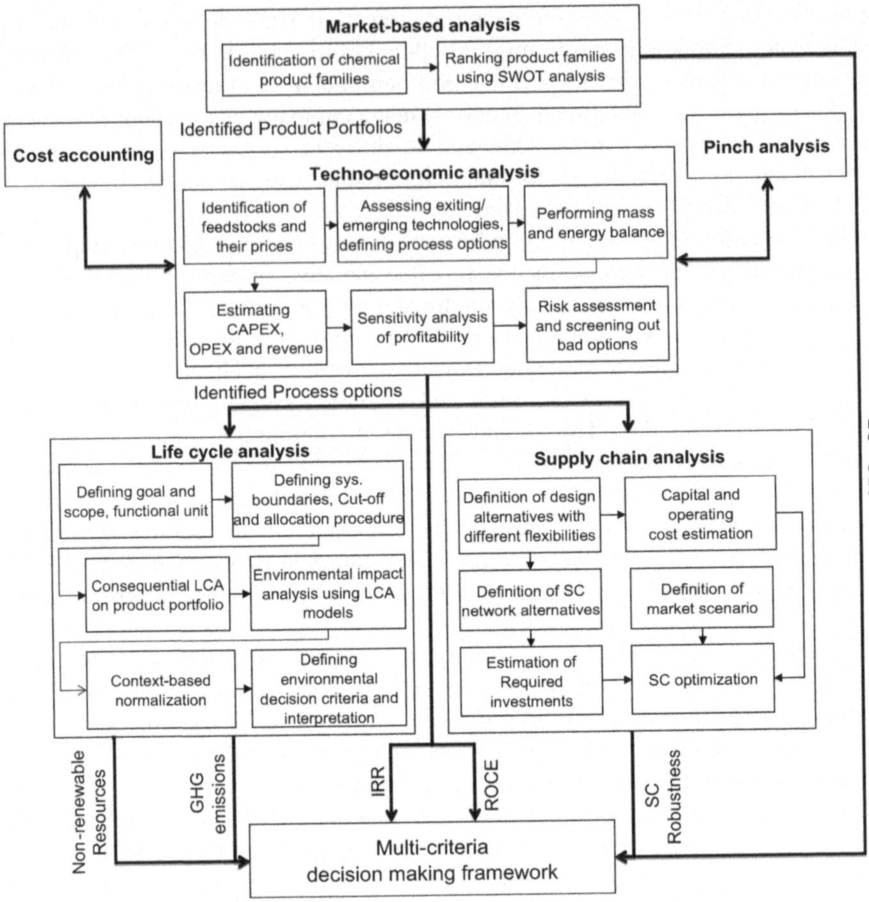

Fig. 8.1 Hierarchical design methodology

a product portfolio is to include building block chemicals and their value-added derivatives. This increases the flexibility of the company, i.e., its adaptability to market volatility and will improve its competitive position in the long run. In this regard, the financial return should not be the only decision-making factor and long-term market position, implied changes in the existing production system and supply chain, and process and supply chain flexibility must be addressed as well (Chambost and Stuart 2009).

Product portfolio definition identifies sets of product portfolios that can potentially be produced in a biorefinery in retrofit to a pulp and paper mill. Few market criteria will be provided by this methodology to be used in the MCDM. Chambost and Stuart (2009) developed a two-stage methodology for the definition of product portfolios. The goal of this step is to identify a list of potential product families for a market region. Products included in the list comprise replacement and/or substitution chemicals and they are analyzed based on

- Identifying the product/process/partners combinations
  - A list of potential products, processes, and partners (procurement, technology provider, transportation, sales) is defined based on the competitive position of the company, i.e., access to feedstock, potential markets in its vicinity, existing processes, and supply chain assets, and financial position;
  - Then the large list of identified products is refined based on market, economic and product specific information such as product functionalities, market size and growth, market saturation and basic margins;
- Ranking product families.

In the second stage of the methodology, product families are ranked using a SWOT analysis which helps identifying strengths, weaknesses, opportunities, and threats of each product within each family. This step assesses risks associated with each family. The outcome of this methodology is a set of ranked product portfolios and the risks associated with them defined based on company's competitive position. It also provides a set of market-related metrics that can be further used in the MCDM. In the next steps of the hierarchical methodology, product portfolios are evaluated using PSE tools which analyze return on investment (techno-economic analysis), environmental performance (LCA), flexibility, i.e., which set of products introduces a better potential for flexibility (SC analysis), and sustainable partnership models (SC analysis).

8.2.2 *Techno-Economic Analysis: Large-Block Analysis*

Techno-economic analysis is carried out through a large-block analysis (LBA) method, which considers processes and their mass and energy balances as large blocks (input–output model). It identifies feasible technologies that can be employed for producing products targeted by product portfolio definition methodology (Dieudonné et al. 2013; Hytönen and Stuart 2010). The LBA constitutes four steps

- In the first step, the available feedstocks in the mill region and their prices are identified;
- Then, the existing and emerging technologies that can use the identified feedstock and produce the targeted products are identified and their mass and energy balance models are developed. This way, the process options are defined;
- Next, capital and operating costs, as well as revenues and profitability are estimated as a function of plant capacity;
- Finally, risk assessment is carried out. The uncertain economic variables which will probably have major impact on the economic performance of the project are

identified. For this purpose, sensitivity analysis of internal rate of return (IRR) is performed by varying economic variables in order to identify the sensitive factors. The probability distribution of risky variables is defined and the IRR and its probability are calculated using the defined distributions.

Through this methodology, the unprofitable process options are screened out. Moreover, apart from IRR, other relevant criteria such as return on capital employed (ROCE) are generated in this step to be used in the final MCDM. Cost accounting techniques are employed to reflect the cost components of total production cost into profitability estimation so that an accurate estimation is achieved. The combination of product portfolios, process technologies, and their phased implementation, i.e., the definition of phases in which the strategies are implemented, form the biorefinery strategies and are further analyzed by other tools that are introduced in the following sections.

8.2.3 Life-Cycle Assessment

Sustainability has recently gained attention in biorefinery process design and implementation, including environmental, health, and safety hazard identification, and life-cycle impact assessment (LCIA) in the decision-making towards multi-criteria optimality. Life-cycle assessment (LCA) is considered as an appropriate tool for assessing the environmental sustainability of technological options due to its capability to evaluate the potential effects on the ecosystem as well as on population and human health that might endanger the current and future generations. In fact, LCA is a technique that assesses potential environmental impacts associated with all stages of a service or a product's life cycle, from "cradle to grave". LCA is preferable for the biorefinery projects and can be used as a tool to evaluate options and projects of replacing fossil-based fuels and products with by bio-based fuels and bio-based products. Moreover, LCA can be employed to incorporate environmental objectives and criteria in decision-making process to help decision-makers with making more informed decisions. Nevertheless, in this project, the LCA is conducted on a "cradle-to-gate" basis since it is assumed that bioproducts and competing products will have the same end-use application, same end-of-life treatment, and same final disposal. LCA studies are categorized into two main types: attributional (ALCA) and consequential (CLCA). Modeling principles of ALCA and CLCA are relatively the same, but what mainly distinguishes the two methods is the choice of processes to be included in the system boundary, which in turn, results from a clear definition of the study goal. Actually, ALCA assesses the environmental impacts of a given product or service through current (existing) and established process or service while CLCA generates information that describe the consequences of future decisions (e.g., implementation of new process or modification of existing process). ALCA is useful for hot spot identification, which at this moment is not possible with CLCA. The main characteristic of CLCA approach is the inclusion of processes to the extent of their expected changes due to a new or future decisions as well as the application of system expansion to include the co-products in multiproduct systems.

A practical LCA methodology is required for the evaluation of biorefinery processes more broadly considering the product diversity portfolio. Particularly for integrated biorefinery projects with multiple bioproducts, CLCA provides more detailed information for decision-making in cases where most of the consequences occur outside the life cycle of biorefinery processes and products under investigation. LCA assesses the environmental impacts of strategies by translating mass and energy quantities into local, regional, and global environmental impact categories, and provides several criteria for decision-making. The general LCA approach is depicted in Fig. 8.2.

LCA methodology includes data collection for the biorefinery processes along with the definition of goal and scope, functional unit, and system boundaries. The functional unit refers to the output or product of a process or a system. The life-cycle inventory and life-cycle impacts are calculated for a given reference flow stream. Then, depending on the context of the analysis, system-based cut-off procedures and system-based allocation procedures can be applied to exclude the impacts of similar process steps, and as a result to properly allocate the environmental burden (consequences due to changes) to the appropriate products (or product portfolios). CLCA is performed on product portfolios to study the incremental environmental consequences of future decisions related to the integrated biorefinery projects. Main steps in the life cycle of a biorefinery process consists of: raw material acquisition and extraction from natural resource including biomass harvesting and preparation, bioproducts production through biorefinery processes including biochemical and thermochemical pathways, product transportation and distribution and ultimately product recycling, reuse or final disposal. Following the methodology of consequential LCA, environmental impacts through the life cycle is assessed using a "cradle-to-gate" perspective for production pathways, which are defined for valorization of the biorefinery products in this study.

Modeling of processes and impact assessment are carried out using SimaPro 8.0 Multiuser LCA software and IMPACT 2002+ (version 2.15), respectively. Regarding the Life-Cycle Inventory (LCI) database, Ecoinvent AmN is employed. This database is developed by Interuniversity Research Centre for the Life Cycle of

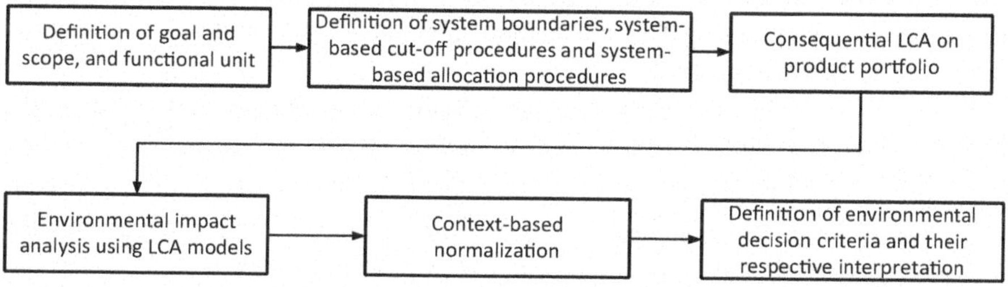

Fig. 8.2 General view of LCA approach

Products, Processes and Services (CIRAIG), to adapt the international Ecoinvent database to the North-American context. The life-cycle impact assessment (LCIA) is performed using specialized software such as SimaPro, which translates life-cycle inventory to environmental impacts. SimaPro is the world's most widely used LCA software. Finally, the results are normalized to be used in the MCDM. In this way, environmental aspects of a strategy are quantified in terms of several criteria such as greenhouse gas (GHG) emissions and nonrenewable energy resources (NRR) provided in kilo-ton CO_2 equivalent (translated in terms of light vehicles in the Canadian context) and MJ of fossil-based energy consumption per MJ of coal-based available reserves, respectively. Sources of data for the life-cycle inventory include mass and energy balances data, and data from biorefinery technology providers.

The proposed methodology also enables incorporating policy instruments as inputs for both economic modeling and MCDM analysis.

8.2.4 Supply Chain Analysis

The SC analysis, in general, can be used at both operational and strategic level decision-making. At the operational level, SC analysis carries out product planning over different time horizons and identifies trade-offs between product orders and anticipated supply and demand. It calculates the profit across the entire SC and accounts for cost contributors that are typically ignored in economic analyses, e.g., detailed variable inventory cost, changeover cost, etc. It can also be used to take into consideration market volatility, and to determine how the flexibility of the production system can be exploited to mitigate market risks in order to maximize profit. Moreover, for the long term, companies should design their SC based on the effect of the design on operational activities. As mentioned earlier, SC analysis can help identifying how the flexibility of a system must be exploited to maximize the profit in a volatile market. Using this capability, SC analysis can be used, at the strategic level, to target the level of flexibility of a production system needed to mitigate the risk of market volatility. Mansoornejad et al. (2013) demonstrate how process design and SC design decisions are interrelated and decision-making in each domain must be executed with respect to the other. Furthermore, SC analysis provides a better insight into the costs and profit incurred by an implemented strategy. Thus, an SC analysis can be used, not only for making short-term, operational decisions related to the management of the SC, but also for making long-term, strategic decisions.

In this regard, the SC analysis proposed in this work has two major parts; designing the production flexibility and designing the SC network. The objectives of the SC analysis are

- Targeting the level of process flexibility, including determining the production capacity as well as the operating window as a design target;

- Designing SC network, including determining the number, the location and the capacity of facilities, as well as selecting partners.

Supply chain analysis includes a SC optimization formulation that aims at maximizing SC profit by identifying the trade-offs between demand and production capabilities, and by finding the optimal alignment of production capacity and market demand. The SC optimization model, proposed by Mansoornejad et al. (2013), is a multi-period, multi-product, multi-echelon model that considers feedstock price and availability, production costs, inventory and delivery costs, as well as product price and demand. Taking this information into account, the SC optimization framework exploits the potential for flexibility and determines which orders must be fulfilled, and therefore, how much of each product must be produced, how they should be stored, and how they should be delivered to maximize SC profit.

The SC optimization model is a tactical model that has some operational components. The model divides each time period into several hours that can be dedicated to production, changeover, and maintenance. In this way, a better cost representation can be made by the model. Different types of biomass provided by several suppliers are defined. Processes are either dedicated, i.e., they produce only one product, or flexible from a product perspective, i.e., they are able to produce several products through different production modes or "recipes". In other words, a flexible process can use different recipes to produce different products in different production modes. Changing from one recipe to another incurs changeover cost and time. Processes can be idled or temporarily shut down for scheduled maintenance. Warehouses can receive material, either feedstock or product, from different sources and plants, and supply different markets. Each market places order in two ways: by contract, i.e., for the long term, and on the spot market, i.e., for the short term. In case of a contract, specific quantities of products must be delivered to the customer in specific time periods. In other words, the contractual orders must be either fulfilled up to a certain prespecified level, or be declined. The spot demand can be partially/completely fulfilled or declined. Transportation routes link suppliers, facilities, and customers together. The model is formulated as an MILP problem with a discrete time horizon of 48 weeks. Each time period is broken down into hours. Several subsets have been created to link parameters and variables to each other. For instance, processes can only produce certain materials. Each supplier may provide a specific type of feedstock and each customer may need a specific product. Production plants may be able to link with some specific suppliers and markets. This, in general, will reduce the possible options and thus, the complexity of the problem. The main decision variables of the model include

- Binary variable for contract selection.
- Binary variable representing the recipe used on a process. Another binary variable is used to determine whether or not a recipe is used consecutively on a process. This binary variable helps considering the changeovers.
- Amount of material processed and produced by each process.

- Flow of materials, i.e., feedstock and product, to and from the mill.
- Number of hours taken by each recipe in a specific process.
- Amount of energy produced and/or consumed by each process.

In this methodology, first, process design alternatives representing different potentials of flexibility are defined. In the second part, SC network alternatives are defined based on the assets of the existing SC and resources that are needed for new products. Then the process and the SC network alternatives are combined to create a set of process-SC network alternatives, called combined alternatives. In the third part, the SC model is run for different levels of volume flexibility of each combined alternative, in case of several market scenarios. The SC profit of each combined alternative is calculated at the operational level over a year, and by considering profit and capital costs, profitability associated with each level of flexibility of each alternative is estimated. Additionally, the robustness of each alternative against all market scenarios is determined using a relevant metric.

The level of flexibility of the alternative that has the best performance is set as the targeted level of flexibility. In the fourth part, several implementation strategies are defined based on the targeted level of flexibility. The advantages and constraints of the mill must be considered when defining strategies through discussions involving the executive board of the mill. SC model is run for each strategy in case of several market scenarios and the performance of each strategy at the operational level is evaluated. Performance metrics, i.e., profitability, robustness, and flexibility, are used to evaluate the performance of the alternatives. The SC analysis consists of four main steps

- Process design: Designing possible levels of flexibility (process alternatives);
- SC network design: Designing possible SC networks (SC network alternatives);
- Targeting the level of flexibility for each alternative using SC optimization;
- Generating implementation strategies based on defined process/SC network alternatives and evaluating them for making the final decision.

In the first step of the methodology, process design alternatives representing different potentials of flexibility are defined. The maximum possible capacities for each process are identified considering three major factors: market demand, feedstock availability, and technical barriers, e.g., maximum possible equipment size, capacity bottlenecks. Then the production system is characterized in terms of

- Overall process configuration, i.e., whether the targeted products are produced via processes in series or in parallel;
- Product flexibility, i.e., whether each process is dedicated to one product or whether several products can be produced by a single process in different production modes;
- Volume flexibility, i.e., whether the process can handle a range of production rates and whether the inherent flexibility of the process is enough or not.

Based on this characterization, process alternatives are defined and their base-case capital cost and operating cost are estimated.

In the second step, SC network alternatives are defined based on the assets of the existing SC and resources that are needed for new products. In this step, the specifications of the new SC must first be identified with respect to targeted product/process options taking into account the potential synergies between the new SC and company's existing SC. More specifically, it must be verified whether the inventory and the transportation system of the existing SC have enough capacity for more biomass as well as new products and whether they are consistent with the properties of the new products. Then, SC network alternatives are defined, representing different options in terms of partnership, location and capacity of warehouses and distribution centers, and transportation network. Finally, the capital cost associated with each alternative is estimated.

In the third step, the flexibility of each design alternative is targeted based on its performance in different market conditions. The process and the SC network alternatives defined in the two previous steps are combined to create a set of process-SC network alternatives, called combined alternatives. Several market scenarios are generated to represent market volatility in terms of product price and demand changes. The SC model is run for different levels of volume flexibility of each combined alternative, in case of several market scenarios are considered. The SC profit of each combined alternative is calculated at the operational level over a year, and by considering profit and capital costs, profitability associated with each level of flexibility of each alternative is estimated. Additionally, the robustness of each alternative against all market scenarios is determined using a relevant metric, which is presented further in detail. The level of flexibility of the alternative that has the best performance is set as the targeted level of flexibility.

In the fourth step, several implementation strategies are defined based on the targeted level of flexibility. The advantages and constraints of the mill must be considered when defining strategies through discussions involving the executive board of the mill. SC model is run for each strategy in case of several market scenarios and the performance of each strategy at the operational level is evaluated. Performance metrics, i.e., profitability and robustness, will be later used in the MCDM along with other metrics.

SC metrics, showing the SC performance in a dynamic market, have been developed to quantify the SC robustness and flexibility against market volatility (Mansoornejad et al. 2013). A metric of robustness (MR) (Eq. 8.1) is defined to quantify the robustness of each alternative under volatile market conditions. The metric represents an aggregate deviation of the downside profits from the base-case profit. The lower the deviation is, the more robust the system will be. Moreover, a metric of flexibility (MF) (Eq. 8.2) has been used, which shows the extent to which the production volume diverges from the nominal production rate. For further details on the metrics, the reader is referred to Mansoornejad et al. (2013).

$$\mathrm{MR} = \left(\sum_{\mathrm{Sc}} (\mathrm{Pr_B} - \mathrm{Pr_{Sc}})/\mathrm{Pr_B}\right)^{-1} \tag{8.1}$$

$$\mathrm{MF} = \sum_{p} \sum_{m} \left(\sum_{t} \left(C_{mpt} - C^{N}_{mp} \right) / \sum_{t} C^{N}_{mp} \right) \quad (8.2)$$

$\mathrm{Pr_B}$ is the base-case profit, $\mathrm{Pr_{Sc}}$ is the profit for scenario Sc, C_{mpt} is the amount of product m produced on process p in a time period t and C^{N}_{mp} is the amount of product m produced on process p by the nominal production rate.

8.2.5 *Multi-criteria Decision-Making*

Multi-criteria decision-making (MCDM) is a systematic tool which facilitates obtaining the right decision by a multidisciplinary panel through identifying the promising biorefinery strategies using a set of conflicting decision criteria. MCDM is performed following several steps within two phases including a pre-panel and a panel phase (Munda et al. 1994; Munda 1995). The pre-panel activity is to review the decision context, define the objectives of the decision-making, define the alternatives and introduce the decision criteria. The panel activity is mainly to interpret the evaluated criteria by decision-makers, to identify the preference of the panel, to quantify the relative importance of the decision criteria, to calculate the final scores representing the performance of the design alternatives, and finally to rank the alternatives and interpret the results in order to make a final decision. At the most important part of the panel activity, that is for quantifying the relative importance (weight) of the decision criteria, first a most important criteria is selected by the panel and a target value is attributed to it. Then, the importance of each criterion is compared with it using the trade-off method and the weights of the criteria are calculated.

In the context of biorefinery, the use of MCDM to weigh the criteria by conducting a panel has received attention only a few years ago (Cohen et al. 2010; Hytönen and Stuart 2011). In this chapter, in addition to considering market, economic, and environmental dimensions for strategic decision-making introduced previously by Sanaei (2014), supply chain criteria has been also incorporated into a single MCDM framework which ultimately attributes a unique sustainability score to each defined biorefinery strategy and ranks them from a sustainability perspective. In the next section, a case study, which focuses on SC analysis, is presented.

8.3 Case Study

The methodology explained in the previous section is applied in a retrofit case study to identify the promising biorefinery strategies to be implemented in a Kraft pulp mill in Canada. To identify biorefinery technologies which can fit well with the studied mill, unique characteristics of the mill were identified. Considering that the recovery boiler is the bottleneck of pulp production processes, lignin precipitation was chosen as a candidate strategy (strategy A) which could ultimately result in lowering costs of the core business and potentially increasing pulp production capacity. Moreover, due to limited availability of the excess heat and power at the studied mill, another strategy would be the less energy-intensive technologies that can be integrated with the existing mill, among which fast pyrolysis, as an energy self-sufficient process, was selected as the second candidate strategy (strategy B).

Two phases of implementation were applied for each candidate strategy. Strategy A includes lignin precipitation in the first phase and converting precipitated lignin into two grades of phenol formaldehyde resins (PF resins) in the second phase. Strategy B consists of bio-oil production by fast pyrolysis in the first phase and lignin extraction from the produced bio-oil followed by converting the extracted lignin into the same grades of PF resins in the second phase. Given the two strategies and the two implementation phases, two process/SC network alternatives, i.e., four alternatives, are defined for each strategy (Fig. 8.3).

In strategies A-1 and B-1, the mill sends its products to a partner which processes lignin to produce resins and delivers them to the market. However, in strategies A-2 and B-2, the mill produces PF resins at the mill site and delivers the resins to the market. To have more flexibility, in strategies A-2 and B-2, two small lines of resin production are installed in parallel, so that the mill can produce both grades of resins at the same time according to market demand. The SC optimization model is run for different levels of process flexibility in case of several market scenarios in order to identify the optimum level of flexibility for each defined strategy. Market scenarios are defined in terms of product price and demand over one year (Fig. 8.4). The scenarios have been defined for a cycle of one year with

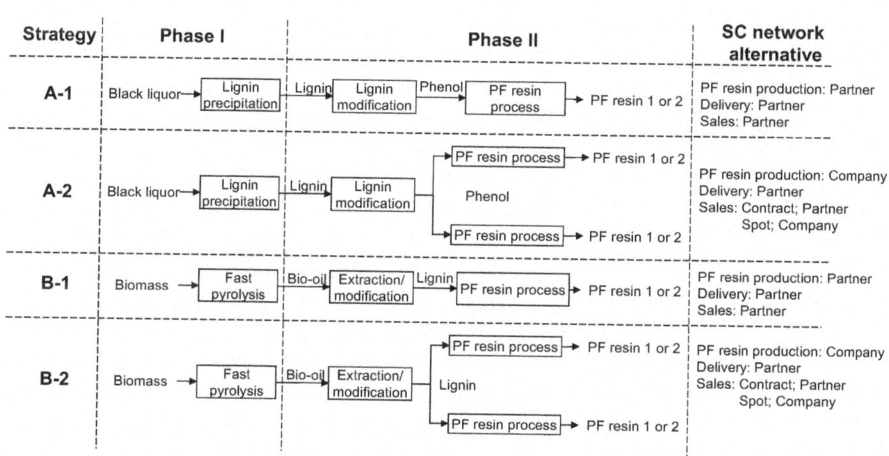

Fig. 8.3 Biorefinery strategies

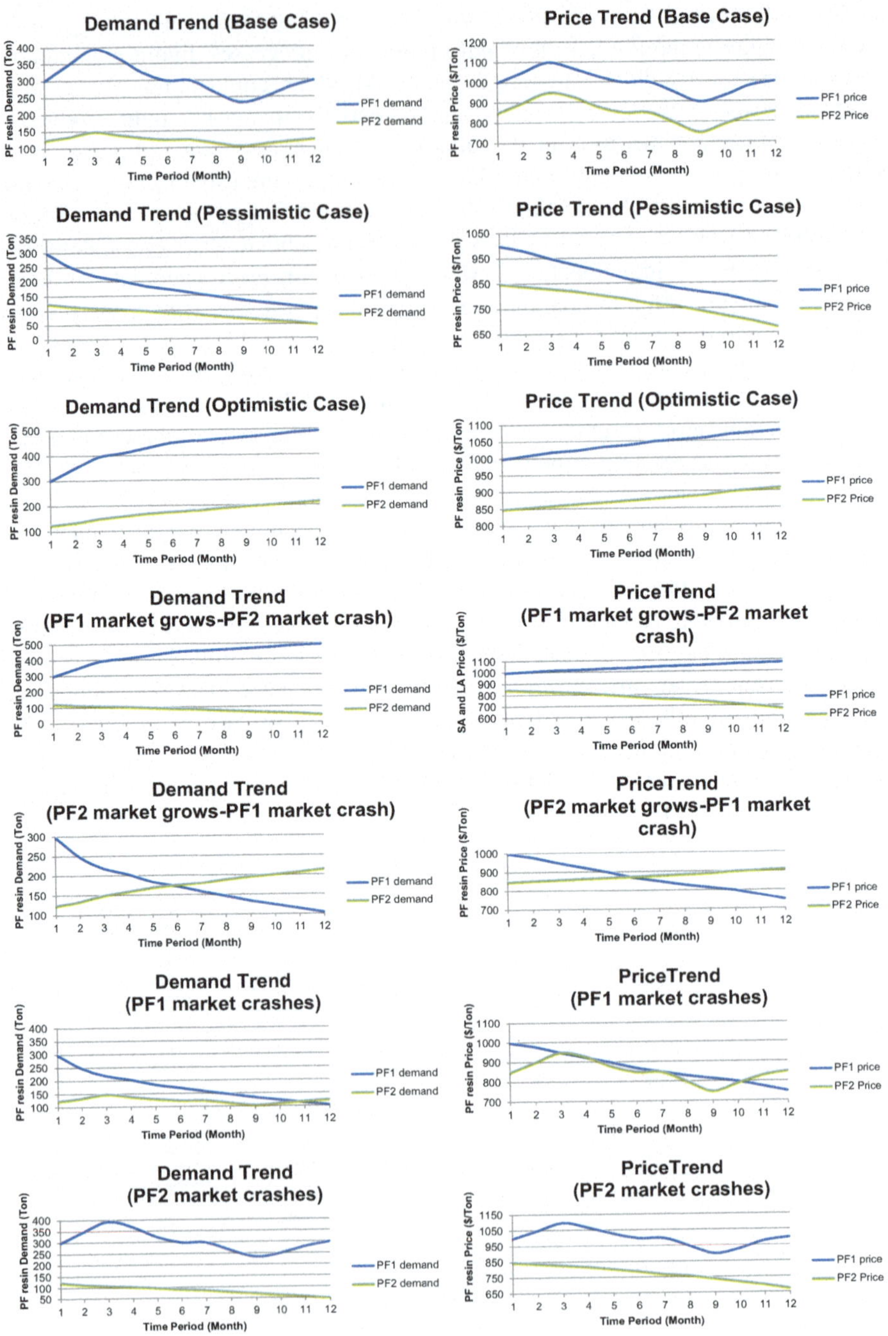

Fig. 8.4 Market scenarios

12 intervals of one month. In each scenario, there is a trend for price and demand that has a certain behavior given the type of the scenario, i.e., optimistic, pessimistic, etc.

The results of flexibility analysis is presented for strategy A-1 as an example. To increase the flexibility of a given process, the process bottlenecks must be identified and modifications in terms of increasing the capacity of equipment or adding new equipment ought to be carried out to increase the operating window of the process. However, this incurs higher capital costs. The flexibility analysis investigates if this increase in capital cost can be compensated by the profit increase that may result from increased flexibility. Profit is calculated by SC model for all market scenarios and for different levels of flexibility.

Figure 8.5a shows that as flexibility increases, profit increases as well. However, after a certain level of flexibility profit increase flattens, because there is not enough demand in the market to be fulfilled by the extra level of flexibility, thus having more flexibility does not change the profit. Figure 8.5b illustrates the capital cost required for each level of flexibility. From 0 to 20% flexibility, the increase in the capital cost is not significant, because with some slight process modifications the level of flexibility can be improved. However, in order to go beyond this level, major modifications are required to be done on the process, which incur more cost. Moreover, with more flexibility, more capacity will be available and the capital cost required for the SC will grow. As a result, the capital cost increases more sharply after 20% flexibility. Therefore, for higher levels of flexibility more capital cost is required (Fig. 8.4b). Thus, it should be verified up to what level of flexibility the rise in capital cost can be compensated by profit increase. For this purpose, profitability is calculated. The result of profitability analysis is illustrated in Fig. 8.4c. Internal Rate of Return (IRR) was chosen as the profitability metric. As shown in Fig. 8.4c, up to 20% flexibility, the increase in capital cost can be compensated by the profit increase. However, for levels of flexibility higher than 20%, the capital cost rise plays the major role and profit increase in that range is not enough to pay off the extra capital cost. Hence, 20% can be targeted as the optimum level of flexibility for this alternative. Same analysis is carried out for all strategies to determine the optimum level of flexibility.

Afterwards, SC model is run once again for the optimum flexibility level of each strategy in case of all market scenarios, and the average profitability and robustness of each strategy is calculated (Fig. 8.6). Strategy A-2 has a higher profit and capital cost compared to strategy A-1, though the higher profit cannot pay off the higher capital, thus it has a lower IRR (about 2% less). However, due to having more flexibility, it is more robust to market volatility. Strategy B-2's profit is higher than that of strategy B-1, but its capital cost is slightly higher. Hence, B-2 has a higher IRR. Moreover, because of being more flexible, strategy B-2 is almost 70% more robust than strategy B-1. Overall, strategy B-2 with higher profitability and more robustness is a better strategy than strategy B-1. In addition, strategy A-2 has a better robustness (two times more) and insignificantly lower IRR compared to

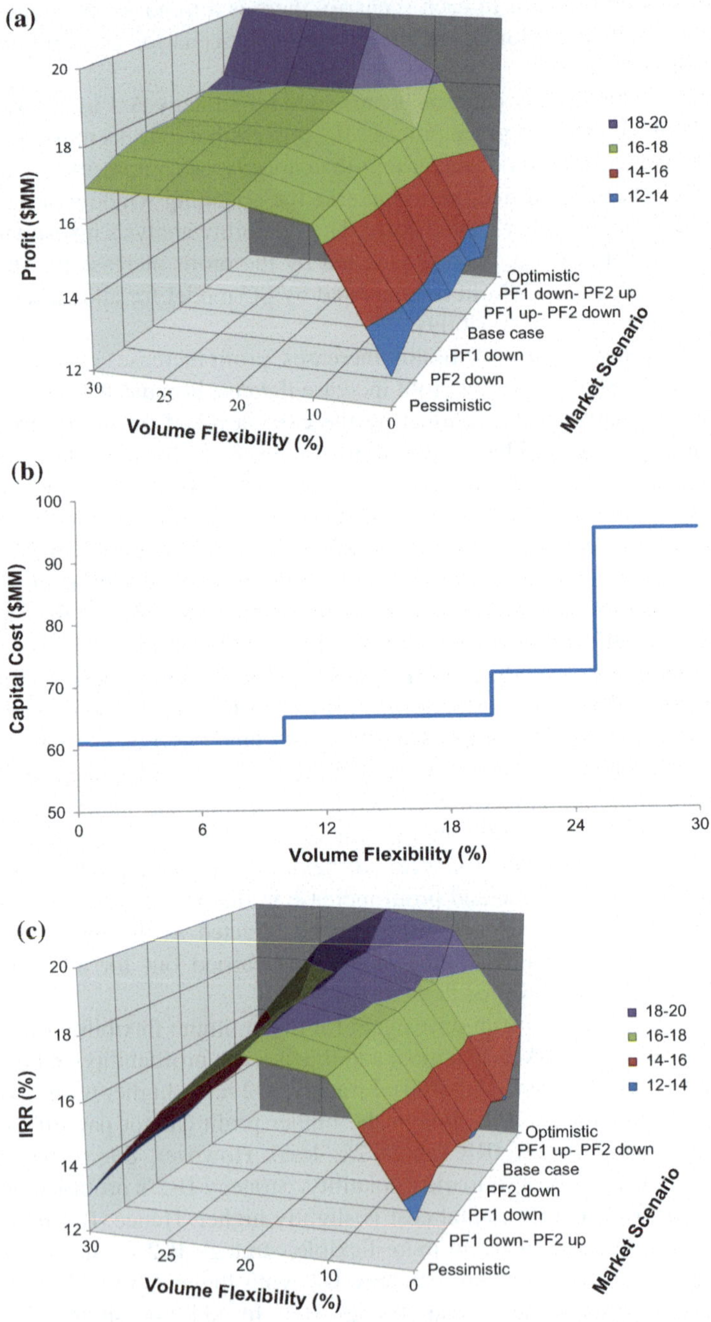

Fig. 8.5 Targeting level of flexibility: **a** profit-levels of flexibility-market scenarios; **b** capital cost versus level of volume flexibility; **c** profitability-levels of flexibility-market scenarios

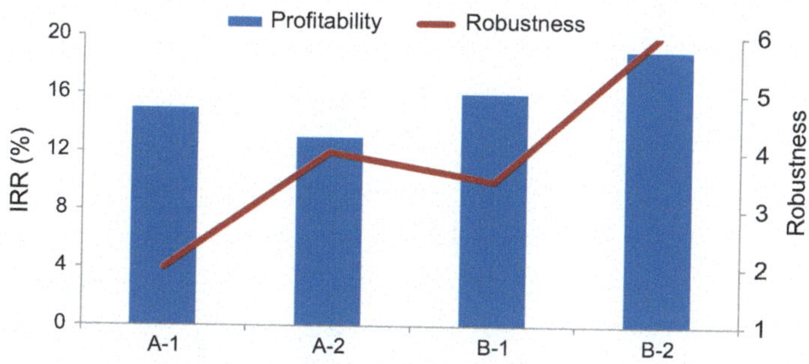

Fig. 8.6 Average profitability and robustness of strategies

strategy A-1. Thus, strategies A-2 and B-2 seem to be more promising in terms of profitability and robustness. In order to identify the most promising strategy between these two, they should be compared in other aspects of their performance. The considered decision criteria (Table 8.1) include two economic criteria representing profitability (IRR) and efficiency of investment (ROCE) of the strategies; one supply chain criterion (MR) reflecting the response of strategies to market conditions; two environmental criteria (GHG and NRR) as the most significant end-point impacts in LCA analyses; and two market criteria (QR and CPC) representing competitiveness of each strategy.

The evaluated criteria for these two strategies show conflicting results which brings complexity to decision-making. For instance, although strategy A-2 has less profitability (IRR), it shows better performance in market perspective (higher QR). Such conflictions confirm the necessity of applying the MCDM tool. A weighting factor (relative importance) is attributed to each decision criterion so that the criteria can ultimately be aggregated into a unique score for each candidate strategy, enabling decision-makers to make the final decision. Not surprisingly IRR was selected as the most important criterion, and using trade-off method the panel gave a weight to each criterion representing its relative importance compared to the most important criterion. The quantified weighting factors are presented in Fig. 8.6a. As shown in Fig. 8.6a, in the context of this study IRR and GHG are the first-ranked criteria with higher than 20% relative importance. In contrast, NRR got the lowest importance among the decision criteria. The low weighting factor of NRR is because both candidate strategies A-2 and B-2 (Fig. 8.6b) show much better performance than the acceptable level of NRR for environment, and thus this criterion could not help panelists to distinguish between the candidate strategies. Based on the normalized value of each criterion, an aggregated sustainability score was calculated for each of the two candidate strategies. This approach resulted in identifying strategy B-2 as the most sustainable strategy in this work with two times higher sustainability score (Fig. 8.7).

Table 8.1 Set of sustainability criteria for decision making

Criteria	Interpretation	Metric
IRR: internal rate of return	Measures the profit/risk ratio under normal market conditions. This ratio should normally be greater than 20%, the minimum target for profitability over the short-term, before considering probable future technology improvements	$NPV = \sum_i CF_i/(1+IRR)^i = 0$
ROCE: return on capital employed	Measures the cash flow generated relative to the invested capital. Higher ROCE is preferred since it indicates better return on invested capital.	$ROCE = EBIT/CE$
MR: metric of robustness	Measures the capability of the system in not deviating from the base-case performance in a volatile market	$MR = \left(\sum_{Sc} (Pr_B - Pr_{Sc})/Pr_B\right)^{-1}$
GHG: greenhouse gas emissions	Measures the potential contribution of a biorefinery project to climate change through GHG emissions in terms of the number of cars equivalent	$\frac{\text{GHG emissions(kg } CO_2 \text{ equivalent)}}{\text{Yearly GHG emissions per average light car(kg } CO_2 \text{ equivalent)}}$
NRR: non-renewable resources	Measures the amount of non-renewable resources consumed or saved by the project in terms of energy compared to the reference case	$\frac{\text{Fossil based energy consumption(MJ)}}{\text{Coal based available energy(MJ)}}$
CPC: competitiveness on production costs	Shows how competitive the biorefinery alternative is compared to the current benchmarked production	$\frac{\text{Production cost benchmark–Current production cost}}{\text{Production cost bencmark}} \times 100$
QR: quality of revenue	Measures the ability of biorefinery strategy to maintain strong margins due to added value in a diversified product portfolio. The greater the value of QR the stronger the biorefinery strategy	$QR = \frac{\text{Revenue from added value products}}{\text{Total revenue}} \times 100$

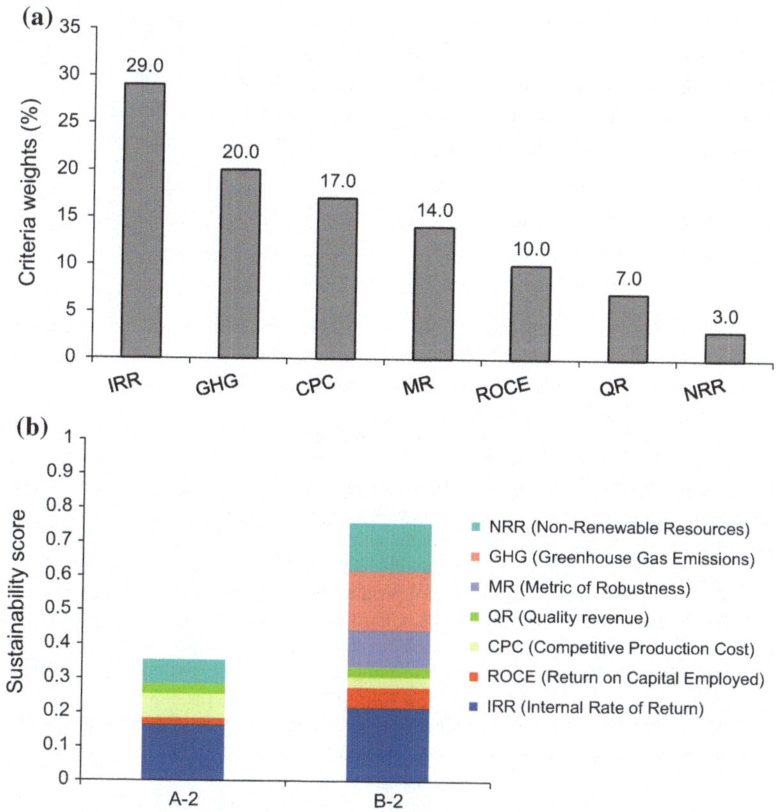

Fig. 8.7 **a** Weights of sustainability criteria; **b** sustainability scores of strategies

8.4 Concluding Remarks

Biorefinery processes need to be designed and optimized toward the sustainable competitiveness goal. They can be integrated into different forest industry processing facilities. Given the characteristics of the host facility, multiple design and implementation strategies are possible. The proposed methodology considers systematically all conflicting criteria that can enable decision-makers to choose the right strategy. Each criterion is calculated by a specific, separate methodology, e.g., techno-economic analysis, SC analysis, LCA. Evaluation criteria are weighted according to their perceived importance. A set of metrics and equations is proposed for the evaluation of GHG emissions in a biorefinery value chain. The result is a set of scores measuring the extent to which each biorefinery strategy achieves the desired objectives such as lowest capital cost, highest production performance, and best LCA-based environmental footprint. From the prescreening stage, it was concluded that two biorefinery strategies (A1 and B1) should be screened out. Despite the promising market perspective of strategy A2, when environmental performance is considered, this strategy becomes less attractive as compared to the strategy B2. Therefore, when combining market-based and techno-economic assessments with life-cycle and supply chain analyses, strategy B2, that is based on fast pyrolysis of lignocellulosic material with lignin extraction and its functional modification to produce biosourced phenol formaldehyde resin, is the most preferred option.

References

Andiappan V, Ko ASY, Lau VWS, Ng LY, Ng RTL, Chemmangattuvalappil NG, Ng DKS (2015) Synthesis of sustainable integrated biorefinery via reaction pathway synthesis: economic, incremental enviromental burden and energy assessment with multiobjective optimization. AIChE J 61:132–146. doi:10.1002/aic.14616

Chambost V, Stuart PR (2007) Selecting the most appropriate products for the forest biorefinery. Ind Biotechnol 3:112–119

Chambost V, Stuart PR (2009) Product portfolio design for forest biorefinery implementation at an existing pulp and paper mill. In: El-Halwlagi MM, Linninger AA (eds) Proceedings of the 7th international on the foundations of computer-aided process design, pp 907–916

Cohen J, Janssen M, Chambost V, Stuart PR (2010) Critical analysis of emerging forest biorefinery (FBR) technologies for ethanol production. Pulp Pap Can 111:24–30

Dieudonné RB, Solvason CC, Sammons NE, Chambost V, Bilhartz DL, Eden MR, El-Halwagi MM, Stuart PR (2013) Product portfolio selection and process design for the forest biorefinery. In: Stuart PR, El-Halwagi MM (eds) Integrated biorefineries: design, analysis, and optimization. CRC Press, Taylor & Francis Group, New York, pp 3–35

Hytönen E, Stuart PR (2010) Biofuel production in an integrated forest biorefinery—technology identification under uncertainty. J Biobased Mater Bioenergy 4:58–67. doi:10.1166/jbmb.2010.1066

Hytönen E, Stuart PR (2011) Techno-economic assessment and risk analysis of biorefinery processes. In: Pistikopoulos EN, Georgiadis MC, Kokossis AC (eds) Proceedings of the 21st European symposium on computer aided process engineering, pp 1376–1380

Kasivisvanathan H, Ubando AT, Ng DKS, Tan RR (2014) Robust optimization for process synthesis and design of multifunctional energy systems with uncertainties. Ind Eng Chem Res 53:3196–3209. doi:10.1021/ie401824j

Kokossis AC (2014) Design of integrated biorefineries. Comput Aided Chem Eng 34:173–185. doi:10.1016/B978-0-444-63433-7.50018-3

Kokossis AC, Yang A (2010) On the use of systems technologies and a systematic approach for the synthesis and the design of future biorefineries. Comput Chem Eng 34:1397–1405. doi:10.1016/j.compchemeng.2010.02.021

Mansoornejad B, Chambost V, Stuart PR (2010) Integrating product portfolio design and supply chain design for forest biorefinery. Comput Chem Eng 34:1497–1506. doi:10.1016/j.compchemeng.2010.02.004

Mansoornejad B, Pistikopoulos EN, Stuart PR (2013) Metrics for evaluating the forest biorefinery supply chain performance. Comput Chem Eng 54:125–139. doi:10.1016/j.compchemeng.2013.03.031

Munda G (1995) Multicriteria evaluation in a fuzzy environment: theory and applications in ecological economics. Physical Verlag, Heidelberg, Germany

Munda G, Nijkamp P, Rietveld P (1994) Qualitative multicriteria evaluation for environmental management. Ecol Econ 10:97–112

Othman MR, Repke JU, Wozny G, Huang Y (2010) A modular approach to sustainability assessment and decision support in chemical process design. Ind Eng Chem Res 49:7870–7881. doi:10.1021/ie901943d

Sammons NE Jr, Yuan W, Eden MR, Aksoy B, Cullinan HT (2008) Optimal biorefinery product allocation by combining process and economic modeling. Chem Eng Res Des 86:800–808. doi:10.1016/j.cherd.2008.03.004

Sanaei S (2014) Sustainability assessment of biorefinery strategies under uncertainty and risk using multi-criteria decision-making (MCDM) approach, Ecole Polytechnique de Montréal. Ph.D. thesis, available on https://publications.polymtl.ca/1423/ (Downloaded date: December 10th, 2015)

Sandin G, Røyne F, Berlin J, Peters GM, Svanstrom M (2015) Allocation in LCAs of biorefinery products: implications for results and decision-making. J Clean Prod 93:213–221. doi:10.1016/j.jclepro.2015.01.013

Santibañez-Aguilar JE, González-Campos JB, Ponce-Ortega JM, Serna-González M, El-Halwagi MM (2014) Optimal planning and site selection for distributed multiproduct biorefineries involving economic, environmental and social objectives. J Clean Prod 65:270–294. doi:10.1016/j.jclepro.2013.08.004

Schaidle JA, Moline CJ, Savage PE (2010) Biorefinery sustainability assessment. Environ Prog Sustain 30:743–753. doi:10.1002/ep.10516

Sharma P, Sarker BR, Romagnoli JA (2011) Decision support tool for strategic planning of sustainable biorefineries. Comput Chem Eng 35:1767–1781. doi:10.1016/j.compchemeng.2011.05.011

Sharma P, Vlosky R, Romagnoli JA (2013) Strategic value optimization and analysis of multi-product biomass refineries with multiple stakeholder considerations. Comput Chem Eng 50:105–129. doi:10.1016/j.compchemeng.2012.10.010

Sugiyama H, Fischer U et al (2008) Decision framework for chemical process design including different stages of environmental, health, and safety assessment. AIChE J 54:1037–1053. doi:10.1002/aic.11430

Sukumara S, Amundson J, Faulkner W, Badurdeen F, Seay J (2012) Multidisciplinary approach in developing region specific optimization tool for sustainable biorefining. Comput Aided Chem Eng 30:157–161. doi:10.1016/B978-0-444-59519-5.50032-0

Zhang Q, Gong J, Skwarczek M, Yue D, You F (2014) Sustainable process design and synthesis of hydrocarbon biorefinery through fast pyrolysis and hydroprocessing. AIChE J 60:980–994. doi:10.1002/aic.14344

Author Biographies

Dr. Behrang Mansoornejad, Ph.D. joined the Industrial Systems Optimization group at CanmetENERGY-Varennes as a process engineer in 2013. He obtained his Bachelor degree in Chemical Engineering and his Master of Applied Science degree in Process Engineering from the University of Tehran, Iran. He completed his Ph.D. thesis in Chemical engineering at Polytechnique Montréal. He has multiple peer-reviewed publications in prestigious journals including Computers and Chemical Engineering and International Journal of Production Economics, and has presented in several international conferences. He has more than 3 years of experience as a process engineer in petrochemical industry as well as more than 3 years of experience in forest biorefinery projects. His key fields of expertise are process design and modeling, techno-economic assessment, supply chain analysis, and strategic decision-making. Dr. Mansoornejad's current professional interests are: modeling and simulation of biorefinery and bioenergy technologies; techno-economic and environmental assessment of biorefinery projects, and long-term strategic decision-making for forest biorefinery.

Dr. Shabnam Sanaei, Ph.D. is working at the Biomaterials Business Unit of the Domtar Inc. as biorefinery specialist, project manager. She has Ph.D. from Chemical Engineering Department Polytechnique Montreal, working in an NSERC Environmental Design Engineering Chair entitled Process Integration in the Pulp and Paper Industry. During her Ph.D. studies, she developed a systematic approach to assess sustainability of biorefinery strategies and to make strategic decisions in the forestry sector under uncertainty and risk, using a set of process systems engineering tools such as techno-economic, Monte Carlo analysis, lottery-based risk attitude measurement and multi-criteria decision-making (MCDM). Before joining Domtar, Shabnam had worked as a consultant in EnVertis Consulting Inc., and as a process engineer in MAPNA Group. Dr. Sanaei's major contributions have been related to the evaluation of integrated biorefinery opportunities and strategic decision-making under uncertainty and risk considering their sustainability performance.

Mrs. Banafsheh Gilani, M.Sc. is a process engineer with the Industrial Systems Optimization group at CanmetENERGY-Varennes. She joined the group in 2013 and has been on secondment to FPInnovations, Vancouver lab since 2014. Mrs. Gilani holds a Master of Applied Science degree in Chemical Engineering from Polytechnique Montreal and she completed her bachelors in chemical engineering at the University of Tehran, in Iran. She has more than 8 years of continuous experience as a process engineer and project coordinator in industrial engineering, procurement, and construction services, as well as 3 years' experience in forest biorefinery projects. Mrs. Gilani's current professional interests are: analyzing different biorefinery technologies to identify the critical steps and operational constrains; and evaluating technical, economic, and environmental performance of technology candidates to select the most viable biorefinery technology which can create new values from biomass. In the past five years, Mrs. Gilani has performed the sustainability assessment of several biorefinery technologies. The focus of her work includes process design of forestry-based biorefinery processes; development of process flow diagrams and mass and energy balances; techno-economic assessment and life-cycle analysis of biorefinery technologies.

Dieudonné R. Batsy, B.Sc. is a Ph.D. candidate at Polytechnique Montréal. He is working with NSERC Chair and Professor Paul Stuart. He is currently working on the implications of climate change and energy policy in the selection of sustainable forest biorefinery strategies. Prior to joining NSERC Chair as a graduate student, Dieudonné received his Bachelor degree in Chemical Engineering from Polytechnique Montreal, and briefly worked as an intern with the NSERC Chair in Environmental Design Engineering on process design and process simulation of a biorefinery. He was the recipient of the best student project award, of the PBR/High Standards and Audacity in Engineering in 2009 of *Prix Excellence Jeunesse of CCQ* (Quebec Community Crossroad) in 2010. The Honorable Pierre Duchesne Lieutenant Governor of Québec awarded him a "bronze medal for youth" to acknowledge his academic excellence and his tangible commitment within his academic institution and community in 2010. Ranked in the top 5% of students in chemical engineering and showing a great potential in research, Polytechnique Montreal awarded him the Da Vinci Profile Award for excellence, which honors balance between studies, social commitment, and student participation.

Dr. Marzouk Benali, Ph.D. is a senior research scientist and manager of the biorefinery R&D program at CanmetENERGY research center of Natural Resources Canada. He is a dedicated and innovative leader with 28 years of experience as a research scientist and project manager in industrial process design, engineering, and optimization. He has co-supervised more than 40 graduate and undergraduate students and more than 20 junior and senior researchers. He has published over 80 peer-reviewed papers in international journals and conference proceedings, 60 technical reports, four book chapters, and has also delivered lectures at more than 50 conferences. He has been granted five patents. He has received numerous awards and honors from the academia, industry, and government. Dr. Benali obtained his Master and Ph.D. degrees in Chemical Engineering from Université de Technologie de Compiègne, in France. Dr. Benali's current research areas cover: design and process engineering, industrial process optimization, and process intensification, with special focus on forest biorefinery and bioenergy developments. This includes: design, modeling, and optimization of biorefinery and bioenergy processes; pilot-scale development of wood biomass pretreatment and fractionation processes; pilot-scale development of sugar technology platform; techno-economic assessment, and LCA-based environmental evaluation, as well as development of multi-criteria decision-making tools.

Paul R. Stuart, Ph.D. After a career in engineering design and consulting, Paul joined the Chemical Engineering Department at Polytechnique-Montréal to become an NSERC Chair in Design Engineering—Process Integration in the Pulp and Paper Industry. In his research program, Paul addresses industry-driven problems using product and process design methodologies and systems analysis tools, targeting the forest products industry and its transformation to new business models such as the biorefinery. He works closely with the CanmetENERGY Laboratory of Natural Resources Canada to develop industrially relevant systems engineering approaches. In 2010, Paul cofounded and is today Principal Consultant of EnVertis Inc, a consultancy seeking to assist forest product companies with business transformation. Paul is also the Scientific Director of a Canada-wide network of researchers conducting Value Chain Optimization (VCO) research for the forest products sector. Paul is a founding Fellow of PAPTAC, a past President of the Canadian Society for Chemical Engineering (CSChE), and a Fellow of the Canadian Academy of Engineering (CAE).

Chapter 9
Biorefineries in the World

Francisco Gírio, Susana Marques, Filomena Pinto,
Ana Cristina Oliveira, Paula Costa, Alberto Reis and Patrícia Moura

Abstract This chapter intends to give a brief overview of current conventional and advanced biomass-based biorefineries in the World. While the conventional biorefineries use mature and commercial technology, the advanced biorefineries (e.g., lignocellulosic-based biofuel biorefineries, microalgae-based biorefineries) have different degrees of technology-readiness level and regardless the process technology, only a few of them have reached the commercial scale although the profitability remains a quest. The most representative's examples of biorefineries in the World are reviewed in this chapter with special emphasis on thermochemical- and biochemical-based biomass processing technologies for advanced biofuel biorefineries at pilot, demo or commercial stage. Few examples of product (non-energetic)-driven biorefineries are also discussed, such as pulp and paper biorefineries and lactic acid-producing biorefineries, mainly because only a limited number are in operation because their key technologies are still in the R&D, pilot or demo stage.

9.1 Introduction

Similarly, to an oil refinery, a biomass-based refinery or "biorefinery" is an industrial plant that produces a variety of products and as a result of the whole utilization of complex raw materials, namely the lignocellulosic biomass. Some of these biorefinery products, e.g., ethanol, butanol, lignin are more oxidized molecules than the hydrocarbons molecules obtained from fossil fuels (Cherubini and Stromman 2010).

Conversely to other more conventional uses of biomass (e.g., the use of wood in pulp and paper mills or the use of oleaginous seeds to biodiesel), a biorefinery industrial plant aims for a greater utilization of the biomass feedstock, for an enhanced mitigation of greenhouse gas (GHG) emissions, for producing fewer

---

F. Gírio (✉) · S. Marques · F. Pinto · A.C. Oliveira · P. Costa · A. Reis · P. Moura
LNEG, Estrada do Paço do Lumiar, 22, 1649-038 Lisboa, Portugal
e-mail: francisco.girio@lneg.pt

M. Rabaçal et al. (eds.), *Biorefineries*, Lecture Notes in Energy 57,
DOI 10.1007/978-3-319-48288-0_9

wastes and residues, and for greater energy efficiency and product income. Other advantages claimed for the biorefinery concept have already been cited in Chap. 1 (Biorefinery concept).

The simplest (i.e., less complex in terms of design) biorefineriesuse sugar-based crops (e.g., sugar beet, sugarcane), starch crops (e.g., cereals, grains, such as corn, cassava, or wheat) or oleaginous crops (e.g., rape, soy), called "first generation" feedstocks, as raw materials. These biorefineries are already maturely established, and the majority of the commercial facilities belonging to this category produce first-generation (1G) biofuels, i.e., biodiesel or bioethanol. In addition, traditional pulp and paper mills can also be categorized as conventional biorefineries, since they produce multiple products from biomass (wood).

Conversely, advanced biorefineries are usually related with the use of more sustainable nonfood crop feedstocks, in particular lignocellulosic residues from forestry, agricultural, agro-industrial, the organic fraction of urban wastes, cellulose wastes (e.g., paper and sludge) and dedicated energy crops. Indeed, lignocellulosic biomass residues represent a promising option as feedstock for bio-based production, under a biorefinery concept, considering their output/input energy ratio, their great availability both in tropical and temperate countries, their moderate cost (primarily related to their transport), and the fact that they do not compete for food and feed production (Kheshgi et al. 2000; Cardona and Sanchez 2007; Lynd et al. 1991; Sanchez and Cardona 2008).

A classical classification of biorefineries has been proposed by Kamm and Kamm (2004), Kamm et al. (2006). They consider four types of biorefineries: whole crop biorefinery (e.g., integrating the whole plant harvested for food and nonfood purposes), green biorefinery (e.g., grass and other immature green biomass as feedstocks), lignocellulosic feedstock biorefinery (e.g., forestry-based lignocellulosic feedstocks) and the two-platform concept, based on main types of technology involved (e.g., sugar platform and the thermochemical platform). This classification has several limitations, e.g., it does not consider the concept of multi-platform biorefineries neither the existence of current algae-based biorefineries.

A more recent classification of biorefineries has been postulated by IEA-Bioenergy Task 42 based on energy-driven (or biofuel-driven) biorefineries and product (non-energetic)-driven biorefineries (IEA-Bioenergy 2014). In the former, the main target is to produce large amounts of energetic products (biofuels, heat, and power) while the latter are mainly focused for production of bio-based products out of biomass while their by-products or residues can be used for energy purposes.

We shall review in this chapter many examples of biofuel-driven biorefineries some of which that have reached the commercial scale although the profitability remains a quest, but not many examples of product-driven biorefineries since only a limited number are in operation, mainly because some key technologies are still in the R&D, pilot or, demo stage.

9.2 Conventional Biorefineries

There are in the World many examples of the so-called conventional biorefineries since they use only a fraction (usually a minor part, in weight) of the raw material as input. These biorefineries are already maturely established, and the majority of the commercial facilities belonging to this category produce first-generation (1G) biofuels, i.e., biodiesel (fatty acid methyl ester) or bioethanol. Besides these above-mentioned energy-based biorefineries, other non-energetic-based biorefineries can be classified as conventional if they use sugar-based crops, starch crops or oleaginous crops, i.e., edible feedstocks. Since all general rules might have exception, traditional pulp and paper mills can also be categorized as conventional biorefineries, since they produce multiple products from biomass, i.e., wood (Ragauskas et al. 2006).

The next sections illustrate some of the most important examples of conventional biorefineries already in the World market.

9.2.1 First-Generation Biodiesel

Biodiesel (fatty acid methyl esters—FAME) is a biofuel that can be used as pure biodiesel fuel or blended with diesel, due to the similarity in their physical and fuel properties. The 1G biodiesel production process is defined as a conventional oleaginous biorefinery since it integrates the production of an energy-based product (biodiesel) and non-energetic based co-products (animal feedstock and glycerin).

Oleaginous crops are the most used raw material for biodiesel production. In Europe, rapeseed is the main feedstock, accounting for 55% of total biodiesel production in 2014 (EU Biofuels Annual 2015), while in the USA, Argentina, and Brazil soybean is the largest biodiesel feedstock. Palm oil from Indonesia and Malaysia has also an important market in this biofuel industry. Besides these prominent oil crops, many other sources such as sunflower, canola, jatropha, and oil-based residues (e.g., used cooking oil, animal fat) can be used for the same purpose.

Figure 9.1 presents a simplified diagram of a 1G biodiesel production process. Using oilseeds as starting raw material, the first processing step corresponds to the oil separation, by pressing (cold or hot) and/or by solvent extraction (usually hexane). Through this process, crude vegetable oil is obtained and meal extract/seed cake incurs as co-products. The latter are usually used for animal feedstock due to their protein content while the crude vegetable oil, mainly composed of triglycerides, is used to produce biodiesel.

Fatty acid methyl esters (biodiesel) are obtained through the esterification and transesterification reactions of free fatty acids and triglycerides, respectively. In the transesterification process, glycerides react with an alcohol (e.g., methanol) in the presence of a catalyst (e.g., an alkali such as sodium methylate) to deliver FAME

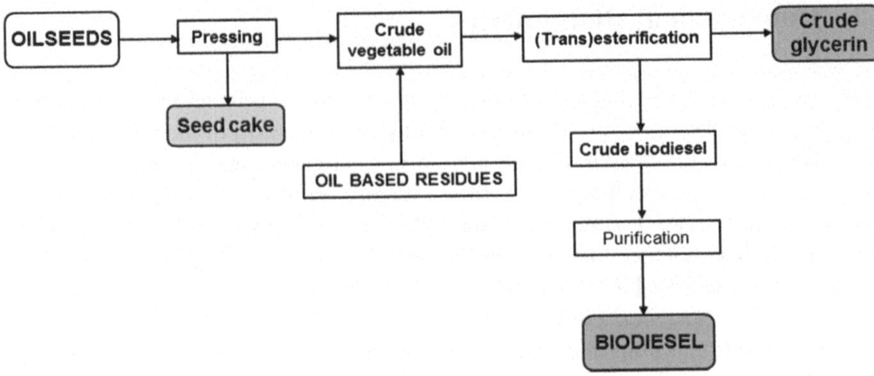

Fig. 9.1 Simplified flowsheet for biodiesel production from oilseeds and/or oil-based residues, co-producing crude glycerin and seed cake

and glycerin. Currently, this conversion process is used at a large scale to convert edible oils into FAME. For raw materials with high acidity, as nonedible oils and oil-based residues, a two-stage process is the most suitable approach to be used. In this case, the alkaline catalyzed transesterification is preceded by an acid catalyzed esterification that allows converting the free fatty acids (FFA) into FAME.

Crude glycerin from biodiesel production can be refined into a pure form and then be used in food, cosmetic and pharmaceutical industries. Nowadays, several applications for non-refined glycerin have been proposed, as anaerobic digestion, animal feeds, and thermochemical/biological conversions to value-added products. However, plants for the production of biodiesel are either stand-alone or integrated in oil mills, and are not yet connected to additional secondary refining of, e.g., crude glycerin into higher added value non-energetic based products shifting these first-generation biodiesel plants into more advanced biorefineries.

Biodiesel has been produced at an industrial scale in the EU since 1992, and nowadays there are approximately 250 plants with an installed production capacity of 23 million tons of biodiesel (European Biodiesel Board). These plants are mainly located in Germany, Italy, Austria, France, and Sweden. Table 9.1 presents the main 1G biodiesel producers in EU in 2014. The available data estimate that EU biodiesel consumption reached 11.1 Mtoe (12.9 million tons) in 2014 (Biofuels barometer, Eurobserv'er 2015).

Recent data from USA report a total of 166 biodiesel producers and an annual production capacity of 9.1 million tons (10.431 billion liters). The main 1G biodiesel producers in USA, with installed production capacity higher than 378 MM liters/year, are RBF Port Neches LLC in Texas (681.3 MM liters/year) and REG Grays Harbor LLC in Washington (378 MM liters/year). Otherwise, biodiesel production from the Midwest region (Petroleum Administration for Defense District 2) was about 71% of the United States total being Louis Dreyfus Agricultural Industries LLC in Indiana (341 MM liters/year) the largest plant (Biodiesel Magazine 2015).

Table 9.1 Main 1G biodiesel producers in EU in 2014[a] (biofuels barometer, Eurobserv'er 2015)

Company	No. of plants	Total production capacity (t/year)
Avril (formely Sofipetrol)	13	2,700,000
Neste Oil	3	1,180,000
ADM Biodiesel	3	975,000
Infinita (Musim Mas)	2	600,000
Marseglia Group (Ital Green oil and Ital Bi Oil)	2	560,000
Verbio AG	2	450,000
Eni	1	300,000
Petrotec	3	185,000

[a]Only units in Europe are considered

9.2.2 First-Generation Bioethanol

Nowadays the most common renewable fuel is 1G bioethanol, i.e., ethanol produced from sugar/starch crops, which is typically blended with petrol in most of the countries as low blends (e.g. E5, E10).

Sugar-based crops require minimal processing since the process only requires direct fermentation of sucrose (obtained via simple sugar extraction) into ethanol by *Saccharomyces* yeasts. Raw sugar is derived from both sugarcane and sugar beet.

Brazil and India are the World's two largest sugar producers, accounting together for over half the World's sugarcane production for the past 40 years (E4tech, RE-CORD and WUR 2015). Brazil is the largest producer of ethanol from sugarcane, which is the cheapest, and the World's second largest ethanol producer, behind the United States, and it is also a pioneer in using ethanol as a motor fuel (Sánchez and Cardona 2008). In 2013/14, Brazil produced 653.5 million tons of sugarcane, which yielded 37.7 million tons of sugar and 27.5 billion liters (7.3 billion gallons) of ethanol. Most of this production is absorbed by the domestic market where it is sold as either pure ethanol fuel or blended with gasoline. All common gasoline sold in Brazil includes superior blends with anhydrous bioethanol ranging from 18 to 27 %vol/vol, being currently 25% (Unica and ApexBrasil 2016). There are 391 ethanol plants operating in Brazil in 2016 (NovaCana 2016), being 126 dedicated to ethanol production and 252 producing both sugar and ethanol. There are 15 additional plants dedicated exclusively to sugar production (Ibeto et al. 2011).

Despite sugar beet is a demanding crop in terms of soil conditions, fertilizer and irrigation, in Europe 188.4 million tons of sugar beet were produced in 2012 (FAO 2014) and sugar beet has long been used to produce fuel ethanol, with France, Germany, Belgium, and Greece leading production (European Biofuels Technology Platform). Although North America has not processed beets for commercial ethanol, nonfood-grade "energy" beets have recently been used to produce commercial-quality ethanol. Following the closure of sugar plants, former sugar beet farmers decided to produce "energy" beets for conversion to ethanol instead of table sugar. That ultimately led to the implementation in North America of the first demonstration plant (one million gallon-capacity) in California (Fresno County), operated by Mendota Bioenergy LLC, using the whole beet as feedstock for ethanol production (Renewable Energy World.com).

Starch processing also involves a fairly mature technology involving enzymatic hydrolysis—liquefaction and saccharification—to break the starch into its constituent sugar, producing a relatively clean glucose stream that is easily fermented to ethanol by *Saccharomyces* yeasts (Gray et al. 2006). There are two production processes: wet milling and dry milling. The main difference between the two lies in the initial treatment of the grain. In traditional dry milling, as shown in Fig. 9.2, the entire corn kernel or other starchy grain is first ground into flour, and processed without separating out the various component parts of the grain. The components of the kernel not intended for fermentation include the germ, fiber ,and protein, are concentrated in the distillers dried grains that are produced as co-products. These dried distillers grains with solubles (DDGS), a protein rich product, are used for animal feedstock. While dry milling is less capital intensive, it also yields less ethanol per bushel of corn than wet milling (Rajagopalan et al. 2005). In wet milling, the grain is soaked or "steeped" in water and dilute sulfurous acid for up to 48 h, to assist in separating the grain into its many component parts. Slurry processing separates the germ from the rest of the kernel, which is further processed to separate the fiber, starch, and gluten, as shown in Fig. 9.3. The fiber and corn gluten are sold as components of animal feedstock while the starch is fermented to ethanol, or commercialized as corn starch or corn syrup (Renewable Fuels Association 2005). Thereby, these ethanol plants shall be categorized as biorefineries since they produce multiple products from the sugar and starch components of biomass.

Starch processing for ethanol is dominated by USA using corn starch. Indeed, 40% of the USA corn harvest already is used for bioethanol production, mainly through dry-grind technology (E4tech, RE-CORD and WUR 2015). Europe leads on wheat, with modest production of barley and other coarse grains. Indeed, wheat is the main crop grown for bioethanol production in Europe, accounting for 0.7% of

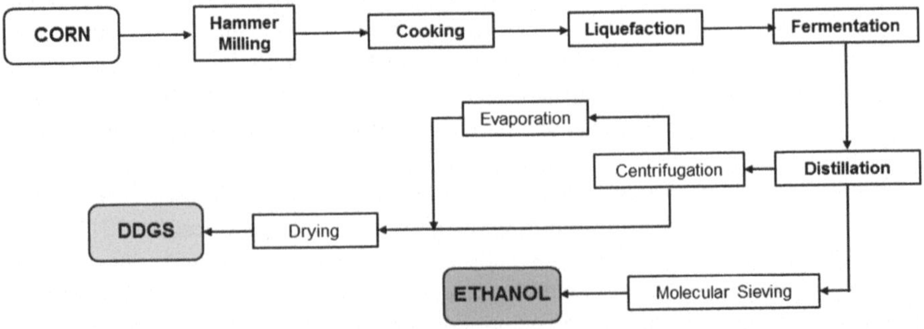

Fig. 9.2 Simplified flowsheet for ethanol production from corn via dry milling, co-producing dried distiller's grains with solubles (DDGS) (adapted from Renewable Fuels Association 2005)

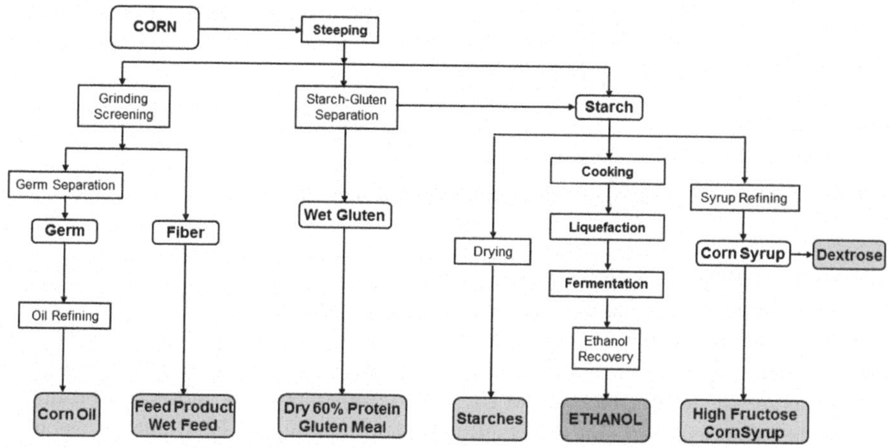

Fig. 9.3 Simplified flowsheet for ethanol production from corn via wet milling (adapted from Renewable Fuels Association 2005)

Table 9.2 Largest commercial producers of 1G bioethanol in the USA (according to data in Ethanol Producer Magazine 2016)

Company	No. of plants	Total capacity (million liters/year)
Archer Daniels Midland Co.	8	>5970
Big River Resources LLC	4	1475
Flint Hills Resources LLC	7	3050
Green Plains	14	4600
Marquis Energy LLC	2	1420
Pacific Ethanol	8	1950
Poet Biorefining	26	6315
Valero Renewable Fuels	11	4960

EU agricultural land and 2% of Europe's grain supply (European Biofuels Technology Platform). China is the global leader for rice production, along with significant production levels of wheat, coarse grains (corn), and sugarcane (E4tech, RE-CORD and WUR 2015).

There are 216 plants producing 59.5 billion liters of ethanol per year from sugar/starch in the United States. Seventy-four of the largest commercial 1G bioethanol facilities in operation in USA, belonging to 8 companies, produce 50% of this capacity (Table 9.2), using corn as raw material. Besides corn, sorghum is also used as a relevant starchy feedstock in the USA, and less than 0.2% of 1G ethanol produced in USA is obtained from wastes (e.g., beverage waste or wheat screenings).

In Canada, there are 17 plants producing 7.3 billion liters of 1G ethanol per year, with 54% of this capacity assured by two companies using corn as raw material, Green Field Ethanol (4 plants with a total yearly capacity of 2.49 billion liters) and Suncor (one plant with a total yearly capacity of 1.5 billion liters). Besides corn, Canada is producing 1G ethanol from wheat, barley and waste beverage (Ethanol Producer Magazine 2016).

9.2.3 Other 1G Sugar Platform Bioproducts

Although bioethanol is the dominant sugar platform product, there are a few commercial biorefinery facilities producing other multiple sugar platform products, such as succinic acid, lactic acid, acetic acid, *n*-butanol, iso-butanol, 1,3-propanediol (PDO), or itaconic acid, from sugar or starch components of plant biomass (E4tech, RE-CORD and WUR 2015). Other primary bio-based products, i.e., made as a first step (by direct microbial fermentation) from sugars, such as 1,4-butanediol (BDO), farnesene, poly-hydroxyalkanoates (PHAs), are currently produced still at R&D, pilot, or demonstration scale. Indeed, in terms of bio-based markets volumes, bioethanol dominates (with 58 billions of dollars of sales per year), followed by *n*-butanol (1.1 billion USD/year), acetic acid (0.8 billion USD/year), and lactic acid (0.7 billion USD/year), with much smaller but still significant markets. Some of these established bio-based products already dominate global production, e.g., ethanol and lactic acid (E4tech, RE-CORD and WUR 2015).

For instance, worldwide demand for lactic acid is expanding, not only for the more conventional use in cosmetic, pharmaceutical, and food preservatives markets, with approx. 45% of market share, but also for industrial applications, mainly for polylactic acid (PLA) production.

PLA, or polylactide, is a biodegradable and recyclable thermoplastic polyester resin (Vaidya et al. 2005), suitable for packaging materials, insulation foam, automotive parts, and fibers (textile and nonwoven) (E4tech, RE-CORD and WUR 2015), and more recently, due do its biocompatibility, for biomedical applications (Vijayakumar et al. 2008). Packaging is likely to remain the key market for PLA, with the expected increase on demand for environmentally friendly starch-based plastics for this application. Indeed, most of PLA producers also manufacture lactic acid. The global annual demand for lactic acid (including PLA) is estimated at 472 ktons, and the global production capacity is estimated to be around 750 ktons per year, with a strong presence in China (E4tech, RE-CORD and WUR 2015).

Lactic acid can be manufactured either by chemical synthesis or by fermentative processes but the latter have been preferentially used in industrial production, accounting for approximately 90% of the total worldwide production (Hofvendahl and Hahn-Hägerdal 2000). The fermentative route allows the selective production of the desired L-lactic acid stereoisomer, with additional advantages in terms of energy efficiency and yield. Moreover, renewable carbon sources are used for microbial fermentation, and actually lactic acid is mostly produced from corn starch (in the USA), tapioca roots, chips or starch (in Asia), or sugarcane and sugar beets (in the rest of the World).

The largest global commercial producer of PLA is USA-based NatureWorks (former Cargill Dow), producing PLA resins under the Ingeo brand, with a commercial production plant in Nebraska (with an annual 150 ktonnes-capacity) and plans for a new plant in Thailand (E4tech, RE-CORD and WUR 2015). NatureWorks' Ingeo PLA is also foreseen to be used in filaments formulations for 3D printing. In short term, they plan the transition for the use of residual lignocellulosic biomass (e.g., corn stover, wood chips, switch grass, or straw) as inherently more sustainable feedstock, replacing the currently used corn starch (NatureWorks LLC website). The largest global lactic acid producer is Corbion Purac (Netherlands). The later produces lactic acid, lactic acid derivatives, and lactides (including lactide resins for high-performance PLA bioplastics), operating 5 production plants, in the USA, the Netherlands, Spain, Brazil, and Thailand (the largest plant, with an annual 100 ktonnes-capacity) (E4tech, RE-CORD and WUR 2015).

In Europe, there are other PLA (and lactic acid) producers, including Synbra Technology, which operates a commercial (annual 5 ktonnes-capacity) plant in Netherlands and a pilot production plant in Switzerland (annual 1 ktonnes-capacity). Cellulac announced the future conversion of a brewery in Ireland into a lactic acid and PLA plant with 2G feedstock. Besides Corbion Purac, Galactic and Jungbunzlauer are operating commercial lactic acid production plants in Europe.

Other PLA producers include the Chinese Zhejiang Hisun Biomaterial (with an annual 5.5 ktonnes-capacity to be expanded to 50 ktonnes; using cassava instead of corn). Other lactic acid producers include Henan Jindan Lactic Acid Technology (with an annual 100 ktonnes-capacity, the largest in Asia) (E4tech, RE-CORD and WUR 2015).

9.2.4 Pulp and Paper Mills

In fact, the existing technology in pulp production can already be denominated as a conventional biorefinery, as long as in addition to pulp a wide range of chemical by-products and energy are produced from biomass, i.e., wood, which is processed in a sustainable manner (Davenport 2008). The paper industry is based on the exploitation of a natural resource, renewable, and featuring carbon sequestration—the forest; it generates products with high carbon content, which gives it a sequestration status (proportional to the lifetime of the product); it generates products several times-recyclable, which allows to extend its lifetime; and it generates biodegradable products, which reduces the impact of its final disposal in the environment. In addition, a very significant and increasing energy amount (corresponding to 69% of its energy needs, in 2010, according to CELPA 2011) is obtained from biomass, e.g., wood wastes, bark and mainly from cooking liquor containing the lignin removed from wood during the manufacturing process (this latter representing 85% of the total biofuels consumed in 2010) (Kheshgi et al. 2000).

The pulp and paper are ideal hosts for the combined production of heat and electricity called cogeneration because a large amount of steam is used for several purposes in production processes (Kheshgi et al. 2000). In fact, the proportion of electricity produced by cogeneration has reached high levels and the paper industry is currently the largest producer of electricity through cogeneration among all the industrial sectors of the European Union. The pulp and paper industry is, in fact, the largest user and producer of renewable energy in the European Union: it produces about 17% of the EU renewable energy and consumes about 13%.

Despite all these improvements in product and process efficiency that have been implemented in the pulp and paper industry, the core business of the majority of mills and infrastructures remains the same. However, this situation should change in the near future, since the pulp and paper industry is in a very favorable position to become the next generation of advanced biorefineries, taking advantage from its great experience on processing large streams of biomass (wood). Implementation of the modern and advanced biorefinery concept by pulp and paper manufacturers will benefit from biomass resource already collected, available, and centralized in one location, eliminating the need for implementation of new collection circuits (Ragauskas et al. 2006). Thereby, many of the companies are already evaluating its potential as a modern and advanced biorefinery, further extending the biorefinery concept, as reported in Sect. 9.6.4.

9.3 Thermochemical-Based Advanced Biorefineries

9.3.1 Introduction

Nowadays the main thermochemical processes for biofuel production are still gasification and pyrolysis. Pyrolysis of biomass produces gases, liquids, and solids (carbonaceous residue). The gases can be used as fuel and the major constituents are hydrogen (H_2), carbon monoxide (CO), carbon dioxide (CO_2), nitrogen (N_2), and gaseous hydrocarbons. The liquids are used as raw material or as fuel either directly or after being converted into secondary fuels (more valuable). The proportion of the obtained products depends on the technology used, type of reactor and of operational conditions, such as the type of gas, pressure, temperature, reaction time, type of solvent and catalyst. The type and composition of the biomass is also of utmost importance. Currently the main purpose of pyrolysis is to obtain bio-liquids to be used as fuel or as raw material in the industry.

In Fig. 9.4 a simplified diagram of biomass pyrolysis' main products is presented.

The gasification process is more mature than pyrolysis. During gasification, biomass is converted into a gas, usually referred as syngas, whose major

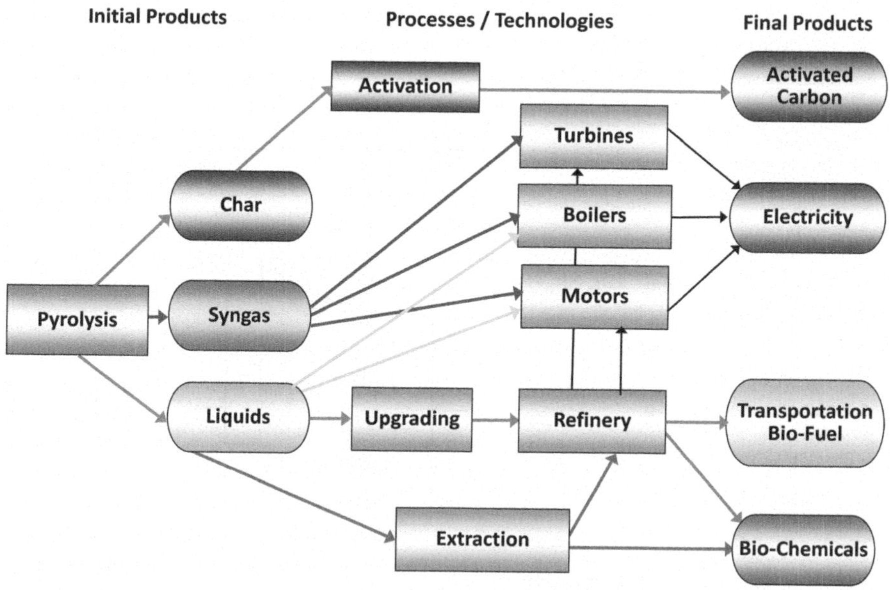

Fig. 9.4 Scheme of the main products of biomass pyrolysis

constituents are CO, H_2, and CO_2, methane (CH_4) and other gaseous hydrocarbons from C_2 to C_4, usually referred as (C_nH_m). Gasification may occur in the presence of different gasification agents, which determine the type of gas produced. The most common gasification agents are oxygen (or air) and steam, but CO_2 or mixtures of any of these components may also be used. The choice of the gasification agent depends on the application of gasification gas. The use of air leads to the production of a gas that is diluted with N_2, thus having a lower calorific value, not suitable for chemical synthesis. The use of oxygen solves the problem of N_2 dilution, but increases the operative costs because of the cost associated with oxygen production.

Gasification gas may be used in chemical synthesis to produce liquid or gaseous fuels. For such use, syngas cleaning and upgrading is required to reduce the contents in tar, alkali metal, and sulfur, N_2, and chlorine compounds. In addition, the H_2/CO ratio in the gasification gas must be between 1.5 and 3.0 depending on the chemical synthesis: Fischer–Tropsch, synthetic natural gas (SNG), alcohol syntheses (methanol, ethanol, and propanol) or synthesis of dimethyl ether (DME). As all these chemical syntheses are already well known, the main challenge is the production of syngas with the suitable gas composition to be further used in these processes (Rapagn et al. 2002; Lv et al. 2007).

Figure 9.5 shows the main products obtained from biomass gasification. Lately, research and development of thermochemical technologies has been focused on the production of alcohol mixtures and DME by chemical syntheses, instead of diesel like biofuels by Fischer-Tropsch synthesis, mainly due to economic reasons and absence of biomass supply chain at large scale and at competitive prices.

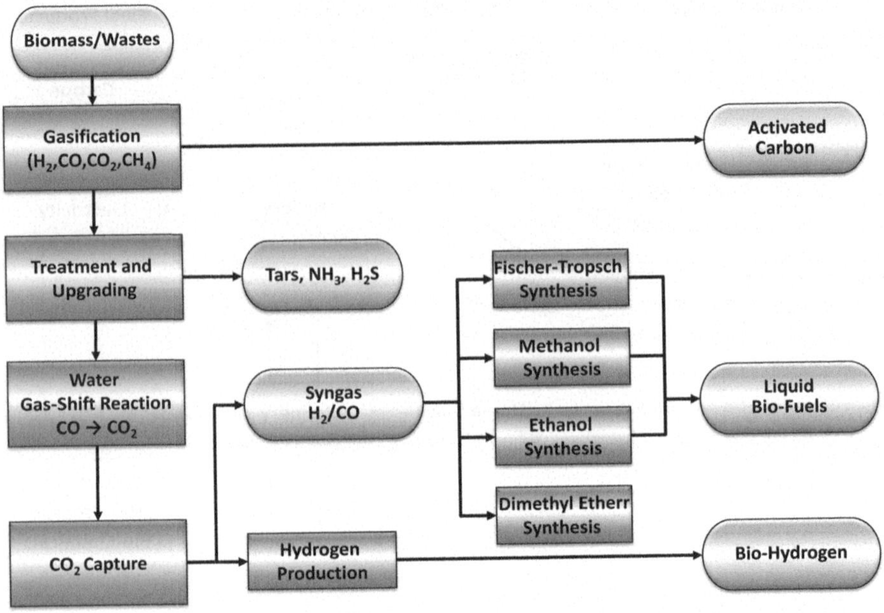

Fig. 9.5 Biomass gasification main products

The basis of a thermochemical-based biorefinery is the production of different biofuels and bio-chemicals using thermochemical processes, mainly gasification and pyrolysis. The main advantage is the production at regional or decentralized level, where biomass is available. Afterwards the high-energy bioproducts can be cost-effectively transported to existing centralized refinery facilities for further processing.

9.3.2 *Most Relevant Thermochemical-Based Processes*

A selection of the most important processes for biofuels production by thermochemical processes in biorefinery concept is presented (IEA Bioenergy 2014; Bacovsky et al. 2013). The criteria for the selection were the maturity of the process and the existence of demonstration or commercialization units. To facilitate the reading and comprehension of the main information available, the data are organized in the below format.

SNG demo

Process/Technology	SNG (synthetic natural gas)
Project owner	Biomassekraftwerk Güssing
Web page	http://www.eee-info.net

(continued)

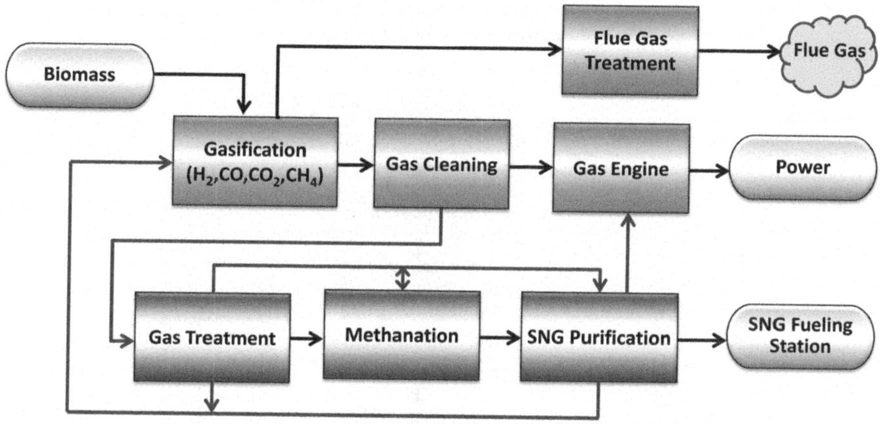

Fig. 9.6 Flow diagram of bioC Biomassekraftwerk Güssing

Fig. 9.7 Diagram block of bioCRACK process, Vienna

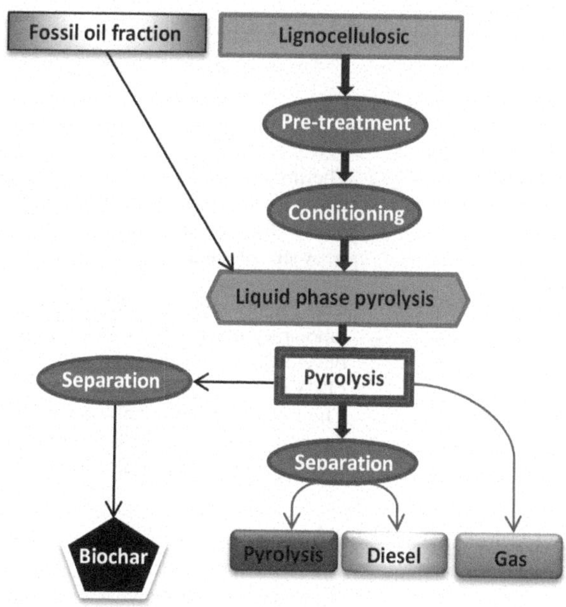

(continued)

Facility type	Demonstration
Operation	Since 2009
Feedstock	350 Nm^3/h of syngas
Outputs	576 t/year; 100 Nm^3/h of SNG
Status	Operation

(continued)

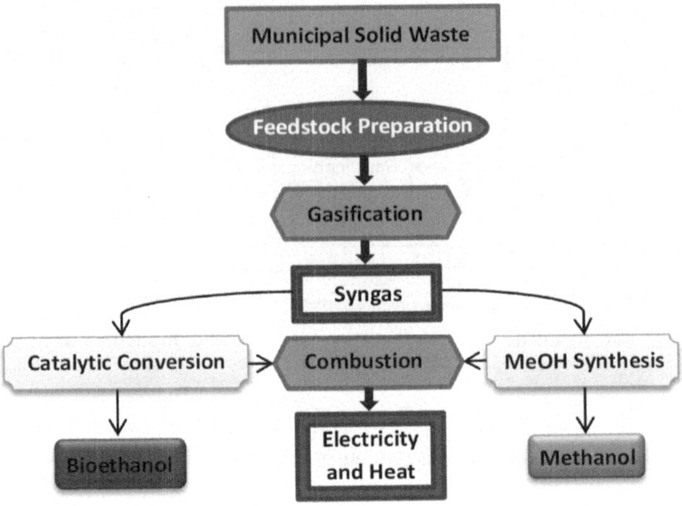

Fig. 9.8 Diagram of Enerkem process, Alberta

(continued)

Location	Güssing, Austria
Summary	The installation uses a gaseous stream from the existing gasifier in Güssing, which is purified before being introduced to the catalytic reactor for conversion to methane, which operates at a temperature between 300 and 360 °C and a pressure range from 1 to 10 bar (Fig. 9.6) This facility with 10 MW started in June 2009, after tests in a laboratory plant of 10 kW for a few years. The technology was developed in collaboration with Vienna University of Technology (TUV) The facility was originally owned by CTU—Conzepte Umwelt Technik AG and was subsequently acquired by Biomassekraftwerk Güssing (Bacovsky et al. 2013)

BDI bioCRACK pilot Plant

Process/Technology	A one-platform (pyrolysis-oil) refinery for the production of diesel fuel, pyrolysis-oil, and bio-char from solid biomass
Project owner	BDI—Bioenergy International AG
Web page	http://www.bdi-bioenergy.com/de-biomass_to_liquid-24.html
Facility type	Pilot plant
Operation	Since 2009
Feedstock	2.4 t/day Lignocellulosic biomass (wood chips, straw)
Outputs	5 t/a raw diesel fuel, pyrolysis-oil, char
Location	OMV refinery Schwechat/Vienna, Austria

(continued)

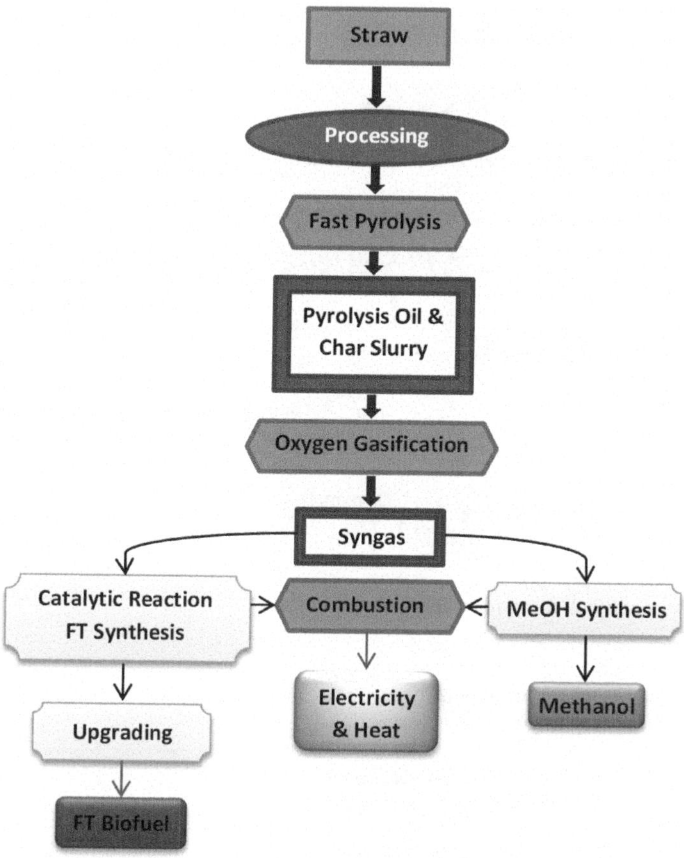

Fig. 9.9 Process block diagram of BIOLIQ plant, Karlsruhe

(continued)

Summary	This process uses the bioCRACK technology to produce mineral diesel blended with biofuels obtained from renewable sources which can be easily upgraded to EN590 quality, using current refinery units The bioCRACK process (Fig. 9.7) converts solid biomass like wood chips and straw directly into raw diesel fuel. The OMV refinery Schwechat was selected for the implementation of bioCRACK pilot plant that was the first worldwide to be implemented in a refinery (IEA Bioenergy 2014) This concept accomplishes two important tasks: production of diesel to meet the growing demand of liquid fuels and increase of biogenic share

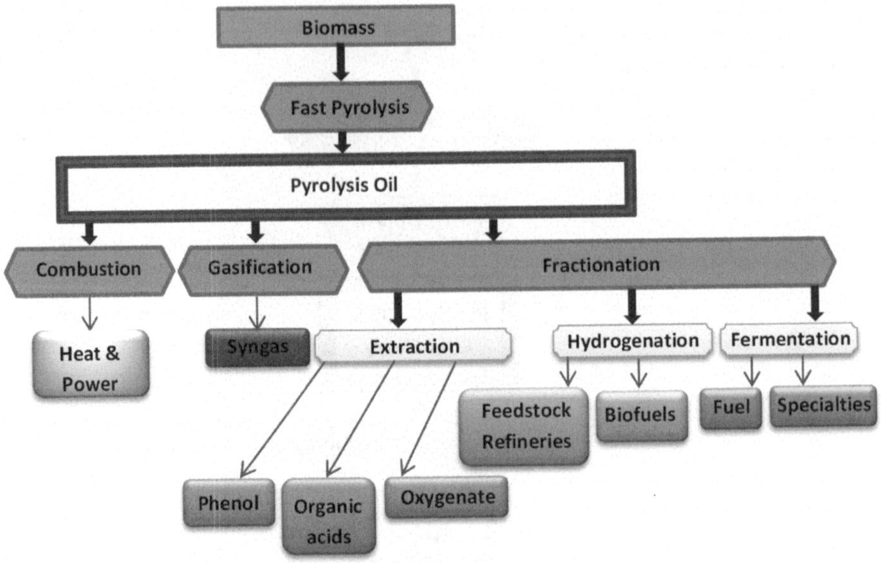

Fig. 9.10 Process block diagram of BTG plant, Hengelo

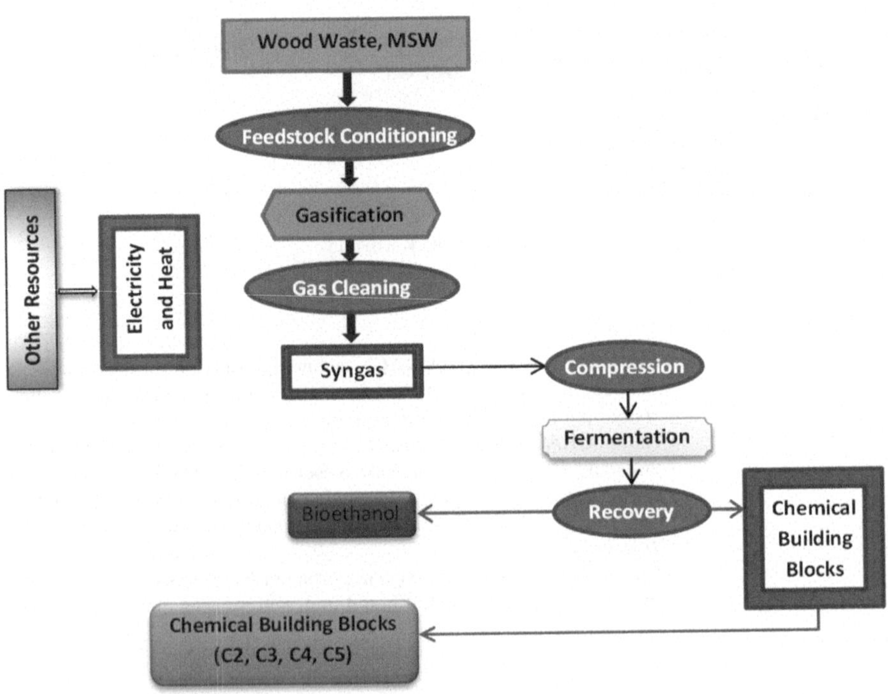

Fig. 9.11 Process block diagram of LanzaTech plant, Glenbrook

Fig. 9.12 Process block diagram of INEOS plant, Florida

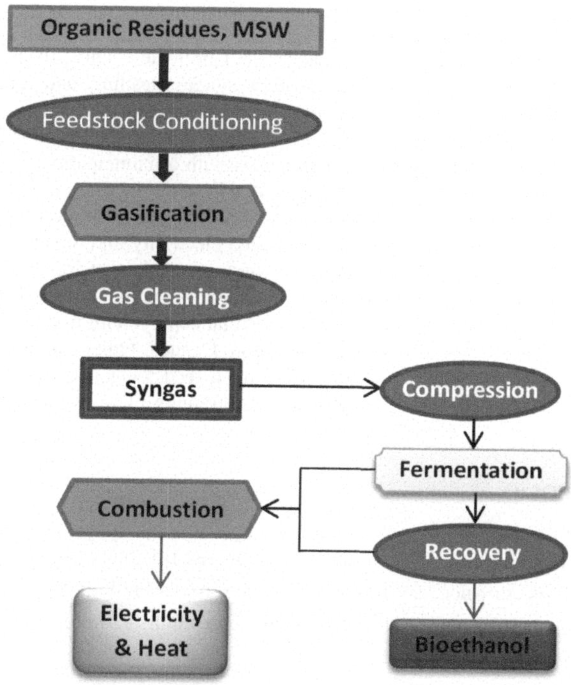

Enerkem waste-to-biofuels facility (Alberta, Canada)

Process/Technology	1-platform (syngas) biorefinery producing biofuels, chemicals, and bioenergy from municipal solid waste
Project owner	Enerkem Alberta Biofuels L.P.
Web page	www.enerkem.com/www.edmontonbiofuels.ca
Facility type	Commercial Scale, under commissioning (Alberta) Demonstration (Westbury)
Operation	Since 2014 (Alberta) In operation since 2009 (syngas), 2010 (methanol), 2012 (ethanol) (Westbury)
Feedstock	100,000 dry tonnes of sorted MSW per year (Alberta) Decommissioned telephone poles and sorted municipal solid waste (Westbury)
Outputs	38×10^6 l/year of biofuels 5 million liters/year (methanol), syngas, ethanol
Location	Edmonton, Alberta (on the integrated waste management site of the City of Edmonton) Westbury facility (Canada)
Summary	Enerkem Alberta Biofuels made an agreement with the City of Edmonton to build, operate a plant and produce 38×10^6 l/year of next-generation biofuels, using around 100,000 of dry tonnes of nonrecyclable and noncompostable municipal solid waste, which will

(continued)

(continued)

	be otherwise landfielded. This process includes feedstock preparation, gasification, syngas cleaning, and catalytic conversion to produce ethanol for transportation use and methanol for chemical applications. The plant started operation on June 4, 2014 and is the result of a fruitful collaboration with a municipality In Fig. 9.8, a diagram of Enerkem process is shown Enerkem Westbury demonstration facility is placed in a rural area and recycles used electricity and telephone poles and railway ties. It converts the non-usable parts of these poles and other waste materials into biofuels and green chemicals. It allowed the process scale-up to full commercial deployment, like the Enerkem Alberta Biofuels facility in Edmonton, Canada. Other important tasks include testing of unconventional feedstocks and raw materials, development of new products and training plant technicians and operators (IEA Bioenergy 2014)

BIOLIQ

Process/ Technology	A 3-platform (pyrolysis-oil, syngas, electricity, and heat) biorefinery producing customized BtL-fuels from (straw)
Project owner	Karlsruhe Institute of Technology (KIT)
Web page	http://www.bioliq.de/english/24.php
Facility type	Pilot Plant
Operation	Since 2008
Feedstock	500 kg/h of Residual biomass
Outputs	BioliqSynCrude®, Synthesis gas, BtL-fuels (Dimethyl ether, Gasoline)
Location	Karlsruhe, Baden-Württemberg
Summary	The Bioliq® pilot plant produces BTL (biomass to liquid) fuels from biomass wastes, mainly dry straw and wood. The integrating process chain also produces synthesis gas and chemicals (Fig. 9.9) The Karlsruhe BTL-concept combines decentralized production of BTL-fuels by rapid pyrolysis and central processing with final industrial-scale refinement. This process is based in energy densification, which increases its efficiency, save carbon dioxide by decreasing routes of transport to refineries and ensures the full exploitation of biomass that is transformed into material, and energy (IEA Bioenergy 2014)

BTG bioliquids refinery

Process/ Technology	1-platform (pyrolysis-oil) biorefinery for the production of chemicals, fuels, power, and heat from lignocellulosic biomass
Project owner	Empyro BV
Web page	www.btgworld.com; www.btg-btl.com
Facility type	Commercial Plant
Operation	Since 2015
Feedstock	5 t/h of wood residues
Outputs	Pyrolysis-oil, power, heat

(continued)

(continued)

Location	Hengelo, the Netherlands
Summary	The pyrolysis facility of Empyro in AkzoNobel site in Hengelo, the Netherlands, was implemented in 2015. This pyrolysis-oil production facility was the first step of a full-scale biorefinery. It is planned that the production capacity will increase gradually till 20 million liters per year of pyrolysis-oil In Fig. 9.10 a process block diagram of BTG plant is shown The pyrolysis process used was created at the University of Twente and has been further developed for 20 years by BTG in Enschede, the Netherlands. Empyro BV was founded by BTG Bioliquids BV (owner of the technology) and Tree Power (long-term investor in renewables) and the main aim was to demonstrate BTG technology on a commercial-scale Pyrolysis plant processes 5t/h of biomass (wood chips) which is mixed with hot sand to be converted in a few seconds into 3.5 ton pyrolysis-oil, char and gas. Enough electricity will be produced for plant own use and steam will be supplied to the salt production of AkzoNobel, located near Empyro. The company FrieslandCampina will use the oil in its facilities located in Borculo to substitute annually 10 million cubic meters of natural gas (IEA Bioenergy 2014)

LanzaTech Plant

Process/ Technology	A 2-platform (syngas, chemical building blocks) biorefinery producing bioethanol and chemical building blocks from wood waste and MSW by syngas fermentation
Project owner	LanzaTech NZ Ltd
Web page	http://www.lanzatech.com/
Facility type	Pilot Plant (New Zealand) Mobile Plant (USA)
Operation	Since 2015 (New Zealand) Since 2014 (USA)
Feedstock	Wood waste and MSW (New Zealand) Woody biomass syngas produced from 125 t/day dry wood residues (USA)
Outputs	Fuel ethanol 250 l/day Ethanol, Chemicals (USA)
Location	BlueScope Steel mill, Glenbrook, New Zealand Soperton, Georgia USA
Summary	LanzaTech was founded in New Zealand in 2005 with the aim of developing and commercializing gas fermentation technologies to produce ethanol to be used as fuel, using carbon monoxide in low-hydrogen waste gases produced by the steel industry. Afterwards, LanzaTech decided to include syngas derived from waste woody biomass and municipal waste LanzaTech has used the pilot plant at the BlueScope Steel mill, Glenbrook, New Zealand to demonstrate its process using real waste gases. New organisms capable of producing a range of chemical building blocks from CO, CO_2, and H_2 have been developed LanzaTech has been developing other projects, two 378,000 l demonstration facilities, with Baosteel and Capital Steel in China and an

(continued)

(continued)

	integrated biorefinery in the Soperton, Georgia, USA using forestry waste (Freedom Pines Biorefinery) (IEA Bioenergy 2014) In Fig. 9.11 a process block diagram of LanzaTech plant is presented

INEOS

Process/ Technology	A 2-platform (syngas, power and heat) Biorefinery producing bioethanol and power from vegetative, yard, and MSW by syngas fermentation
Project owner	INEOS New Planet BioEnergy
Web page	www.ineosbio.com
Facility type	Commercial scale
Operation	Since 2012
Feedstock	Vegetative, yard, and MSW
Outputs	Bioethanol, power
Location	Indian River County BioEnergy Center, Florida
Summary	INEOS Bio and its joint venture partner, New Planet Energy, have invested millions of dollars to build an innovative bioethanol facility in the Indian River County BioEnergy Center, near Vero Beach, Florida. The main objective is the production of eight million gallons per year of third-generation bioethanol from renewable biomass, such as: vegetative, yard, and municipal solid waste (MSW). The combination of gasification and fermentation processes to treat these wastes is used in this process Six megawatts (gross) of clean renewable power will also be produced to be used in the facility and the excess will be exported to the local community. Other objectives are the demonstration of key equipment at full commercial-scale The project started in February 2011 and construction was accomplished in the second quarter of 2012, with commissioning and start-up till the end of the year Figure 9.12 shows a process block diagram of INEOS plant (IEA Bioenergy 2014)

9.4 Biochemical-Based Advanced Biorefineries

In opposition to conventional first-generation biochemical-based technologies, which are well-established processes that are already operational on a commercial scale, most of advanced technologies, commonly referred to as second- or third-generation, are still in the research and development (R&D), pilot or demonstration stage. A few examples of advanced biorefineries, in particular biofuel plants, at commercial stage already exist although struggling to compete with the equivalent and cheaper first-generation plants. In this chapter, the terminology "advanced" biorefineries includes production technologies based on the use of nonfood crops as feedstocks.

9.4.1 *Biochemical-Based Advanced Lignocellulosic Biorefineries*

Although there are many possible ways for the transformation of lignocellulosic biomass into biofuels and bioproducts, many products that are familiar today, including citric acid, ethanol, and lactic acid, are mainly fermentation-based (Zwart 2006). Indeed, carbohydrates can be converted through fermentation, by bacteria, fungi ,or yeast (genetically modified or not), into alcohols, organic acids, alkenes, lipids ,and other chemicals, under diverse process conditions (e.g., low/high pH, anaerobic/aerobic, nutrient rich/deprived). Monosaccharides and disaccharides are very widely converted whereas larger molecules such as oligosaccharides and polysaccharides (e.g., starch) are not easily metabolized by a wide range of organisms, and thus the latter are not included within the so-called "sugar platform". This terminology addresses any combination of C5 carbohydrates (pentoses, e.g., xylose, arabinose, ribose), C6 carbohydrates (hexoses, e.g., glucose, fructose, galactose), and/or C12 carbohydrates (e.g., saccharose, lactose, maltose) that exists within a pathway from biomass feedstock toward final biofuel or biochemical products (E4tech, RE-CORD and WUR 2015).

Thereby, under a "sugar platform" pathway, the achievement of carbon sources metabolically active from complex lignocellulosic substrates requires the prior degradation of its constituent polysaccharides, cellulose, and hemicellulose fractions, into their building blocks (Gírio et al. 2010). Therefore, there is a need of a mechanical comminution followed by a biomass pretreatment before application of an enzymatic process (van Wyk 2011), as shown in Fig. 9.13. The main purpose of any pretreatment is thus to free cellulose and hemicellulose fractions from the

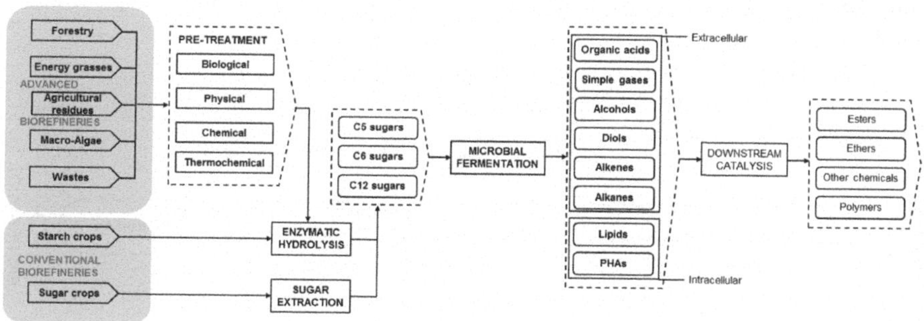

Fig. 9.13 Biochemical-based biorefineries—pathways via the sugar platform (adapted from E4tech, RE-CORD and WUR 2015)

lignin, reduce cellulose crystallinity, and increase the porosity of the material (Cheng and Timilsina 2011; Geddes et al. 2011; Zhang 2008).

Pretreatment has been recognized as the main bottleneck and one of the most expensive processing steps in cellulosic biomass-to-fermentable sugars conversion. In addition, improved bioprocesses through a better integration of the engineering and biology operation units (e.g., integration of pretreatment, enzymatic hydrolysis, fermentation and downstream processing technologies) should also be achieved to bring a bio-based process using lignocellulose feedstocks to commercial scale (e.g., Proethanol2G EU 7th FWP project website). Some of commercial lignocellulosic pretreatment technologies (e.g., PROESA™ Technology, Beta-Renewables, Italy; Liberty™ Technology, POET/DSM, USA; Abengoa, USA) are already available for the production of bioethanol and co-products from a wide variety of woody material, wastes, and other residues, contributing to the deployment of advanced biorefineries using raw materials other than readily available sugar and starch feedstocks. However, these advanced biorefineries are not yet mature but still under development.

These biochemical-based advanced bioethanol biorefineries in the World might be distinguished by the different technologies applied as pretreatment, which necessarily promotes a distinct biomass fractionation, as represented in Fig. 9.14 (E4tech, RE-CORD and WUR 2015).

The pretreatment technologies are grouped by process type, existing biological (microbial/fungi), physical, chemical, and thermochemical process options. Often, they differ according to the way how they extract/solubilize the lignin fraction during pretreatment or in a later stage in value chain (corresponding to the two distinct dashed arrows in Fig. 9.14). Indeed, lignin separation can be promoted by applying strong alkaline conditions (e.g., with sodium hydroxide and sodium sulphite with high temperature, such as used in kraft pulping), dilute and concentrated acids, organic solvents (as organosolv) or ionic liquids (Bacovsky et al. 2013).

Europe is leading research on lignin removal, with several companies conducting pilot/demo activities on this step, such as GreenValue, Switzerland or Borregaard Industries AS that owns and operates the World's most advanced biorefinery, producing advanced biochemicals (e.g., vanilla flavor vanillin), biomaterials (e.g. specialty celluloses) and bioethanol, using all the components of wood.

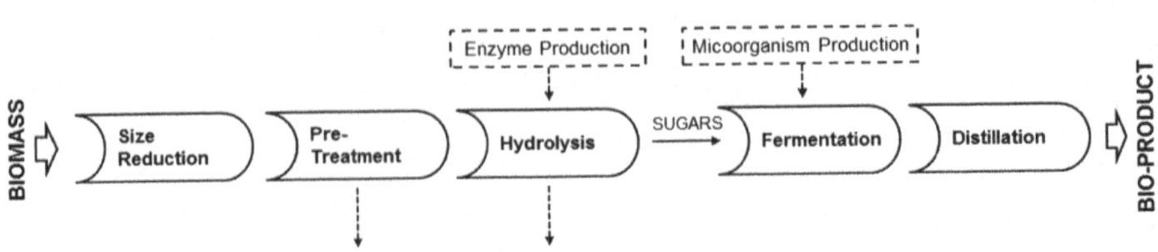

Fig. 9.14 Processing steps in a general biochemical-based lignocellulosic biorefinery (adapted from IEA-Bioenergy 2014)

The more mature pretreatment technologies include acid hydrolysis and hydrothermal (e.g., hot liquid water, steam explosion) pretreatments, the latter typically do not liberate lignin but solubilizes and hydrolyze hemicellulose component to different extents (Carvalheiro et al. 2004). Although steam explosion can be performed as uncatalyzed treatment (Glasser and Wright 1998), it is often applied as acid-catalyzed steam explosion (Sassner et al. 2008) or ammonia fibre explosion (AFEX) (Sendich et al. 2008). Wet oxidation treatment is also a form of hydrothermal treatment involving the use of oxygen and water at elevated temperatures and pressure, promoting the oxidation of lignin and its conversion into CO_2, H_2O, and carboxylic acids.

There are other emerging pretreatment technologies, such as supercritical (CO_2) pretreatment, ionic liquids and microbial (with fungi and bacteria) treatments (E4tech, RE-CORD and WUR 2015; Bacovsky et al. 2013), but these are still at research level, i.e., at a Technology Readiness Level (TRL) below 5, and thus commercial, or even demonstration or pilot processes are inexistent, and these will be out of the scope of this text.

Indeed, this chapter deals with projects involving the conversion of sugars to bioproducts (fuels, materials, or chemicals) via novel pathways, having commercialization as target following applied research and developments from pilot-scale (and above) within companies (not academia). Given this, once bioethanol is the sugar-based bioproduct at higher TRL in terms of commercialization, with first-of-a-kind commercial plants already operational, priority focus will be given to lignocellulosic bioethanol plants, rather than different fuels and chemicals.

Significant advances have been achieved in the past several years in all aspects of lignocellulose conversion into ethanol making it potentially competitive in economic terms, particularly for niche markets (Gray et al. 2006). Thus, whereas starch-based ethanol and sugarcane-based ethanol are now mature industries, companies such as Beta-Renewables/Biochemtex (Crescentino, Italy), Inbicon/Dong Energy (Kalundborg Denmark), Abengoa (Babilafuente, Spain), Clariant (Straubing, Germany) in Europe; Abengoa (Hugoton, Kansas), Blue Sugars Corporation (formerly KL Energy Corp.), BP Biofuels (formerly Verenium), Enchi Corporation (formerly Mascoma Corporation); Dupont (Nevada, Iowa) and POET-DSM (Emmetsburg, Iowa) in USA; Iogen Corporation in Canada; GranBio (Alagoas) and Raízen/Iogen (Piracicaba) in Brazil, are giving the first steps as major players to commercialize cellulosic ethanol. The existent pilot and demonstration plants will serve as platforms to identify bottlenecks and potential barriers to full commercialization of 2G bioethanol in the future.

Given this, a brief overview of the current worldwide most relevant (preferentially closest to commercial as possible) industrial activities on 2G bioethanol manufacturing, based on different pretreatment technologies, will be presented (Table 9.3). In addition, records containing more detailed process information were built for eight of these facilities, selected as the most representative (in terms of maturity) of each technological approach.

Table 9.3 Most relevant (commercial, demo or pilot) cellulosic ethanol production facilities in operation in the World

Project owner/Name	Location	Technology (pretreatment)	Typology	Raw material	Product(s) (total capacity)	Start-up link
POET/DSM Advanced Biofuels/ Project LIBERTY	Emmetsburg, Iowa, USA	Two-stage acid-catalyzed steam explosion	Commercial	Corncobs and corn stover	Ethanol (60,000 t/year), biogas and CHP (from lignin)	2014 http://www.projectliberty.com/
GranBio	São Miguel dos Campos, Alagoas, Brazil	PROESA™—uncatalyzed steam explosion	Commercial	Sugarcane straw	Ethanol (65,000 t/year) and 50 MWe (from lignin and bagasse)	2014 http://www.granbio.com.br/
Biochemtex/ Beta Renewables	Crescentino, Italy	PROESA™—uncatalyzed steam explosion	Commercial	Wheat straw (currently). Later on, Arundo donax.	Ethanol (60,000 t/year), biogas, 13 MW electricity and animal feed	2013 www.betarenewables.com/
Abengoa Bioenergy Biomass of Kansas	Hugoton, Kansas, USA	Sulfuric acid-catalyzed steam explosion	Commercial	Corn stover	Ethanol (75,000 t/year) and 18 MWe electricity	2015 http://www.abengoabioenergy.com/web/en/2g_hugoton_project/
Raízen Energia Participações S/A	Piracicaba, S. Paulo, Brazil	Iogen's Technology-acid-catalyzed steam explosion	Commercial	Sugarcane bagasse and straw	Ethanol (32,000 t/year) + electricity	2015 http://www.raizen.com/
Dupont	Nevada, Iowa, USA	Dilute ammonia hydrolysis with heat	Commercial	Corn stover	Ethanol (90,000 t/year) + CHP (from lignin)	2015 http://www.dupont.com/
Inbicon/DONG Energy	Kalundborg, Denmark	Autohydrolysis	Demonstration	Straw, corn stover, and bagasse	Ethanol (4300 t/year), C5 molasses (13,900 t/year) and lignin pellets (11,400 t/year)	2009 http://www.inbicon.com/en

(continued)

Table 9.3 (continued)

Project owner/Name	Location	Technology (pretreatment)	Typology	Raw material	Product(s) (total capacity)	Start-up link
Blue Sugars Corp. (former KL Energy) and Petrobras	Uton Wyoming, USA	Dilute-acid impregnation combined with thermal–mechanical	Demonstration	Sugarcane bagasse and other lignocellulosic biomass	Ethanol (4500 t/year) and lignin	2009 http://www.bluesugars.com/
Abengoa Bioenergía	Babilafuente Salamanca, Spain	Sulfuric acid-catalyzed steam explosion	Demonstration	Cereal straw (mostly barley and wheat)	Ethanol (4000 t/year)	2009 http://www.abengoa-bioenergy.com/web/es/
BP Biofuels/ Jennings Demonstration Facility	Jennings, Louisiana, USA	Dilute-acid impregnation combined with thermo-mechanical	Demonstration	Dedicated energy crops	Ethanol (4200 t/year)	2009 http://www.bp.com/en_br/brazil/o-que-fazemos/biocombustiveis.html
Borregaard AS/BALI Biorefinery Demo	Sarpsborg, Norway	BALI™ process—sulphite pretreatment (continuous technology, with patent applied in 2009)	Demonstration	Sugarcane bagasse, straw, wood, energy crops, other lignocellulosic biomass	Ethanol (110 t/year) or C5/C6 sugars (220 t/year) and lignin specialty chemicals (200 t/year)	2013 http://www.borregaard.com/
Biogasol APS/ BornBioFuel 2	Aakirkeby, Bornholm, Denmark	Carbofrac™—catalyzed “wet explosion”	Demonstration	Straw, various grasses and garden waste	Ethanol (4000 t/year), biogas, lignin and hydrogen	2015/2016 http://www.biogasol.com/
Iogen Corporation	Ottawa, Canada	Modified acid-catalyzed steam explosion	Demonstration	Cereals straw, corn stover,	Ethanol (1600 t/year)	2007

(continued)

Table 9.3 (continued)

Project owner/Name	Location	Technology (pretreatment)	Typology	Raw material	Product(s) (total capacity)	Start-up link
				sugarcane bagasse		http://www.iogen.ca/
Chempolis Ltd./Chempolis Biorefining Plant	Oulu, Finland	Formico™—organosolv with formic and acetic acids	Demonstration	Non-wood biomass—e.g., straw, reed, empty fruit branch, bagasse, corn and stalks	Ethanol (5000 t/year) and paper pulp from cellulose, various chemicals (e.g., acetic acid and furfural as chemical intermediates) from hemicellulose	2008 http://www.chempolis.com/
Clariant (former Süd-Chemie)	Straubing, München, Germany	Sunliquid™ process –with mechanical and thermal pretreatment	Pilot	Wheat straw and other agricultural residues	Ethanol (1000 t/year)	2012 http://www.sunliquid.com/
Enchi Corporation (former Mascoma Corp.)	Rome, New York, USA	CBP with little or no pretreatment —only mechanical disruption or hydrothermal	Pilot	Wood chips, switchgrass and other lignocellulosic biomass	Ethanol (500 t/year) and lignin	2003 http://www.enchicorp.com/
Abengoa Bioenergía	York, Nebraska, USA	Sulfuric acid-catalyzed steam explosion	Pilot	Corn stover	Ethanol (75 t/year)	2007 http://www.abengoa-bioenergy.com/web/es/
SP Plant (former SEKAB)/	Örnsköldsvik, Sweden	Two-stage Dilute-acid hydrolysis with/without SO_2 impregnation	Pilot	Softwood, sugarcane straw and bagasse	Ethanol (160 t/year), biogas and lignin	2005 http://www.sekab.com/biorefinery/demo-plant/

(continued)

Table 9.3 (continued)

Project owner/Name	Location	Technology (pretreatment)	Typology	Raw material	Product(s) (total capacity)	Start-up link
FUTUROL/ Procethol2G	Pomacle, France	Acetic acid and formic acid-based organosolv pretreatment	Pilot	Wheat and barley straw, corn stover, wood, sugarcane bagasse	Ethanol (700 t/year); C5 syrup (650 t/year); Lignin (750 t/year)	2011 www.projet-futurol.com
Weyland AS	Bergen, Norway	Concentrated acid hydrolysis	Pilot	Mostly wood (spruce and pine) and agricultural residues	Ethanol (158 t/year) and lignin	2010 http://www.weyland.no/
Virdia (former HCL Clean Tech)	Danville, VA, USA	CASE™ process (with concentrated HCl)	Pilot	Mostly softwood (pine) and hardwood (eucalyptus)	Cellulosic sugars and lignin	2012 http://biomaterials.storaenso.com/about-us/
Lignol Innovations Ltd.	Burnaby, BC, Canada	Alcell™ process—organosolv with ethanol (unique fractionation into cellulose, hemicellulose and lignin)	Pilot	Woody biomass—hardwood and softwood residues	Ethanol, pure lignin and other valuable co-products	2009 https://www.sdtc.ca/en/organizations/lignol-innovations-ltd
CIMV	Pomacle, Marne, France	Organosolv with formic acid and acetic acid	Pilot	Wheat straw	Paper pulp and glucose from cellulose, C5-sugars from hemicellulose, and lignin	2006 http://www.cimv.fr/

BETA-RENEWABLES

Process/ Technology	PROESA™—uncatalyzed steam explosion
Project owner	Biochemtex/Novozymes
Web page	http://www.betarenewables.com/
Facility type	Commercial
Operation	Since 2013
Feedstock	Started with wheat straw and they are planning to use energy crops (*Arundo donax* (giant reed), miscan-thus, and switch grass) in the future
Outputs	Ethanol (60,000 t/year) and biogas
Location	Crescentino, Italy
Summary	Beta Renewables is a joint venture established at the end of 2011 between Biochemtex, a company of the Mossi Ghisolfi Group, and the USA fund TPG (Texas Pacific Group), and Nozozymes, World leader in the enzymes industry that became a shareholder of the company in 2012. PROESA™ technology, basically consisting of an uncatalyzed steam explosion pretreatment followed by simultaneous saccharification and fermentation using a strain engineered to convert both C5 and C6 sugars to ethanol. PROESA has been already licensed to other World biofuel companies (e.g., GranBio, Brazil). Residual lignin is only used for power

POET/DSM (LIBERTY project)

Process/ Technology	Two-stage dilute-acid hydrolysis
Project owner	POET/DSM Advanced Biofuels
Web page	http://www.projectliberty.com/
Facility type	Commercial
Operation	Since 2013
Feedstock	Corncobs and corn stover
Outputs	Ethanol (75,000 t/year) and biogas
Location	Emmetsburg, Iowa, USA
Summary	POET-DSM Advanced Biofuels, is a joint venture between POET and Royal DSM. The new commercial cellulosic ethanol plant is based on dilute-acid or acid catalyzed steam explosion followed by enzymatic hydrolysis with enzymes provided by DSM. They plan to license this integrated technology package, converting corn crop residue to cellulosic bioethanol, to third parties and also to the other 26 existing corn ethanol plants in POET's network (Bacovsky et al. 2013)

ABENGOA

Process/ Technology	Steam explosion
Project owner	Abengoa USA
Web page	http://www.abengoa-bioenergy.com/web/es/
Facility type	Commercial

(continued)

(continued)

Operation	Since 2015
Feedstock	Corncobs and corn stover
Outputs	Ethanol (75,000 t/year) and biogas
Location	Hugoton, Kansas, USA
Summary	This Spanish company did build an advanced biofuel plant at US benefiting from the generous governmental incentives for deployment advanced biofuel plants. It uses a steam explosion technology as biomass pretreatment and a property in-house enzyme production for enzymatic hydrolysis step

BIOFLEX

Process/ Technology	Two-stage steam explosion
Project owner	GranBio
Web page	http://www.granbio.com.br/
Facility type	Commercial
Operation	Since 2015
Feedstock	Sugarcane straw
Outputs	Ethanol (65,000 t/year) and 50 MWe electricity
Location	Alagoas, Brazil
Summary	Bioflex plant is a sugarcane straw-based 2G bioethanol plant and integrated in the same site of 1G ethanol and sugar plant. They bought the license to operate with PROESA™ Technology from the Italian Beta-renewables which is an uncatalyzed two-stage steam explosion technology. Besides ethanol, they co-produce 70 MWe, having a surplus of 50 MWe which they sells to the grid

Enchi Corporation

Process/ Technology	CBP with little or nopretreatment—only mechanical disruption or hydrothermal
Project owner	Enchi Corporation (former Mascoma Corp.)
Web page	http://www.enchicorp.com/
Facility type	Demonstration
Operation	Since 2003
Feedstock	Wood chips, switchgrass, and other lignocellulosic biomass
Outputs	Ethanol (500 t/year) and lignin
Location	Rome, New York, USA
Summary	Enchi Corporation (former Mascoma Corp.) combines enzyme production, hydrolysis and fermentation in a single step (consolidated bioprocessing or CBP). It uses thermophilic bacteria that are engineered to produce large amounts of the enzymes promoting cellulose hydrolysis and ferment the resulting sugars into ethanol (Bacovsky et al. 2013). Biomass is processed without external enzymes and with little or no pretreatment, thus avoiding the two main barriers to cost-effective processing today. Mascoma Corporation was a USA biofuel company founded in 2005 by Robert Johnsen (CEO), Lee Lynd and Charles Wyman, two professors from Dartmouth College, to produce cellulosic ethanol made from wood and switchgrass (Mascoma 2007). In November 2014, the yeast-related business assets including the Mascoma name were purchased by Lallemand, Inc. of

(continued)

(continued)

	Montreal, Canada, and renamed Mascoma LLC which is now a subsidiary of Lallemand. The remaining business assets of the former Mascoma Corp., including the thermophilic bacteria technology and the pilot plant in Rome (NY) were renamed Enchi Corporation (Wikipedia 2016). Enchi Corporation, together with J.M. Longyear, under the joint venture Frontier Renewable Resources, is planning to build a commercial 2G ethanol facility (with 60,000 t-annual capacity) in Kincheloe (Michigan, USA) (Bacovsky et al. 2013)

INBICON

Process/ Technology	Autohydrolysis
Project owner	Inbicon (DONG Energy)
Web page	http://www.inbicon.com/
Facility type	Demonstration
Operation	Since 2009
Feedstock	Straw, corn stover, and bagasse
Outputs	Ethanol (4300 t/year), C5 molasses (13,900 t/year), and lignin pellets (11,400 t/year)
Location	Kalundborg, Denmark
Summary	DONG Energy is aiming to license their Inbicon technology and know-how to partners around the World to build commercial scale Inbicon Biomass Refineries. Inbicon's core technology consists of a three-stage continuous process: mechanical conditioning of the biomass; hydrothermal pretreatment (at 180–200 °C); followed by enzymatic hydrolysis at a high dry matter level (up to 30% of dry matter), providing continuous liquefaction. Hydrothermal pretreatment at Inbicon consists of an autohydrolysis by using compressed hot water processes with no added chemical catalyst. This pretreatment is economically and environmentally attractive, generating low levels of inhibitory products, with no significant effect on lignin and cellulose. However only solubilizes hemicellulose in oligomeric form (Carvalheiro et al. 2004) In Inbicon process—version 1—C5 sugars derived from hemicellulose, are separated into a liquid fraction that is further concentrated, and marketed as C5 molasses for biogas or animal feed. Since mid-2013, Inbicon did implement the version 2 technology using C5/C6 mixed sugar fermentation to ethanol. Thereby, in addition to ethanol, the demo plant only co-produces lignin pellets used as feedstock for power production (integrated with Asnæs power station in Kalundborg (Dong Energy power plant), allowing the highest possible energy efficiency

CLARIANT

Process/ Technology	Sunliquid™ process—with mechanical and thermal pretreatment
Project owner	Clariant
Web page	http://www.sunliquid.com/

(continued)

(continued)

Facility type	Demonstration
Operation	Since 2012
Feedstock	Wheat straw and other agricultural residues
Outputs	Ethanol (1000 t/year)
Location	Straubing, München, Germany
Summary	Clariant (former Süd-Chemie) has developed the Sunliquid™ process for converting agricultural residues into biofuel, based on a mechanical and thermal pretreatment and an integrated process including enzyme production (using pretreated feedstock as substrate), optimized enzymes, simultaneous conversion of C5 and C6 into ethanol and innovative (based on adsorption technology) highly energy saving purification method (as alternative to conventional distillation) (Bacovsky et al. 2013). Based on this Sunliquid™ process, Clariant has been successfully operating a first pilot plant at its research facility [Clariant's Biotech & Renewables Center (BRC)] in Munich since 2009, producing up to two tons of ethanol per year. In July 2012, this Germany's largest plant to date started into operation in Straubing. Once designed for industrial plants with a production capacity of 50,000–150,000 t of cellulosic ethanol per year, Clariant intends to license out Sunliquid™ technology

IOGEN

Process/ Technology	Modified acid-catalyzed steam explosion
Project owner	Iogen Corporation
Web page	http://www.iogen.ca/
Facility type	Demonstration
Operation	Since 2007
Feedstock	Cereals (wheat, oat, or barley) straw, corn stover, sugarcane bagasse ,and other agricultural residues
Outputs	Ethanol (1600 t/year)
Location	Ottawa, Canada
Summary	Iogen's cellulosic ethanol process is based on separate hydrolysis and fermentation using a multi-stage hydrolysis process. The process involves C6 and C5 fermentation, and includes enzymes production once Iogen Corporation is also a manufacturer and marketer of enzyme products for application involving fiber modification or hydrolysis (e.g., for pulp and paper, grain processing, brewing, textile, and animal feed industries). In 2009, Iogen became the first cellulosic ethanol producer to sell its advanced biofuel at a retail service station (Bacovsky et al. 2013). For a one-month period, a 10% cellulosic ethanol blend was available for sale to the general consumers at an Ottawa Shell station. More than 60,000 liters of Iogen's cellulosic ethanol was sold (Iogen 2016). Iogen is seeking partners and licensees to commercialize its technology, and the Brazilian ethanol company Raízen Energia Participações S/A, the nation's leading manufacturer of sugarcane ethanol, is operating its commercial biomass-to-ethanol facility (adjacent to Raízen's Costa Pinto sugarcane mill) using Iogen Energy's advanced cellulosic biofuel technology

BornBioFuel 2

Process/ Technology	Carbofrac™—catalyzed 'wet explosion'
Project owner	Biogasol APS
Web page	http://www.biogasol.com/
Facility type	Demonstration
Operation	Since 2013
Feedstock	Straw, various grasses and garden waste
Outputs	Ethanol (4000 t/year), biogas, lignin, and hydrogen
Location	Aakirkeby, Bornholm, Denmark
Summary	BioGasol APS, company founded in 2006 as a spin-out of the Technical University of Denmark (DTU), combines wet oxidation with steam explosion as core technology. The process is called 'wet explosion' and the use of oxygen and pressure release at high temperature (170–200 °C) are combined, as a variant of conventional hydrothermal treatment. Additives such as acid, base, or solvents act as catalysts in the process, and dilute acid-catalyzed steam explosion pretreatment is BioGasol's reference process. All by-products are further converted to energy carriers (ethanol, hydrogen, methane and solid biofuel). The pretreatment technology developed, Carbofrac™, is licensed by the company, together with high-yield C5/C6 co-fermentation technology Pentoferm™ using the unique thermophilic Petocrobe™ (a recombinant derived from *Thermoanaerobacter italicus*) (Bacovsky et al. 2013)

DuPont cellulosic ethanol demonstration plant

Process/ Technology	Dilute ammonia hydrolysis
Project owner	DuPont
Web page	http://www.dupont.com/
Facility type	Demonstration
Operation	Since 2010
Feedstock	Corn stover, switchgrass, and sugarcane bagasse
Outputs	Ethanol (750 t/year), biogas, lignin and hydrogen
Location	Vonore, Tennessee, USA
Summary	DuPont Genencor Science, a joint venture between Danisco, Genencor and DuPont, is the only company applying alkaline processing for biomass pretreatment on industrial scale (Bacovsky et al. 2013). The developed conversion process, including novel enzymes and fermentation organisms, will be deployed at the first commercial DuPont Cellulosic Ethanol plant, which will be nearing operation in Nevada (Iowa) and it will produce ethanol from corn stover (corncobs, leaves, and stalks). DuPont offers cellulosic ethanol licenses to other companies, supplying enzymes and ethanologens for production at a licensed plant

9.4.2 Biochemical-Based Advanced Algae Biorefineries

Despite the intense effort in R&D in microalgae in recent years, not only in the food, cosmetic, and pharmaceutical fields but also in the production of conventional biofuels (biodiesel, bioethanol) and other more advanced than not even commercial phase (long-chain alcohols, hydrogen, hydrocarbons, and jet-fuel), the development and the transition to the desired scale of demonstration have not been so accelerated as expected in Europe (Gírio et al. 2013) and worldwide. The theoretical maximum productivity in oil from microalgae (354 m^3 ha^{-1} $year^{-1}$) is several orders of magnitude higher than the most productive terrestrial oleaginous culture so far (palm) but still very remote and unlikely from real values obtained according with the present state-of-the-art and technical knowledge. This gap is mainly due to technological barriers (Nurra et al. 2014; González et al. 2015; Jacob-Lopes et al. 2015) and problems of scale-up (Lopes da Silva and Reis 2015) that should be overcome at a short-term timeframe, in particular with regard to mild, simple, sustainable, and inexpensive methods of biomass pre-concentration and harvesting, cell disruption, and extraction of target components for a wide range of applications, in particular for bio-energy. The still high cost of producing microalgal biomass makes it unattractive for "high volume-low cost" applications as in the case of biofuels. The present costs of producing microalgal-based oils (for biodiesel FAME) are one order of magnitude higher than the FAME biodiesel made from vegetable oils (food). On the other hand, the concept of biorefinery applied to microalgal biomass which would lower production costs of biofuels through the integrated co-production of high value-added products to pay the technology is facing difficulties to be an expanded reality (Gao et al. 2012; Gírio et al. 2013; Hariskos and Posten 2014; Lopes da Silva and Reis 2015). As a result, it is not surprising that, to date, the number of demonstration facilities for the production of bioenergy vectors from micro and macro-algae, solar radiation and CO_2 as well as for upgrading to biofuels for use in carrier sector is limited worldwide and the available data are scarce. There have been some large-scale facilities for the production of microalgae in the Netherlands (AlgaePARC, Wageningen), southern Europe (Portugal and Spain), USA (Sapphire energy), Brazil (Solazyme), and India with operational flexibility that can take them to shift their conversion for bioenergy production if the conditions and markets are favorable. Among them is the greatest pilot installation that exists in Europe (Allmicroalgae, Portugal) which has 1200 m^2 of photobioreactors operation to a 1300 m^3 microalgae production volume, based in Leiria-Pataias, Portugal. This facility will allow accelerating and to implement the R&D necessary for the cultivation of microalgae in order to demonstrate the real potential as a sustainable and economically competitive technology for the production of advanced biofuels. In this sense, it is to be hoped that new flagship projects arise in a short-term timeframe. The use of microalgae coupled to wastewater treatment for energy emerges as a very promising possibility for its lower cost and the environmental benefits associated with. A selection of the most important processes about biofuels production by microalgae-based processes in the frame of the biorefinery concept is presented next with special emphasis on selected cases in which the selection criteria was the maturity of the process and the existence of demonstration or commercialization units, if available (Petrick et al. 2013). The main information available is organized in table format for easy reading and understanding.

Allmicroalgae/AlgaFarm

Process/ Technology	Photoautotrophic microalgal production with CO_2 fixation
Project owner	Cecil/A4F
Web page	www.a4f.pt
Facility type	Demo/Commercial
Operation	Since 2013
Feedstock	CO_2, fertilizers
Outputs	Microalgal biomass (*Chlorella*), Pigments (possibility of reconversion to biofuels)
Location	Pataias, Portugal
Summary	The cement Portuguese company Secil and the biotech Portuguese company A4F, founded AlgaFarm, a joint venture to develop a process that initially used the combustion gases (CO_2) from the cement industry in the production of microalgae. The industrial-scale production unit resulted from the expansion of a pilot plant that has been in operation for two years. It is currently in commercial operation and produces *Chlorella vulgaris* directed for the food segment The AlgaFarm was designed with four main sectors: resources, production, processing, and management. Each sector covers several systems that together constitute the largest production facilities of microalgae Worldwide with closed photobioreactors, reaching 1300 m^3 of total volume of production The downstream processing of biomass includes harvesting through ultrafiltration to a biomass concentration between 5 and 10% on dry weight basis. The treatment includes pasteurization, spray drying ,and final packaging under protective atmosphere free from O_2. The AlgaFarm is currently a microalgae plant equipped with data acquisition system and fully automated control Allmicroalgae it is the new supplier of Allma Chlorella—and has unveiled plans for a significant new phase of investment in its production facilities Allmicroalgae is now selling Chlorella ingredients under the Allma brand, with production continuing at the existing Algafarm plant in Leiria, 100 km north of Lisbon. This site will be revamped—according to the company announcement—with new, state-of-the-art production technology that will boost output and increase quality levels

A4F

Process/ Technology	Direct Bioethanol production from autotrophic microalgae
Project owner	A4F
Web page	www.a4f.pt
Facility type	Pilot

(continued)

(continued)

Operation	Since 2014
Feedstock	CO_2, fertilizers, microalgae
Outputs	Bioethanol (production culture volume: 3 m^3)
Location	Lisbon, Portugal
Summary	A4F AlgaFuel, S.A. is a spin-out from Necton S.A. company, dedicated to the development and implementation of bioengineering projects for the industrial production of microalgae. This unit is the first pilot unit of microalgae "GMO-compliance" in Europe, which allows the production and R&D of microalgae genetically recombined without risk of release into the environment

Algaenergy

Process/ Technology	Photoautotrophic microalgal production with CO_2 fixation
Project owner	Various companies among which are Repsol and Iberdrola
Web page	http://www.algaenergy.es
Facility type	Pilot
Operation	Since 2012
Feedstock	CO_2, fertilizers
Outputs	Microalgae, biodiesel (production culture volume: 40 m^3)
Location	Adolfo Suárez Airport, Barajas, Spain
Summary	The Spanish company AlgaEnergy, owned by various companies among which are Repsol and Iberdrola, leads the project named CO_2Algaefix which consists in carbon dioxide capture to the preindustrial scale on the basis of the development and initial experience obtained in the pilot installation

Buggypower S.L.

Process/ Technology	Photoautotrophic microalgal production with CO_2 fixation
Project owner	Partnership between Empresa de Electricidade da Madeira (EEM) and BioFuel Systems (http://www.biopetroleo.com/)
Web page	http://www.buggypower.eu
Facility type	Demo
Operation	Since 2011
Feedstock	CO_2, fertilizers
Outputs	Microalgae, biodiesel (culture volume capacity: 900 m^3 photobioreactors)
Location	Porto Santo, Portugal
Summary	Buggypower, has built and currently operates a facility for the production of biofuels with microalgae, on the island of Porto Santo, Madeira, Portugal. This innovative project has as main objective to replace the fuel from fossil resources, currently used in EEM thermal power station with biofuel produced by microalgae. When optimized this project, Porto Santo will become a fully sustainable island and "eco-friendly", in terms of energy supply and use

Kalundborg symbiosis

Process/ Technology	Mixotrophic microalgal production with wastewater treatment
Project owner	Public and private enterprises (industrial cluster)
Web page	www.symbiosis.dk/en
Facility type	Demo
Operation	Since 2013
Feedstock	Wastewater
Outputs	Microalgae; Treated wastewater (40 m^3 photobioreactors)
Location	Kalundborg, Dennmark
Summary	"Kalundborg Symbiosis" is an industrial ecosystem, where a residue or by-product of an industrial company is used as a resource by another company, in a closed loop. This industrial symbiosis is a local collaboration with several public and private companies that buy and sell their waste products one to each other, resulting in mutual benefits want to either environmental economic. The demo unit of microalgae is located within an industrial cluster and benefit from synergies, including energy efficiency

Bäckhammars Algbruk

Process/ Technology	Mixotrophic microalgal production for biodiesel with wastewater treatment
Project owner	SP Technical Research Institute of Sweden
Web page	Not available
Facility type	Pilot
Operation	Since 2014
Feedstock	Pulp and paper wastewater
Outputs	Biodiesel
Location	Bäckhammars, Sweden
Summary	The SP Technical Research Institute of Sweden leads a project named "Bäckhammars Algbruk" in partnership with 11 other partners such as Perstrop (expert in specialty chemicals). The project aims to study the use of microalgae in the production of biofuels

TNO-Valorie

Process/ Technology	Mixotrophic microalgal production for biodiesel with wastewater treatment
Project owner	GAIA consortium created by TNO (Netherlands Organization for Applied Scientific Research)
Web page	Not available
Facility type	Pilot
Operation	Since 2014
Feedstock	Microalgae
Outputs	Specialized compounds and biodiesel
Location	Mobile plant/factory, The Netherlands

(continued)

(continued)

Summary	The project T-Valorie (VALORIE: "Versatile Algae on-site Raw Ingredient Extractor") aims at the construction of mobile factories used in the extraction, fractionation and specialized compounds and biodiesel production from microalgae. The objective is to adapt the mobile plant according to the type of raw material and existing technologies as well as the desired products. Another objective is to generate knowledge about the needs of downstream processing of biomass collected and business opportunities for the refined products so that the producers can assess the market value and the economic viability of the process. This project is owned by the industrial consortium called GAIA, created by TNO ("Netherlands Organization for Applied Scientific Research"). The consortium comprises the Algae Food and Fuel B.V., TNO, and a group of companies including Sabic, De Wit Oils and Van Varese Paint

AlgaePARC

Process/ Technology	Autotrophic microalgal production for biofuels and other products
Project owner	WUR (Wageningen University)
Web page	http://www.wageningenur.nl/en/Expertise-Services/Facilities/AlgaePARC.htm
Facility type	Pilot
Operation	Since 2011
Feedstock	CO_2
Outputs	Microalgae, other products (7 m^3 outdoors photobioreactors; 2 m^3 indoors photobioreactors
Location	Wageningen, The Netherlands
Summary	The AlgaePARC is a research facility located at the Wageningen University dedicated to the development of methods of outdoors sustainable production of microalgae with a competitive cost

Algenol biofuels

Process/ Technology	Autotrophic microalgal production for bioethanol and other products
Project owner	Algenol Biotech LLC
Web page	http://www.algenol.com
Facility type	Demo
Operation	Since 2013
Feedstock	CO_2, saltwater
Outputs	Ethanol, spent microalgae to biodiesel, gasoline ,and jet fuel (3 acre area with outdoor photobioreactors)
Location	Fort Myers, Florida, USA
Summary	Algenol is a global, industrial biotechnology company that is commercializing its patented algae technology platform for production of ethanol using proprietary algae, sunlight, carbon dioxide, and saltwater. Algenol takes advantage of enhancing a natural ability found in many

(continued)

(continued)

	strains of cyanobacteria to produce ethanol by over expressing fermentation pathway enzymes channeling the majority of photosynthetically fixed carbon into ethanol production rather than routine cell maintenance. It is claimed that it is the only renewable fuel production process that can convert more than 85% of its CO_2 feedstock into fuel

BioProcess Algae, LLC

Process/ Technology	Autotrophic microalgal production for animal feeds, fish feeds, nutraceuticals, chemicals, and transportation fuels
Project owner	BioProcess Algae LLC [a joint venture between Clarcor (NYSE: CLC), BioHoldings, Ltd. and Green Plains Inc. (NASDQ: GPRE)]
Web page	http://www.bioprocessalgae.com
Facility type	Demo
Operation	Since 2009
Feedstock	Ethanol plant's CO_2 exhaust gas; (waste heat from the ethanol plant)
Outputs	Microalgae, animal feeds, fish feeds, nutraceuticals, chemicals and transportation fuels
Location	Shenandoah, Iowa, USA (shared with Green Plains Inc. ethanol plant)
Summary	BioProcess Algae, LLC designs, builds, and operates commercial scale bioreactors that enable efficient conversion of light and CO_2 into high-value microbial feedstock. The competitive advantage of BioProcess Algae cultivation technology involves the use of thin biofilms. The technology at the heart of BioProcess Algae cultivators is a unique high surface area, biofilm-based approach to enhance light penetration, productivity, harvest density, and gas transfer—all traditional bottlenecks to low-cost algae cultivation

Sapphire Energy, Inc.

Process/ Technology	Autotrophic microalgal production for omega-3 oils, high-value aquaculture and animal feed ingredients and renewable fuels
Project owner	Sapphire Energy, Inc
Web page	http://www.sapphireenergy.com/
Facility type	Demo
Operation	Since 2009
Feedstock	CO_2, non-potable water
Outputs	Microalgae, animal feeds, fish feeds, nutraceuticals, chemicals, and transportation fuels
Location	Las Cruces, New Mexico, USA (one million liter production ponds)
Facility type	Commercial (Algae Farm)
Operation	Planned to start by 2017
Feedstock	CO_2 (56 metric tons per day); non-potable water
Outputs	Microalgae (1600 tons of biomass per year), omega oils, high-value protein, fuels
Location	Columbus, New Mexico, USA (300 cultivated acres for cultivation)

(continued)

(continued)

Summary	Sapphire's technology uses sunlight, CO_2, non-potable water, nonarable land, nutrients, and novel strains of algae in outdoor ponds to produce algae which then are converted into high-value oils, aquaculture and animal feeds, fuels, and other valuable products. The process is claimed to be economic, scalable, and sustainable as there are not competition for farmable land or fresh water, together with the absorption and reuse of CO_2 emissions. Sapphire claims to operate the largest algae-to-energy facility in the World. In the desert scrub outside of Columbus, New Mexico, Sapphire Energy is commissioning the most advanced, algae production facility in the World. Sapphire Energy's Algae Farm is expected to be the World's first commercial demonstration scale algae conversion farm, integrating the entire value chain of algae-based production, from cultivation, to production, to extraction

Solazyme

Process/ Technology	Heterotrophic microalgal production for oils and fuels
Project owner	Solazyme Inc. (a joint venture with Bunge Global Innovation LLC)
Web page	http://www.solazyme.com/
Facility type	Flagship/Commercial (625,000 L fermentation tanks)
Operation	Since 2014
Feedstock	Sugarcane bagasse
Outputs	Microalgae, oils, fuels, skin and personal care products, encapsulated lubricant (Encapso), bioplastics—total production: 100,000 metric ton renewable oils/year
Location	Orindiúva/Moema, São Paulo, Brazil
Summary	The oil production facility is the Solazyme Bunge Renewable Oils joint venture facility based in Orindiúva-Moema-SP, Brazil. The facility is capable of producing a range of oils and the same inputs are required for the different oils. The only difference is the type of algae strain that is grown in the fermentation tanks. The joint venture facility has a low carbon footprint due to the fact that waste sugarcane material (bagasse) provides all of the energy to power both the sugar mill and the oil production facility. The Brazilian facility can also feed excess energy back to the grid

9.5 Bio-thermo-chemical-based Lignocellulosic Advanced Biorefineries

Biomass gasification produces a gaseous mixture that is called synthesis gas or syngas whose main compounds are H_2, CO, CO_2, CH_4, and other gaseous hydrocarbons from C_2 to C_4. As aforementioned, syngas may have different applications, including its conversion into liquid fuels. Two procedures are possible for the conversion of syngas produced from gasification of biomass into liquid fuels, the chemical catalytic conversion and fermentation. The chemical synthesis of ethanol and higher alcohols constitutes basically a reaction of CO hydrogenation conducted at high temperature and pressure and catalyzed by rhodium-, molybdenum-, copper-based, or modified Fischer–Tropsch catalysts (Lee et al. 2014). In this case, the challenge of finding highly selective, inexpensive, and robust catalysts, capable of preventing the need of a deep purification of the feed syngas currently still persists (Debabov 2013). Another option is syngas fermentation, which is subsequently addressed.

9.5.1 Syngas Fermentation

The fermentation of syngas explores the use of H_2, CO, and CO_2 as substrates for microbial metabolism, for the synthesis of biofuels (e.g., ethanol) and other carbon-based chemical building blocks. This process has the advantage of taking place at near ambient conditions of temperature and pressure, acting the microorganisms as highly specific biocatalysts more tolerant to gas contaminations than chemical catalysts (Kennes et al. 2016). There are also several advantages of ethanol production from syngas over up-to-date biochemical-based technologies of second-generation (2G) biomass conversion. One relates to the possibility of avoiding the complex pretreatments and the costly enzymes necessary for the saccharification of the biomass. Additionally, the gasification enables larger feedstock flexibility and a more complete utilization of the biomass, since it retrieves the chemical energy stored in all parts of the biomass, including in the more recalcitrant fractions (Daniell et al. 2012). This impacts directly in the biomass conversion and ethanol production yields.

9.5.2 Products from Microbial Gas Fermentation

Syngas can be metabolized by acetogenic anaerobic bacteria capable of growing autotrophically in $CO_2 + H_2$, CO, or the mixture of all these, by the Wood–Ljungdahl pathway. In this pathway, CO or CO_2 are reduced and condensed to form acetyl-CoA (Schiel-Bengelsdorf and Dürre 2012; Latif et al. 2014). Acetyl-CoA can be used for the synthesis of cell carbon or can serve as intermediate for the production of several naturally occurring acids and alcohols, e.g., acetate, ethanol, lactate, and 2,3-butanediol (Bertsch and Müller 2015; Kennes et al. 2016). The acetogens *Clostridium autoethanogenum*, *C. ljungdahlii*, *C. carboxidivorans*, *C. acetobutylicum* and *C. ragsdalei* have been object of the most intensive research applied to CO metabolism and syngas fermentation (Abrini et al. 1994; Abubackar et al. 2011; Cotter et al. 2009; Guo et al. 2010; Köpke et al. 2010; Kundiyana et al. 2011; Ramió-Pujol et al. 2015; Younesi et al. 2005). Specific strains of this group have been used in large-scale industrial processes for syngas fermentation and are protected as industrial property. Genetic manipulation has habilitated these microorganisms to increase their tolerance to ethanol and to synthesize a portfolio of products, such as acetone, isopropanol, iso-butanol, terpenes, methyl ethyl ketone, 1-butanol, 2-butanol, propanal, propan-2-one, propan-1-ol and/or propan-2-ol, or its precursors, with potential application within the scope of a syngas fermentation-based biorefinery (Simpson et al. 2012a, b; Köpke and Liew 2012; Chen et al. 2013; Köpke et al. 2013; Gak et al. 2014). The potential of carboxydotrophic bacteria to act as biocatalyzers in the reduction of short-chain carboxylic acids into their corresponding alcohols, such as *n*-propanol, *n*-butanol, *n*-pentanol, *n*-hexanol, and iso-butanol, during syngas fermentation was also demonstrated (Perez et al. 2013).

The synthesis of biopolymers from syngas has been investigated under the frame of an EU collaborative project, SYNPOL, Biopolymers from Syngas Fermentation. This project focuses the optimization of the gasification and pyrolysis processes using biodegradable waste along with microbial strain improvement, particularly of the purple non-sulfur bacteria *Rhodospirillum rubrum*, for the synthesis of poly-hydroxyalkanoate (PHA) and poly-hydroxybutyrate (PHB) from syngas (Drzyzga et al. 2015; Dürre and Eikmanns 2015).

9.5.3 Syngas Composition and Mass Transfer Limitation

The overall fermentation yield is dependent of the syngas composition which, in turn, depends on the gasifier, the biomass type, and the gasification conditions (Liew et al. 2013; Munasinghe and Khanal 2011). A syngas where CO is uniquely used as carbon source whereas H_2 is used as energy source is more suitable for the overall carbon recovery (Bertsch and Müller 2015). In this case, the H_2 in the syngas is used to supply electrons for the reductive conversion of CO into acetyl-CoA, in a process catalyzed by a bifurcating hydrogenase. However, even low concentrations of CO may inhibit the enzyme activity. This shifts CO from carbon source to electron donor in a reaction catalyzed by CO dehydrogenase (CODH), and leads to carbon losses by the formation of CO_2, compromising the process efficiency and undermining the non-polluting aspect of syngas fermentation (Bertsch and Müller 2015). Data from the literature point to suitable H_2/CO ratios of less than 0.5 for air-blown updraft gasifiers, but between 0.5 and 1.0 for air-blown downdraft, air-blown fluidized bed, oxygen-blown, and indirectly heated gasifiers (Wilkins and Atiyeh 2011). Additionally, to maintain the culture stability and the efficiency of carbon conversion, the syngas should be free of other detrimental compounds, such as C_2H_2 or NO_X that inhibit hydrogenase activity, tars that were associated to cell dormancy, HCN that is toxic, and sulfur compounds that may affect cell function (Abubackar et al. 2011; Ahmed et al. 2006; Liew et al. 2013).

Gas–liquid mass transfer limitation is another technical hurdle in the process of syngas fermentation. This is due to the low solubility of CO and H_2 in liquid media, especially at the temperature range at which the fermentation process takes place and aggravated by the stoichiometry of carbon monoxide to ethanol synthesis (6CO: $1CH_3CH_2OH$) (Daniell et al. 2012). To overcome this problem, the bioreactor should be specifically designed and operated to favor the gas–liquid dispersion and solubility, so that microorganisms can have the maximum access to the gaseous substrate (Munasinghe and Khanal 2011).

9.5.4 Industrial Examples of Syngas Fermentation

Two major companies, INEOS Bio and Lanzatech, are currently using syngas fermentation at industrial scale for the production of ethanol, electricity and chemicals (Table 9.4) (Bacovsky et al. 2013). The diagram of Lanzatech and INEOS processes are shown in Sect. 9.3 (Figs. 9.11 and 9.12, respectively).

The activity of INEOS Bio is focused in the use of waste and nonfood crop biomass to produce carbon-neutral bioethanol and renewable energy (INEOS Bio 2012). Since 2012, the company is responsible for 10 patent applications with the aim to increase the yield and productivity of ethanol production by bacterial fermentation of gaseous substrates and improve the gasification process with CO_2 sequestration (Gaddy et al. 2012, 2014).

The process of ethanol production by syngas fermentation by INEOS Bio comprises (INEOS Bio 2012):

- Reception of biomass waste, which may include organic municipal solid waste, agricultural, vegetative ,and forestry residues and products, lignocellulosic energy crops, wood waste, and organic commercial and industrial wastes, for which remote or on-site mechanical, biological, and/or thermal pretreatment may be necessary;
- Biomass drying, using the heat that is recovered from gasification;
- Gasification by a two-step, oxygen-blown process to generate synthesis gas and avoid the formation of dioxins and furans;
- Cooling, cleaning, and compression of the syngas before entering the bioreactor;
- Addition of nutrients and fermentation of the syngas. Here, the bioethanol is synthesized in minutes, at low temperature and pressure and with high yield (maximum theoretical yield: 135 gallons of ethanol per dry ash-free ton of material) and selectivity;
- The liquid from the bioreactor is continuously removed, filtered, distilled, and dehydrated to obtain anhydrous bioethanol;
- Additionally, renewable power is generated by heat recovery from the hot syngas and by combustion of the off gas from the bioreactor.

Table 9.4 Biorefinery facilities based in syngas fermentation

Company, country	Input stream	Biorefinery type	Products	Output capacity	Facility type	Source
Lanzatech, NZ	Various	2-platform biorefinery (syngas, chemical building blocks)	Ethanol 2,3-BDOchemicals	Laboratorial scale	R&D	www.lanzatech.com Bacovsky et al. (2013), IEA Bioenergy (2014)
Lanzatech, NZ	Steel flue gas		Ethanol	90 t/year	Pilot	
Lanzatech, China (1st)	Steel flue gas		Ethanol	300 t/year	Demonstration	
Lanzatech, China (2nd)	Steel flue gas		Ethanol	300 t/year	Pilot	
Lanzatech, Taiwan	Steel flue gas		Ethanol	100 kg/day	Demonstration	
Lanzatech, USA	Woody biomass syngas (125 t/day dry wood residues)		Ethanol chemicals	15,000 t/year	Mobile pilot	
Lanzatech, India	Carbon monoxide containing gas from MSW		Ethanol	300 t/year	Planned	
INEOS Bio plant, USA	Syngas from vegetative yard waste, garden waste, clean woody construction and demolition debris	2-platform biorefinery (syngas, electricity and heat)	Ethanol electricity and heat	24,000 t/year	Commercial	www.ineosbio.com Bacovsky et al. (2013), IEA Bioenergy (2014)
INEOS Bio plant, UK	Syngas from vegetative yard waste, garden waste, clean woody construction and demolition debris		Ethanol electricity and heat	24,000 t/year	Project	

The core of Lanzatech activity in gas-to-liquid bioprocesses focuses mainly the carbon capture of industrial flue gases for the fermentative production of a myriad of low-carbon chemicals and fuels, including ethanol (www.lanzatech.com). The company is responsible for more than 40 patent applications in the past 6 years regarding genetic improvement of proprietary carboxydotrophic acetogenic strains and process optimization, for example to produce recombinant strains incorporating new biosynthesis pathways, to increase the bacterial tolerance to gas contaminants and ethanol, or to improve the carbon capture from the gas stream (Köpke and Liew 2012; Simpson et al. 2012a, b; Chen et al. 2013; Köpke et al. 2013; Schultz and Derek 2013).

The process of gas fermentation in Lanzatech consists of (www.lanzatech.com):

- Reception of the gas steams, from steel mills and processing plants, syngas generated from any biomass resource such as municipal biowaste, organic industrial waste, and agricultural waste, coal-derived syngas, and reformed natural gas;
- Feed of the gas into the bottom of the bioreactor to promote its dispersion by moving upward in the liquid medium. This contact is important to promote a better gas-to-liquid mass transfer, which is one of the major hurdles in the fermentation of gaseous substrates;
- Fermentation by Lanzatech proprietary microbial strains, which are tolerant to contaminants and admit a flexible H_2 content in the gas;
- Withdrawal of the fermentation product to a hybrid separation system, for products and co-products recovery and process water recycling;
- The separated products are used directly as fuel and chemicals or after conversion to common chemicals or drop-in fuels.

Other companies have demonstrated their interest in the process of microbial gas fermentation. Coskata Inc. is an American company with expertize in the production of cellulosic ethanol which in 2012 was seeking to expand its commercial activity to synthesis gas fermentation from diverse feedstock, such as woody biomass, agricultural residues, municipal wastes and natural gas (www.biofuelsdigest.com/bdigest/2014/03/25/coskata-biofuels-digests). The company is responsible for numerous patent applications in the past 5 year regarding the improvement of process efficiency by optimization of the reactor design and operation, protection from HCN and isolation of the novel ethanologenic species *Clostridium coskatii* (Hickey 2013; Hickey et al. 2014; Zahn and Saxena 2012).

Syngas Biofuels Energy Inc. is an American biotechnology company that has developed a "Reverse Global Warming" technology based in engineered biocatalysts for the manufacture of commodity chemicals and fuels from air CO_2 (www.syngasbiofuelsenergy.com). The company advertises the commercialization of scalable fermentation modules of 20,000 gallons for iso-butanol production, by continuous gas blend fermentation using a recombinant *Clostridium* (Gak et al. 2014).

9.6 Future Trends on Advanced Biorefineries

9.6.1 Lignocellulosic-Based Bioethanol Biorefineries

The shift from pilot- and demo-scale production of lignocellulosic ethanol to competitive full-scale production still requires further reduction of the production cost.

Most of the current 2G bioethanol demonstration facilities are only producing cellulosic ethanol, with the hemicellulose fraction (up to 40%) of feedstock being used for other purposes (e.g., animal feeding) (Ferreira 2011). Thereby, one promising strategy will be to integrate the production of ethanol into a biorefinery scheme in which the biomass components of lignin, hemicellulose and extractives are converted into co-value-added products, instead of energy application, in order to overcome the costs associated with pretreatment and enzymes for cellulose hydrolysis (Caesar 2008; Sammons et al. 2007).

Other approach is to integrate 1G plant with a 2G plant on a single site to optimize personnel, utilities, equipments, and other industrial synergies. An attractive option is to integrate cellulosic ethanol production with starch-based ethanol production to use the whole agricultural crop (Hahn-Hägerdal et al. 2006). For instance, in USA, POET/DSM has integrated ethanol production from corncobs into an existing grain ethanol plant and uses part of the collected biomass for power production (CHP integration). Also, cellulosic ethanol process co-produces biogas, which will meet a significant fraction of the adjacent grain ethanol plant's power needs (Bacovsky et al. 2013). The same did occur with GRANBIO plant in Alagoas-Brazil but using sugarcane and straw as 1G and 2G feedstock, respectively.

Moreover, lignocellulosic sugars obtained through enzymatic hydrolysis are diluted with C6 sugars and they are fermented together in the 1G plant (E4tech, RE-CORD and WUR 2015). ICM, Inc. has also co-located its pilot-scale cellulosic biorefinery—using corn fiber, switchgrass and energy sorghum—with the existing grain-to-ethanol facility at LifeLineFoods, LLC in St. Joseph, Missouri (USA) (IEA-Bioenergy 2014). Also, Green Field Specialty Alcohols Inc., Canada's largest producer of fuel ethanol and industrial alcohol, has installed its cellulosic ethanol pilot plant next to 1G corn ethanol facility in Chatham, Ontario. This company produces 2G ethanol, from agricultural crop residues, forestry residues and dedicated energy crops, by applying "Twin Screw Extruder Technology" which allows single-stage or two-stage continuous percolation/hot water/steam explosion pretreatment (IEA-Bioenergy 2014).

9.6.2 Feedstock-Flexible Biorefineries

Conceptually, biorefineries being flexible relatively to feedstock will also be advantageous once the risks associated with feedstock availability will be minimized (E4tech, RE-CORD and WUR 2015). Thereby, modern biorefinery processes should be based on fractionation and pretreatment approaches requiring only relatively small adjustments for application to several biomass feedstocks, allowing the plant to be fed with multiple different feedstocks.

9.6.3 Cluster-Based Biorefineries

Cluster-based biorefineries constituted by different value chains site plants aggregated as a cluster shall be more competitive, such as demonstrated by the successful implementation of the Chemical Cluster (five-site plants) in Stenungsund (Sweden) —Aga, AkzoNobel, Borealis, Ineos and Perstorp—developing a joint strategy of producing sustainable products (Five Clusters 2013). New technologies are being explored for integrating the production of biomass-derived fuels and other products, such as 1,3 propanediol, polylactic acid, and isosorbide, in a single facility (E4tech, RE-CORD and WUR 2015).

9.6.4 Integrated Pulp and Paper Biorefineries

The implementation of biorefinery concept is very attractive for pulp and paper manufacturers, providing high value-added products and pulp production in a closed loop. Manufacturers will continue to have as main objective to produce paper, but the producers have begun to explore how to use the waste streams and by-products—such as bark and wood wastes or the cooking liquor containing lignin and some hemicelluloses (glucomannans from softwood pulp and xylans from hardwood pulp), extractives, including resins and triglycerides, or even the ultimate sludge from their wastewater treatment—for the simultaneous production of biofuels and biochemicals (McElroy 2007; Muffler and Ulber 2008). These new products will be integrated with the existing product lines in the paper industry, while maintaining the properties of pulp and paper and introducing minor modifications in the plant, or alternatively may promote integration with other industries by identifying appropriate synergies (van Ree and Annevelink 2007).

Indeed, there are already some examples of pulpand paper mills that have already implemented this approach. For instance, the concept of the Austrian company Schweighofer Fiber (plans postponed) foresees integration of ethanol production into an existing pulp mill with production of ethanol and energy and recycling of chemicals from the sulphite spent liquor (SSL). Borregaard Industries are successfully producing ethanol from SSL since 1938 (Bacovsky et al. 2013). In their Sarpsborg Biorefinery, spruce chips are treated with acidic calcium bisulfite cooking liquor, promoting hemicellulose hydrolysis to various sugars. After concentration of this SSL obtained, the sugars are fermented into ethanol (Bacovsky et al. 2013). Also, Oji Holdings Corporation (OJI), a Japan's paper manufacturer founded in 1873 that is the sixth largest paper manufacturing company in the World, with support from New Energy and Industrial Development Organization (NEDO), have developed a mechanochemical pulping process for conversion of cellulose to ethanol. This technology is being applied in a pilot plant producing ethanol (65 t-annual capacity) from eucalyptus wood, in Hiroshima (Japan) since 2011 (Bacovsky et al. 2013). The Canadian Alberta-Pacific Forest Industries Inc., one of the largest pulp companies in the World, in simultaneous with kraft pulp production, is manufacturing bio-methanol (4000 t/year) by separating and purifying it from its waste gas stream (steam stripper off-gas) in their commercial demo unit in Alberta (Canada) (Alpac 2016). This product is used for the on-site production of chlorine dioxide, used in pulp whitening operations and the surplus is available for sale to external industries. Domtar Corporation, one of the largest producers of kraft pulp in North America (with 9 pulp and paper mills and 1 paper mill, in USA and in Canada) (Domtar 2016) has created, in partnership with FPInnovations, CelluForce, a company that is producing nanocrystalline cellulose from a fraction of the mill's kraft pulp, based on new patented acid hydrolysis (IEA-Bioenergy 2014), in a demo plant located in Windsor (Québec, Canada) (CelluForce 2016). In addition to pulp, paper, heat, and power, several kraft pulping companies—such as Zellstoff Pöls (in Pöls, Austria) and Carter Holt Harvey Pulp & Paper in its Kinleith mill, New Zealand's largest pulp and paper mill located in the central North Island—are commercially co-producing tall oil and turpentine as attractive marketable intermediate chemicals recovered from black liquor (IEA-Bioenergy 2014).

9.6.5 Higher Added Value Products-Driven Biorefineries

So far biofuel-driven (or energy-driven) biorefineries have been discussed, i.e., with the main goal producing huge volumes of relatively low-value energy (or fuels) out of biomass (IEA-Bioenergy 2014).

There also some product-driven biorefineries, i.e., which have as the main goal produce smaller amounts of relatively higher value-added bio-based products out of biomass and primary (agro) and secondary (process) residues are used to produce energy (power/heat) for internal or external use. Currently, only limited product-driven lignocellulosic biorefineries are in operation, mainly because of the fact that some key technologies are still in the R&D, pilot and demo-phase. However, their potential is huge, and it is generally believed that a refocus will take place concerning optimal sustainable biomass use from mainly energy (fuel) applications to chemical/material applications, and even to biorefineries that use biomass for both "Food" and "Non-food" applications (IEA-Bioenergy 2014).

For instance, in 2013, Cellulac Ltd. Galway (Ireland) announced plans to convert a brewery in Ireland—Great Northern Brewery, Dundalk, Co. Louth, the second largest brewery in Ireland—into a lactic acid and related products (PolyLactic Acid and Ethyl Lactate) plant using 2G feedstocks (together with lactose whey permeate). This company currently has a pilont plant in Postdam, Germany (E4tech, RE-CORD and WUR 2015; IEA-Bioenergy 2014).

Cobalt Technologies, in cooperation with Rhodia and Andritz, are building a demonstration plant in Brazil for the production of butanol from sugarcane bagasse. They combine dilute acid hydrolysis pretreatment with ABE-fermentation and claim that enzymatic hydrolysis is not necessary in their process (E4tech, RE-CORD and WUR 2015).

9.7 Concluding Remarks

A comprehensive overview of the current status of the biorefinery plants all over the World was presented, addressing the benefits, constraints and future challenges of these installations. In general terms, the rational use of the total fraction of the biomass, the broadening of the feedstock sources and the production of a vast product portfolio confer sustainability advantages to the biorefinery pathways and increase their breakthrough chances in the future.

Although it is still not evident that exists a clear winning technology (e.g., biochemical versus thermochemical) for biomass processing, namely lignocellulosic biomass, it rises from this chapter that some commercial cellulosic biofuel plants (biochemical-based) are already fully operational in Europe, USA and Brazil and are leading the current technologies for advanced biofuels in World.

References

Abrini J, Naveau H, Nyns EJ (1994) *Clostridium autoethanogenum*, sp. nov., an anaerobic bacterium that produces ethanol from carbon monoxide. Arch Microbiol 161(4):345–351. doi:10.1007/bf00303591

Abubackar HN, Veiga MC, Kennes C (2011) Biological conversion of carbon monoxide: rich syngas or waste gases to bioethanol. Biofuels, Bioprod Biorefin 5:93–114

Ahmed A, Cateni BG, Huhnke RL, Lewis RS (2006) Effects of biomass-generated producer gas constituents on cell growth, product distribution and hydrogenase activity of *Clostridium carboxidivorans* P7T. Biomass Bioenergy 30:665–672. doi:10.1016/j.biombioe.2006.01.007

Alberta Pacific Forest Industries Inc. (2016) https://alpac.ca. Accessed Feb 2016

Bacovsky D, Ludwiczek N, Ognissanto M, Wörgetter M (2013) Status of advanced biofuels demonstration facilities in 2012. IEA Bioenergy Task 39 Report 1–209

Bertsch J, Müller V (2015) Bioenergetic constraints for conversion of syngas to biofuels in acetogenic bacteria. Biotechnol Bioeng 8:210. doi:10.1186/s13068-015-0393-x

Biodiesel Magazine—The latest news and data about biodiesel, Dec 2015. http://www.biodieselmagazine.com/plants/listplants/. Accessed Feb 2016

Bioenergy2020+ (Renewable Energy Network Austria, Austrian Bio Energy Centre), FJ-BLT (Biomass-Logistics-Technology) IEA Bioenergy Task 39 'Commercializing 1st- and 2nd-Generation Liquid Biofuels from Biomass': Database on facilities for the production of liquid and gaseous biofuels for transport. http://demoplants.bioenergy2020.eu/. Accessed Feb 2016

Caesar B (2008) Industrial biotechnology: more than just ethanol. Factors driving industry growth. Ind Biotechnol 4:50–54

Cardona CA, Sanchez OJ (2007) Fuel ethanol production: process design trends and integration opportunities. Bioresour Technol 98:2415–2457

Carvalheiro F, Duarte LC, Medeiros R, Gírio FM (2004) Optimization of brewery's spent grain dilute-acid hydrolysis for the production of pentose-rich culture media. Appl Biochem Biotechnol 113–116:1059–1072

CelluForce (2016) www.celluforce.com. Accessed Feb 2016

CELPA—Portuguese Association of Paper Industry. www.celpa.pt. Accessed Oct 2011

Chen W, Liew FM, Köpke M (2013) Recombinant microorganisms and uses therefor. World Patent WO 2013/180584

Cheng JJ, Timilsina GR (2011) Status and barriers of advanced biofuel technologies: a review. Renew Energy 36:3541–3549

Cherubini F, Stromman AH (2010) Production of biofuels and biochemicals from lignocellulosic biomass: estimation of maximum theoretical yields and efficiencies using matrix agebra. Energy Fuels 24:2657–2666

Cotter JL, Chinn MS, Grunden AM (2009) Influence of process parameters on growth of *Clostridium ljungdahlii* and *Clostridium autoethanogenum* on synthesis gas. Enzyme Microb Technol 44(5):281–288. doi:10.1016/j.enzmictec.2008.11.002

Daniell J, Köpke M, Simpson S (2012) Commercial biomass syngas fermentation. Energies 5:5372–5417. doi:10.3390/en5125372

Davenport R (2008) Chemicals and polymers from biomass. Ind Biotechnol 4:59–63

Debabov VG (2013) Bioethanol from synthesis gas. Appl Biochem Microbiol 49(7):619–628. doi:10.1134/S000368381307003X

Domtar (2016) www.domtar.com. Accessed Feb 2016

Drzyzga O, Revelles O, Durante-Rodríguez G, Díaz E, García JL, Prieto A (2015) New challenges for syngas fermentation: towards production of biopolymers. J Chem Technol Biotechnol 90 (10):1735–1751. doi:10.1002/jctb.4721

Dürre P, Eikmanns BJ (2015) C1-Carbon sources for chemical and fuel production by microbial gas fermentation. Curr OpinBiotechnol 35:63–72. doi:10.1016/j.copbio.2015.03.008

E4tech, RE-CORD, WUR (2015) From the sugar platform to biofuels and biochemicals. Final report for the European Commission, contract No. ENER/C2/423-2012/SI2.673791

Ethanol Producer Magazine (2016) http://www.ethanolproducer.com/plants/listplants/US/Existing/Sugar-Starch/. Accessed Feb 2016

Eurobserv'er (2015) Biofuels barometer, http://www.eurobserv-er.org/biofuels-barometer-2015. Accessed Feb 2016

European Biodiesel Board. http://www.ebb-eu.org/stats.php. Accessed Jan 2016

European Biofuels Annual (2015) (GAIN Report Number: NL5028). Global Agriculture Information Network, USDA Foreign Agriculture Service

European Biofuels Technology Platform. Accelerating deployment of advanced biofuels in Europe. http://biofuelstp.eu/sugar_crops.html. Accessed Jan 2016

FAO (2014) FAO Statistics - Food and agriculture data. http://www.fao.org/faostat/en/#home. Accessed Feb 2016

Ferreira CSR (2011) Strategies to improve D-xylose and D-glucose co-fermentation for the production of second generation bioethanol. Master Thesis in Biotechnology, Instituto Superior Técnico, Lisboa, Portugal

Five Clusters (2013) Five clusters in west Sweden with strength and potential for the future. http://www.vgregion.se/upload/Regionutveckling/Publikationer/2013/Five_Clusters_2013.pdf

Gaddy JL, Arora DK, Ko C-W, Phillips JR, Basu R, Wilkstrom CV, Clausen EC (2012) Methods for increasing the production of ethanol from microbial fermentation. US Patent 2012/122173

Gaddy JL, Ko C-W, Phillips JR, Slape MS (2014) Methods for sequestering carbon dioxide into alcohols via gasification fermentation. US Patent 2014/45257

Gak E, Tyurin M, Kiriukhin M (2014) Genome tailoring powered production of isobutanol in continuous CO_2/H_2 blend fermentation using engineered acetogen biocatalyst. J Ind Microbiol Biotechnol 41:763–781. doi:10.1007/s10295-014-1416-5

Gao Y, Gregor C, Liang Y, Tang D, Tweed C (2012) Algae biodiesel—a feasibility report. Chem Central J 6(Suppl 1):S1. http://journal.chemistrycentral.com/content/6/S1/S1

Geddes CC, Nieves IU, Ingram LO (2011) Advances in ethanol production. Curr Opin Biotechnol 22:312–319

Gírio FM, Fonseca C, Carvalheiro F, Duarte LC, Marques S, Bogel-Łukasik R (2010) Hemicelluloses for fuel ethanol: a review. Biores Technol 101:4775–4800

Gírio F, Kurkela E, Kiel J, Lankhorst RK (2013) Report EERA-EIBI WORKSHOP "Longer term R&D needs and priorities on bioenergy: bioenergy beyond 2020", p 47

Glasser WG, Wright RS (1998) Steam-assisted biomass fractionation. II. Fractionation behavior of various biomass resources. Biomass Bioenergy 14:219–235

González LE, Díaz GC, Aranda DAG, Cruz YR (2015) Fortes MM biodiesel production based in microalgae: a biorefinery approach. Nat Sci 7:358–369. doi:10.4236/ns.2015.77039

Gray KA, Zhao LS, Emptage M (2006) Bioethanol Curr Opin Chem Biol 10:141–146

Guo Y, Xu J, Zhang Y, Xu H, Yuan Z, Li D (2010) Medium optimization for ethanol production with *Clostridium autoethanogenum* with carbon monoxide as sole carbon source. Bioresour Technol 101(22):8784–8789. doi:10.1016/j.biortech.2010.06.072

Hahn-Hägerdal B, Galbe M, Gorwa-Grauslund M, Lidén G, Zacchi G (2006) Bio-ethanol—the fuel of tomorrow from the residues of today. Trends Biotechnol 24:549–556

Hariskos J, Posten C (2014) Biorefinery of microalgae—opportunities and constraints for different production scenarios. Biotechnol J 9:739–752

Hickey R (2013) Processes for enhancing the performance of large-scale, stirred tank anaerobic fermentors and apparatus therefor. US Patent 2013/0078689

Hickey R, Jianxin D, Reeves A, Richard ET (2014) Processes for the anaerobic bioconversion of syngas to oxygenated organic compound with in situ protection from hydrogen cyanide. US Patent 2014/0273125

Hofvendahl K, Hahn-Hägerdal B (2000) Factors affecting the fermentative lactic acid production from renewable resources. Enzyme Microb Technol 26:87–107

Ibeto CN, Ofoefule AU, Agbo KE (2011) A Global overview of biomass potentials for bioethanol production: a renewable alternative fuel. Trends Appl Sci Res 6:410–425

IEA Bioenergy (2014) BIOREFINING—sustainable and synergetic processing of biomass into marketable food & feed ingredients, products (chemicals, materials) and energy (fuels, power, heat). IEA Bioenergy Task 42 Report 1–66

INEOS Bio, USA (2012) Process technology process description. http://www.ineos.com/globalassets/ineos-group/businesses/ineos-bio/technology/ineos-us-bio-process-description-brochure_april-2012.pdf. INEOS Process Technology Brochure

Iogen (2016) http://www.iogen.ca/cellulosic_ethanol/fuel_demo.html. Accessed Feb 2016

Jacob-Lopes E, Ramírez MLG, Queiroz MI, Zepka LQ (2015) In: Jacob-Lopes E, Zepka LQ (eds) Microalgal biorefineries. Biomass production and uses, pp 81–106. doi:10.5772/59969, ISBN 978-953-51-2181-7

Kamm B, Kamm M (2004) Principles of biorefineries. Appl Microbiol Biotechnol 64:137–145

Kamm B et al (2006) Biorefineries. In: Industrial processes and products. Wiley-VCH

Kennes D, Abubackar HN, Kennes C (2016) Bioethanol production from biomass: carbohydrate vs syngas fermentation. J Chem Technol Biotechnol 91:304–317. doi:10.1002/jctb.4842
Kheshgi HS, Prince RC, Marland G (2000) The potential of biomass fuels in the context of global climate change: focus on transportation fuels. Annu Rev Energy Environ 25:199–244
Köpke M, Liew F (2012) Production of butanol from carbon monoxide by a recombinant microorganism. World Patent WO 2012/053905
Köpke M, Held C, Hujer S, Liesegang H, Wiezer A, Wollherr A et al (2010) *Clostridium ljungdahlii* represents a microbial production platform based on syngas. Proc Natl Acad Sci USA 107(29):13087–13092. doi:10.1073/pnas.1004716107
Köpke M, Liew F, Mueller A (2013) Recombinant microorganisms and uses therefor. World Patent WO 2013/185123
Kundiyana DK, Wilkins MR, Maddipati P, Huhnke RL (2011) Effect of temperature, pH and buffer presence on ethanol production from synthesis gas by *Clostridium ragsdalei*. Bioresour Technol 102(10):5794–5799. doi:10.1016/j.biortech.2011.02.032
Latif H, Zeidan AA, Nielsen AT, Zengler K (2014) Trash to treasure: Production of biofuels and commodity chemicals via syngas fermenting microorganisms. Curr Opin Biotechnol 27:79–87. doi:10.1016/j.copbio.2013.12.001
Lee JH, Reddy KH, Jung JS, Yang EH, Moon DJ (2014) Role of support on higher alcohol synthesis from syngas. Appl Catal A 480:128–133. doi:10.1016/j.apcata.2014.04.026
Liew FM, Köpke M, Simpson SD (2013) Gas fermentation for commercial biofuels production. In: Liquid, gaseous and solid biofuels—Conversion techniques, pp 125–173. doi:10.5772/52164
Lopes da Silva T, Reis A (2015) Algal biorefinery: an integrated approach. In: Debabrata D (eds) Scale up problems for the large scale production of algae. Springer, Berlin, pp 155–186. doi:10.1007/978-3-319-22813-6_6
Lv P, Yuan Z, Wu C, Ma L, Chen Y, Tsubaki N (2007) Bio-syngas production from biomass catalytic gasification. EnergyConvers Manage 48:1132–1139. doi:10.1016/j.enconman.2006.10.014
Lynd LR, Cushman JH, Nichols RJ (1991) Wyman CE Fuel ethanol from cellulosic biomass. Science 251:1318–1323
Mascoma (2007) Mascoma buys Hoosier biofuels startup. Mass high tech. J N Engl Technol
McElroy AK (2007) Not so run of the mill. Biomass Mag 10:38–43
Muffler K, Ulber R (2008) Use of renewable raw materials in the chemical industry—Beyond sugar and starch. Chem Eng Technol 31:638–646
Munasinghe PC, Khanal SM (2011) Biomass-derived syngas fermentation into biofuels. Biofuels 101(13):79–98. doi:10.1016/B978-0-12-385099-7.00004-8
NatureWorks LLC. http://www.natureworksllc.com/#3. Accessed Jan 2016
NovaCana (2016) http://www.novacana.com/usinas-brasil/. Accessed Jan 2016
Nurra C, Torras C, Clavero E, Ríos S, Rey M, Lorente E, Farriol X, Salvadó J (2014) Biorefinery concept in a microalgae pilot plant. Culturing, dynamic filtration and steam explosion fractionation. Bioresour Technol 163:136–142. doi:10.1016/j.biortech.2014.04.009
Perez M, Richter H, Loftus SE, Angenent LT (2013) Biocatalytic reduction of short-chain carboxylic acids into their corresponding Alcohols with syngas fermentation. Biotechnol Bioeng 110(4):1066–1077. doi:10.1002/bit.24786
Petrick I, Dombrowski L, Kröger M, Beckert T, Kuchling T, Kureti S (2013) Algae biorefinery—Material and energy use of algae. DBFZ Report No. 16, p 164
Proethanol2G EU 7th FWP project. www.proethanol2g.org. Accessed Feb 2016
Ragauskas AJ, Nagy M, Kim DH, Eckert CA, Hallett JP, Liotta CL (2006) From wood to fuels. Integrating biofuels and pulp production. Ind Biotechnol 2:55–65
Rajagopalan S, Ponnampalam E, McCalla D, Stowers M (2005) Enhancing profitability of dry mill ethanol plants. Appl Biochem Biotechnol 120(1):37–50
Ramió-Pujol S, Ganigué R, Bañeras L, Colprim J (2015) How can alcohol production be improved in carboxydotrophic *Clostridia*? Process Biochem 50(7):1047–1055. doi:10.1016/j.procbio.2015.03.019

Rapagn S, Provendier H, Petit C, Kiennemann A (2002) Foscolo PU Development of catalysts suitable for hydrogen or syn-gas production from biomass gasi cation. Biomass Bioenergy 22:377–388

Renewable Energy World. http://www.renewableenergyworld.com/articles/print/volume-17/issue-2/bioenergy/ethanol-from-energy-beets-a-viable-option.html. Accessed Jan 2016

Renewable Fuels Association (2005) How ethanol is made? http://www.ethanolrfa.org/how-ethanol-is-made/. Accessed Feb 2016

Sammons N, Eden M, Yuan W, Cullinan H, Aksoy B (2007) A flexible framework for optimal biorefinery product allocation. Environ Prog 26:349–354

Sanchez OJ, Cardona CA (2008) Trends in biotechnological production of fuel ethanol from different feedstocks. Bioresour Technol 99:5270–5295

Sassner P, Martensson CG, Galbe M, Zacchi G (2008) Steam pretreatment of H_2SO_4 impregnated salix for the production of bioethanol. Bioresour Technol 99:137–145

Schiel-Bengelsdorf B, Dürre P (2012) Pathway engineering and synthetic biology using acetogens. FEBS Lett 586(15):2191–2198. doi:10.1016/j.febslet.2012.04.043

Schultz M, Derek G (2013) Improved carbon capture in fermentation. World Patent WO 2013/119866

Sendich E, Laser M, Kim S, Alizadeh H, Laureano-Perez L, Dale B, Lynd L (2008) Recent process improvements for the ammonia fiber expansion (AFEX) Process and resulting reductions in minimum ethanol selling price. Bioresour Technol 99:8429–8435

Simpson SD, Köpke M, Liew FM (2012a) Recombinant microorganisms with increased tolerance to ethanol. World Patent WO 2012/105853

Simpson SD, Köpke M, Liew FM, Chen WY (2012b) Recombinant microorganisms and uses therefor. World Patent WO 2012/115727

Unica (Brazilian Sugarcane Industry Association) and ApexBrasil (Brazilian Trade and Investment Promotion Agency) (2016) http://sugarcane.org/sugarcane-products/ethanol. Accessed Jan 2016

Vaidya AN, Pandey RA, Mudliar S, Kumar MS, Chakrabarti T, Devotta S (2005) Production and recovery of lactic acid for polylactide—An overview. Crit Rev Environ Sci Technol 35:429–467

van Ree R, Annevelink E (2007) Status report biorefinery. Agrotechnology and Food Science Group, Wagninger, NL

van Wyk JPH (2011) Biotechnology and the utilization of biowaste as a resource for bioproduct development. Trends Biotechnol 19:172–177

Vijayakumar J, Aravindan R, Viruthagiri T (2008) Recent trends in the production, purification and application of lactic acid. Chem Biochem Eng Q 22:245–264

Wikipedia (2016) Mascoma Corporation. https://en.wikipedia.org/wiki/Mascoma_Corporation. Accessed Feb 2016

Wilkins MR, Atiyeh HK (2011) Microbial production of ethanol from carbon monoxide. Curr Opin Biotechnol 22(3):326–330. doi:10.1016/j.copbio.2011.03.005

Younesi H, Najafpour G, Mohamed AR (2005) Ethanol and acetate production from synthesis gas via fermentation processes using anaerobic bacterium, *Clostridium ljungdahlii*. Biochem Eng J 27(2):110–119. doi:10.1016/j.bej.2005.08.015

Zahn JA, Saxena J (2012) Novel ethanologenic *Clostridium* species, *Clostridium coskatii*. US Patent 2012/0156747

Zhang YHP (2008) Reviving the carbohydrate economy via multi-product lignocellulose biorefineries. J Ind Microbiol Biotechnol 35:367–375

Zwart RWR (2006) Biorefinery the Worldwide Status at the Beginning of 2006—Including the highlights of the 1st international biorefinery workshop and an overview of markets and prices of biofuels and chemicals, Wageninger, NL

Author Biographies

Francisco Gírio, Ph.D. is Ph.D. in Biochemistry from University of Lisbon (1994), Principal Researcher, and Head of Bioenergy Unit of LNEG - National Laboratory of Energy and Geology, Lisbon, Portugal. In his research career participated as researcher in more than 40 research projects on lignocellulose bioconversion being coordinator of 13, and he is currently member of the Management Board of Bioenergy Joint Program of the European Energy research Alliance, member of the Scientific Council of IMDEA Energy (Spain) since 2012 and president of SIADEB - Iberoamerican Society for the Development of Biorefineries (www.siadeb.org) since 2010. Between 2010 and 2015, he was the national representative at the European Industrial Bioenergy Initiative (EIBI) and he coordinated the European 7th FWP Proethanol2G project (2010–2014) on improving the technology for second-generation bioethanol. He published over than 90 peer-review papers and more than 200 papers and lectures in international conferences. His main focus of research is advanced biofuels pathways, including lignocellulosic bioethanol and biorrefineries sustainability. In 1987, he was awarded the Merit certificate from USDA (USA) for "Contribution in the advance of the scientifically knowledge on xylose fermentation by yeasts and in 2010 he was one of the winners of the "Green Project Awards 2010" in the category "Research and Development."

Susana Marques, Ph.D. is a Researcher at the Bioenergy Unit of the Portuguese National Laboratory of Energy and Geology (LNEG). She graduated in Chemical Engineering (1996) and obtained her Ph.D. in Biotechnology (2012) at Instituto Superior Técnico (IST, Lisboa). She has participated in more than 30 Portuguese and European Projects on lignocellulose bioconversion and co-authored 1 book chapter, 23 papers in International peer-reviewed journals, and more than 100 papers in International conferences. Her main focus of research has been the upgrading of lignocellulosic wastes and by-products into bio-based commercial products (e.g., bioethanol, lactic acid, and biopolymers), working on process intensification including enzymatic hydrolysis and fermentation, preferentially integrated with pretreatment and downstream processing. She was been awarded with *Honour Mention in the Green Project Awards 2009*—category of Research and Development—Project untitled "Upgrading of Wastes from the Pulp and Paper Industry under the Biorefinery Concept."

Filomena Pinto, Ph.D. has Ph.D. in Chemical Engineering, she is a Senior Researcher at Bioenergy Unit of LNEG, and an invited professor at FCUL—Faculty of Sciences of Lisbon. She has co-supervised several M.Sc. thesis (15 already finished) and Ph.D. thesis (5 already finished). Her main research areas cover: thermochemical conversion technologies (gasification, pyrolysis, liquefaction, and hydrothermal liquefaction) for coal, biomass and wastes, syngas cleaning and bio oil upgrading processes, emissions control from thermochemical processes, hydrogenation of oils and animal fats, and CO_2 capture. She is the author of 58 papers in international journals with referee, 41 papers in conference proceedings with referee, 19 chapters in books, and more than 145 papers in international conference proceedings.

Ana Cristina Oliveira, Ph.D. is a Senior Researcher at Bioenergy Unit of National Laboratory of Energy and Geology (LNEG), expert in liquid biofuels production, namely biodiesel, from different raw materials and technologies. She graduated in Chemistry at University of Lisbon (1986) and obtained her M. Sc. in Biotechnology/Biochemical Engineering (1990) and her Ph.D. in Biotechnology (2000) at Technical University of Lisbon (IST). Currently, in addition to her research activity, she is responsible for the liquid biofuels characterization group at LBA (Biofuels and Environment Laboratory) that is an accredited laboratory at LNEG. She also has been integrating the Sustainability Group of biofuels and bioliquids of LNEG's Bioenergy Unit. She is invited professor at Faculty of Sciences of University of Lisbon lecturing the course Energy from Biomass. She has supervised 20 M.Sc. students. She has participated in several national (coordinating 2 of them) and European projects and has published scientific papers in international peer-reviewed journals and in international conferences.

Paula Costa, Ph.D. is a Researcher at Bioenergy Unit of LNEG (National Laboratory of Energy and Geology) working mainly in the area of thermochemical conversion of wastes for alternative fuels production using pyrolysis, gasification, and hydrogenation processes. She obtained Ph.D. in Chemical Engineering at the Faculty of Science and Technology of the New University of Lisbon, titled "Production of liquid and gaseous hydrocarbons by plastic waste pyrolysis" in 2006. She is, also, an invited professor at FCUL (Faculdade de Ciências of the University of Lisbon)/UL (since 2008) and has supervised several MSc thesis (16 already finished). She co-authored 1 book chapter's, 13 papers in international journals, 6 papers in conference proceedings with referee, and more than 35 papers in international conference proceedings.

Alberto Reis, Ph.D. Chemical Engineer, Ph.D. in Biochemical Engineering, Researcher, and currently Deputy Head of the Bioenergy Unit within the National Laboratory of Energy and Geology (LNEG) in Lisbon, Portugal and Coordinator of the Biochemical Engineering Program at the same institution. Team member in six ongoing national and international projects in Bioenergy, and Coordinator of project SIMBIOALGA (New symbiotic approach for a truly sustainable integrated microalgae production directed to a biorefinery platform). Coordinator of the Iberian-Latin-American Thematic Network CYTED P711RT0095 "Iberian-Latin-American Society of Applied Algology SI3A" (162 researchers, 29 groups/institutions, 10 countries). Alberto Reis' scientific interests are in the areas of fermentation and photobioreactor technology, biological CO_2 fixation, and waste treatment. Author or co-author of 65 international peer-review papers.

Patrícia Moura, Ph.D. is a Researcher at the Bioenergy Unit of the National Laboratory of Energy and Geology (LNEG) in Lisbon, Portugal, and the person-in-charge of the Microbial Cell Factories and Enzymes Programme. She graduated in Agro-Industrial Engineering in 1991, obtained her M.Sc. in Biochemistry Engineering in 1996 and her Ph.D. in Agro-Industrial Engineering in 2007, at the Universidade Técnica de Lisboa. Currently, her research interests include anaerobic fermentation processes, biochemical conversion of lignocellulosic by-products, organic waste and microalgae biomass, production of biohydrogen and organic acids, and syngas fermentation. She has participated as team member in several national and international projects and as Executive Coordinator in a national project at LNEG involving anaerobic biochemical conversion processes and valorization of residual biomass. Author/co-author of more than 30 peer-reviewed and conference proceedings papers, 5 national patents, and 15 oral communications and 38 posters in national and international conferences.

Index

M. Rabaçal et al. (eds.), *Biorefineries*, Lecture Notes in Energy 57,
DOI 10.1007/978-3-319-48288-0

F

T

U

The manufacturer's authorised representative in the EU is Springer Nature Customer Service Centre GmbH, Europaplatz 3, 69115 Heidelberg, Germany. If you have any concerns regarding our products, please contact ProductSafety@springernature.com

Printed and bound by CPI Group (UK) Ltd, Croydon, CR0 4YY
22/09/2026
02221059-0007